D1091276

Non-Viral Vectors for Gene Therapy

Second Edition
Part I

Advances in Genetics, Volume 53

Non-Viral Vectors for Gene Therapy

Second Edition
Part I

Edited by

Leaf Huang
Center for Pharmacogenetics
University of Pittsburgh School of Pharmacy
Pittsburgh, Pennsylvania

Mien-Chie Hung
Department of Molecular and Cellular Oncology
The University of Texas
M.D. Anderson Cancer Center
Houston, Texas

Ernst Wagner
Ludwig-Maximilians-Universität München
Munich, Germany

ELSEVIER
ACADEMIC
PRESS

AMSTERDAM • BOSTON • HEIDELBERG • LONDON
NEW YORK • OXFORD • PARIS • SAN DIEGO
SAN FRANCISCO • SINGAPORE • SYDNEY • TOKYO

Elsevier Academic Press
525 B Street, Suite 1900, San Diego, California 92101-4495, USA
84 Theobald's Road, London WC1X 8RR, UK

This book is printed on acid-free paper.

For all information on all Elsevier Academic Press publications
visit our Web site at www.books.elsevier.com

ISBN: 0-12-017653-X

PRINTED IN THE UNITED STATES OF AMERICA
05 06 07 08 09 9 8 7 6 5 4 3 2 1

To Our Families

Contents

7 Toxicity of Cationic Lipid-DNA Complexes   189
Nelson S. Yew and Ronald K. Scheule

3   CATIONIC POLYMERS

8 Polyethylenimine (PEI)   217
Barbara Demeneix and Jean-Paul Behr

9 Pluronic Block Copolymers for Gene Delivery   231
Alexander Kabanov, Jian Zhu, and Valery Alakhov

10   Terplex Gene Delivery System   263
Sung Wan Kim

11   Design of Polyphosphoester-DNA Nanoparticles for Non-Viral Gene Delivery   275
Hai-Quan Mao and Kam W. Leong

Contributors

Numbers in parentheses indicate the pages on which the authors' contributions begin.

Ayesha Ahmad (119) Materials Department, Physics Department, and Molecular, Cellular and Developmental Biology Department, University of California, Santa Barbara, Santa Barbara, California 93106

Valery Alakhov (231) Supratek Pharma Inc, 215 Boul. Bouchard, Suite 1315, Dorval, Quebec H9S 1A9, Canada

Jean-Paul Behr (217) Evolution des Régulations Endocriniennes, Muséum National d'Histoire Naturelle, 7, rue Cuvier, 75231 Paris and Chimie Génétique, Faculté de Pharmacie, route du Rhin, 67401 Illkirch, France

Sabine Boeckle (333) Pharmaceutical Biology - Biotechnology, Department of Pharmacy, Ludwig-Maximilians-Universitaet Muenchen, Butenandtstr. 5-13, D-81377 Munich, Germany

Christine C. Conwell (3) Center for Pharmacogenetics, School of Pharmacy, University of Pittsburgh, Pittsburgh, Pennsylvania 15261

Pieter Cullis (157) Inex Pharmaceuticals Inc., Burnaby, BC, Canada V5J 5J8

Carsten Culmsee (333) Pharmaceutical Biology - Biotechnology, Department of Pharmacy, Ludwig-Maximilians-Universitaet Muenchen, Butenandtstr. 5-13, D-81377 Munich, Germany

Barbara Demeneix (217) Evolution des Régulations Endocriniennes, Muséum National d'Histoire Naturelle, 7, rue Cuvier, 75231 Paris and Chimie Génétique, Faculté de Pharmacie, route du Rhin, 67401 Illkirch, France

Heather M. Evans (119) Materials Department, Physics Department, and Molecular, Cellular and Developmental Biology Department, University of California, Santa Barbara, Santa Barbara, California 93106

Kai Ewert (119) Materials Department, Physics Department, and Molecular, Cellular and Developmental Biology Department, University of California, Santa Barbara, Santa Barbara, California 93106

Mitsuru Hashida (47) Department of Drug Delivery Research, Graduate School of Pharmaceutical Sciences, Kyoto University, Sakyo-ku, Kyoto 606-8501, Japan

Leaf Huang (3) Center for Pharmacogenetics, School of Pharmacy, University of Pittsburgh, Pittsburgh, Pennsylvania 15261

Alexander Kabanov (231) Department of Pharmaceutical Sciences and Center for Drug Delivery and Nanomedicine, College of Pharmacy, University of Nebraska Medical Center, Omaha, Nebraska 68198

Yasufumi Kaneda (307) Division of Gene Therapy Science, Graduate School of Medicine, Osaka University, Suita, Osaka 565-0871, Japan

Sung Wan Kim (263) Department of Pharmaceutics and Pharmaceutical Chemistry, University of Utah, Salt Lake City, Utah 84112

Kostas Kostarelos (71) Imperial College Genetic Therapies Centre, Department of Chemistry, Imperial College London, London, SW7 2AY, United Kingdom

Kam W. Leong (275) Department of Biomedical Engineering, The Johns Hopkins University, Baltimore, Maryland 21205

Alison J. Lin (119) Materials Department, Physics Department, and Molecular, Cellular and Developmental Biology Department, University of California, Santa Barbara, Santa Barbara, California 93106

Ann Logan (19) Molecular Neuroscience Group, Department of Medicine, Wolfson Research Laboratories, University of Birmingham, Birmingham, B15 2TH, United Kingdom

Ian MacLachlan (157) Provita Biotherapeutics Incorporated, Burnaby, BC, Canada V5G 4Y1

Hai-Quan Mao (275) Department of Materials Science and Engineering, The Johns Hopkins University, Baltimore, Maryland 21205

Ana Martin-Herranz (119) Materials Department, Physics Department, and Molecular, Cellular and Developmental Biology Department, University of California, Santa Barbara, Santa Barbara, California 93106

Andrew D. Miller (71) Imperial College Genetic Therapies Centre, Department of Chemistry, Imperial College London, London, SW7 2AY, United Kingdom

Toshihiro Nakajima (307) GenomIdea Inc., 7-7-15 Saito-Asagi, Ibaragi, Osaka 567-0085, Japan

Makiya Nishikawa (47) Department of Biopharmaceutics and Drug Metabolism, Graduate School of Pharmaceutical Sciences, Kyoto University, Sakyo-ku, Kyoto 606-8501, Japan

Martin L. Read (19) Molecular Neuroscience Group, Department of Medicine, Wolfson Research Laboratories, University of Birmingham, Birmingham, B15 2TH, United Kingdom

Cyrus R. Safinya (119) Materials Department, Physics Department, and Molecular, Cellular and Developmental Biology Department, University of California, Santa Barbara, Santa Barbara, California 93106

Ronald K. Scheule (189) Genzyme Corporation, Framingham, Massachusetts 01701

Leonard W. Seymour (19) Department of Clinical Pharmacology, University of Oxford, Oxford, OX2 6HE, United Kingdom

Nelle L. Slack (119) Materials Department, Physics Department, and Molecular, Cellular and Developmental Biology Department, University of California, Santa Barbara, Santa Barbara, California 93106

Yoshinobu Takakura (47) Department of Biopharmaceutics and Drug Metabolism, Graduate School of Pharmaceutical Sciences, Kyoto University, Sakyo-ku, Kyoto 606-8501, Japan

Ernst Wagner (333) Pharmaceutical Biology - Biotechnology, Department of Pharmacy, Ludwig-Maximilians-Universitaet Muenchen, Butenandtstr. 5-13, D-81377 Munich, Germany

Seiji Yamamoto (307) Division of Gene Therapy Science, Graduate School of Medicine, Osaka University, Suita, Osaka 565-0871, Japan

Nelson S. Yew (189) Genzyme Corporation, Framingham, Massachusetts 01701

Jian Zhu (231) Department of Pharmaceutical Sciences and Center for Drug Delivery and Nanomedicine, College of Pharmacy, University of Nebraska Medical Center, Omaha, Nebraska 68198

Preface

Since the pioneering discovery by Felgner *et al.* (1987) that cationic lipid can efficiently transfect cells, there was a surge of research activity in this area. The field received another boost when Nabel *et al.* (1993) successfully completed a small phase I clinical trial in gene therapy of melanoma using cationic liposome as a vector. The data strongly suggest that non-viral vectors may be efficacious and safe in humans. Since then, many different cationic lipids and polymers have been developed as vectors, some of them also entered into clinical trials and others became commercial transfection agents.

An equally important event occurred in 1990 (Wolff *et al.*). The work indicated that naked DNA can transfect muscle cells when injected intramuscularly. This is the beginning of using physical methods to introduce DNA into cells. Since then, all major physical techniques, including pressure, electricity, sound, light, heat, and particle bombardment have been attempted. Some of these methods are quite efficient. For example, the hydrodynamic injection methods developed independently by Zhang *et al.*, 1998 and Liu *et al.*, 1998, is the best method of transfecting liver cells among all viral and non-viral vectors.

There was also much progress made in the molecular biological design of the transgene expression system. For example, site-specific integration of the transgene is now possible for prolonged gene expression without the threat of insertional mutagenesis.

Since the publication of the first edition of *Non-viral Vectors for Gene Therapy* in 1999, the field has experienced significant progress in both chemical and physical vectors. More importantly, many mechanistic studies have appeared to address how the vector works and why the vector produces toxicity. It is safe to state that 18 years after the Felgner's publication, the field of non-viral vector for gene therapy is approaching maturity.

Due to much expansion of the field, it is not possible to include all chapters of this edition in a single book. The two volumes are roughly divided into chemical and physical methods emphasizing mechanistic aspects of the vector. The new edition is then ended with a high note of delivering siRNA for therapeutic purpose. RNA interference is definitely a new dimension in non-viral gene therapy which will attract much attention in the years to come.

We wish to thank Pat Gonzalez of Elsevier and Nicole Sebula of the University of Pittsburgh for excellent editorial assistances throughout various phases of production for these two volumes.

Leaf Huang
Mien-Chie Hung
Ernst Wagner

Section 1

BASIC CONCEPTS

1

Recent Advances in Non-viral Gene Delivery

Christine C. Conwell and Leaf Huang
Center for Pharmacogenetics, School of Pharmacy
University of Pittsburgh
Pittsburgh, Pennsylvania 15261

Advances in Genetics, Vol. 53
Copyright 2005, Elsevier Inc. All rights reserved.

0065-2660/05 $35.00
DOI: 10.1016/S0065-2660(05)53001-3

ABSTRACT

Gene therapy has been deemed the medicine of the future due to its potential to treat many types of diseases. However, many obstacles remain before gene delivery is optimized to specific target cells. Over the last several decades, many approaches to gene delivery have been closely examined. By understanding the factors that determine the efficiency of gene uptake and expression as well as those that influence the toxicity of the vector, we are better able to develop new vector systems. This chapter will provide a brief overview of recent advances in gene delivery, specifically on the development of novel non-viral vectors. The following chapters will provide additional details regarding the evolution of non-viral gene delivery systems.   © 2005, Elsevier Inc.

I. INTRODUCTION

Gene delivery holds great promise as a therapeutic agent for a vast array of medical ailments including cancer, genetic disorders and acquired diseases. The ideal gene delivery vehicle would exhibit cell specificity, minimal immune response, efficient release of DNA into cells, and have a large DNA capacity. Viral vectors such as attenuated viruses, adenoviruses and retroviruses, have thus far proven to have significantly more efficient gene expression than most non-viral vectors. The success of viral vectors has been severely limited due to the potential for a specific immune response to the vector that could hinder gene delivery as well as elicit a severe inflammatory reaction and cause nonspecific gene integration into the host genome. Non-viral vectors are not expected to elicit a specific immune response or randomly integrate DNA into the genomic DNA of the host and therefore are looked to as the future of gene delivery systems. Non-viral vectors include cationic polymer and lipid-based encapsulation of DNA as well as the delivery of naked DNA by physical mechanisms. Although non-viral vectors are considered superior vehicles for gene delivery due to their decreased immunogenicity, their success has been severely limited by inefficient cellular uptake and gene expression. Recent advances made in the field of non-viral gene delivery, specifically strategies to improve target specificity and gene uptake and release, and to further reduce the nonspecific immune response, are addressed briefly below.

II. CATIONIC LIPIDS

Felgner *et al.* reported the use of lipid-based vectors in the late 1980s and since then these vesicles have been considered one of the most promising methods for non-viral gene delivery (Felgner *et al.*, 1987). The cationic head groups make strong electrostatic associations with the DNA, eventually leading to the collapse of the anionic polymer. The length and degree of saturation of the lipid chain is significant in determining the stability and toxicity of the liposome. To form a stable complex, the cationic lipid is often combined with a neutral lipid and/or a helper lipid to form a liposome stable under physiological conditions. Plasmid DNA containing the gene of interest is incorporated into the liposome to form liposome/DNA complexes, or lipoplexes.

A. Ligand-mediated targeting

Significant efforts have been made to improve gene expression by modifying lipoplexes to target-specific cell types. By assuring delivery of the vectors to the specific tissues, the probability of successful delivery is substantially increased. Significant advances have been made towards the targeting of tumor cells *in vivo*. Since cancer cells are known to over-express receptors (e.g., folate and transferrin), lipoplexes have been modified to contain ligands that are recognized by tumor cells. Folate molecules have been tagged to lipids for incorporation into the lipoplexes (Dauty *et al.*, 2002; Zuber *et al.*, 2003). In addition to target-specific molecules, short chain fragments of antibodies have also been used to target tumor cells. Specifically, fragments of short chain antibodies for transferrin have been covalently conjugated to the liposome to form immuno-lipoplexes (Xu *et al.*, 2002). Both systems showed an increased affinity for binding tumor tissues *in vitro* and the immunolipoplex binding was also increased *in vivo* (Dauty *et al.*, 2002; Xu *et al.*, 2002; Zuber *et al.*, 2003). To further enhance the efficiency of these targeted vectors, polyethylene glycol (PEG) was attached to the vectors prior to delivery (i.e., PEGylation). This modification was expected to increase lipoplex solubility as well as improve circulation of the vector (Ogris *et al.*, 1999). The PEG-modified vectors were found to have enhanced binding of lipoplexes to the target cancer cells as compared to unmodified lipoplexes (Yu *et al.*, 2004; Zuber *et al.*, 2003).

B. Toxicity

The toxicity of lipoplexes has been a major limitation for their use as *in vivo* gene delivery systems. As mentioned above, lipid-based vectors do not trigger a cellular immune response (i.e., specific recognition), however these vectors may be recognized as foreign and initiate the production of cytokines such as tumor necrosis factor-α (TNF-α), interferon-γ (INF-γ), interleukin-6 (IL-6), and IL-12. The toxicity of the lipoplexes may be largely attributed to the composition of the liposomes. Optimization of lipid-based vectors (e.g., incorporation of various concentrations of helper lipids) has been relatively exhausted in recent years, as has the development of novel cationic lipids.

Therefore, it has been necessary to address the toxicity of lipoplexes by other methods. Tan *et al.* have shown that the sequential injection of liposome and plasmid DNA can significantly reduce the inflammatory response induced by systemic gene delivery (Tan *et al.*, 2001). By first injecting liposome, then waiting a short time and injecting the plasmid DNA, the authors were able to decrease the levels of the cytokines, TNF-α and IL-12 by greater than 80% as compared to lipoplex delivery (Tan *et al.*, 2001). More recently, Liu *et al.* described the preparation of a nonimmunostimulatory lipid-based vector. In contrast to the standard lipoplex, this vector contains lipid, DNA and an inflammatory suppressor molecule that specifically inhibits the production of the cytokine, NF-κB (Liu *et al.*, 2004b). These "safeplexes" successfully delivered DNA to a number of tissues, all of which displayed a significant decrease in TNF-α as compared to lipoplex delivery. Additionally, delivery of safeplexes did not trigger a significant increase in the levels of IL-12 and INF-γ, which also act as indicators of the severity of the induced immune response. Thus, the addition of an immunosuppressor molecule within the lipoplex can significantly reduce the toxicity associated with lipid-based non-viral vectors. These advances among others will be addressed in detail in Chapter 8.

III. CATIONIC POLYMERS

Cationic polymers condense DNA into compact structures by neutralizing the anionic charge on the DNA. The resulting cationic polymer/DNA complexes, or polyplexes, encapsulate the DNA into small particles for gene delivery. Common polycations include polylysine, polyamines such as polyethylenimine, histone proteins, polyarginine-containing proteins (i.e., protamine, HIV-TAT), and cationic dendrimers. As with lipids, not all cationic polymers are optimal for gene delivery and issues such as efficiency and toxicity must be considered. Many cationic polymers that condense DNA can not withstand the stringent conditions of delivery, such as high ionic strength, therefore it is essential to use polymers that bind tightly, but reversibly, to the DNA.

A. Novel polymer mixtures

Polyethylenimines (PEI) have been shown to have one of the highest transfection efficiencies of all cationic polymers and have become a favorite non-viral gene delivery vector in the last decade (Boussif *et al.*, 1995). PEI is appealing as a delivery vector because it has a high charge density for optimal DNA condensation as well as the ability to act as a "proton sponge", which promotes release from endosomal compartments (Kircheis *et al.*, 2001). Toxicity issues, however, have limited the overall success of this cationic polymer for *in vivo* gene delivery.

In an effort to circumvent toxicity issues while maintaining the transfection efficiency, modified PEI molecules have been designed. These modifications include investigating various molecular weights of branched and linear PEI, conjugation of PEI with PEG (Hong *et al.*, 2004), methylation to a charged quaternary ammonium derivative (Brownlie *et al.*, 2004), and cholesterol–PEI conjugates (Furgeson *et al.*, 2002, 2003), among others (Kichler, 2004). A recent study by Brownlie *et al.* used modular modifications of PEI to create several new variations of the molecule. The presence of PEG decreased toxicity of the PEI vector nearly 10-fold, whereas the quaternary ammonium derivate decreased the toxicity up to 4-fold (Brownlie *et al.*, 2004). By modifying the PEI to create more biocompatible complexes, the toxicity of this vector may be reduced, providing a superior polymer for gene delivery. These alternative molecules will be addressed further in Chapter 9.

Combining lipids and polymers to form new vectors (i.e., lipopoly-plexes) has provided additional options for more efficient cationic polymer-based vector delivery. Specifically, complexes containing lipid-protamine-DNA (LPD) as a modified gene delivery system were investigated for toxicity and efficiency as compared to standard lipid and polymer vectors. LPD was shown to have increased efficiency of gene delivery as compared to cationic liposomes (Li and Huang, 1997; Li *et al.*, 1998). A recent study by Arangoa *et al.* described LPD complexed with asialofetuin (AF), a target ligand-specific for the receptor found on hepatocytes in large numbers (Arangoa *et al.*, 2003). The LPD-AF vectors had enhanced gene expression by approximately 3-fold over lipoplex-AF vectors and a 10-fold increase over the unmodified lipoplex, indicating that the presence of protamine significantly promoted gene delivery under these conditions. Furthermore, modified LPD vectors have been explored for treatment of tumors (Dileo *et al.*, 2003a; Whitmore *et al.*, 2001). Complexes containing CpG oligonucleotides were shown to elicit a strong immune response and increased cytokine production, leading to the reduction in tumor activity *in vivo* (Whitmore *et al.*, 2001). The increased efficiency of LPD vectors as compared with other non-viral vectors, along with recent advances in targeting of the complexes, has provided a promising system for non-viral gene delivery.

Block co-polymers have recently been developed as an adaptation of polymer-based gene delivery systems. Many of the modified vectors contain a frequently used cationic polymer (e.g., PEI, poly-L-lysine (PLL), poly-histidine) that condenses DNA in combination with a stabilizing polymer such as PEG (Ahn *et al.*, 2004; Miyata *et al.*, 2004; Putnam *et al.*, 2003). The production of copolymers allows for combinations of valuable characteristics in individual monomers to be combined to form a novel polymer. Recently, Li and Huang investigated the copolymer, poly(D,L-lactide-co-4-hydroxyl-L-proline) (PHLP). The polymer hydroxyproline is a component of naturally occurring substances such as gelatin and collagen, and therefore, should minimize toxicity of the vector. This biocompatible copolymer was found to have increased gene expression over longer periods of time and lower toxicity as compared to PEI and PLL, making it an excellent candidate for further investigations (Li and Huang, 2004). Additionally, by creating biodegradable linkages between the polymers (i.e., incorporation of esters or thiols), vectors may be designed to be increasingly biocompatible and have a higher propensity to release DNA under specific intracellular conditions (discussed below) (Miyata et al., 2004). Block copolymers have also been designed using molecules that are not cationic in nature. A prominent example is the Pluronic® polymers, which consist of various ratios of ethylene oxide and propylene oxide (Kabanov et al., 2002). These polymers have been shown to have increased gene expression as compared to naked DNA in skeletal muscle tissue. Additionally, efficient delivery of DNA was found with low concentrations of the copolymer, which further reduces toxicity issues (Lemieux et al., 2000). The use of copolymers as non-viral vectors allows for the adjustment of specific characteristics of the vector by manipulating the individual molecules, ratios and linkages incorporated, providing many opportunities to optimize encapsulation and delivery.

The use of natural molecules (e.g., albumin, chitosan, and gelatin) for the encapsulation of DNA for gene delivery is also a promising option for non-viral gene delivery. One advantage to using natural molecules is the reduced toxicity of the vector, partially due to the biodegradability of the polymer. Chitosan, a natural polysaccharide, has been successfully used to deliver DNA both in vitro and in vivo (Chellat et al., 2005; Koping-Hoggard et al., 2004; Kumar et al., 2002; Mansouri et al., 2004). Nanospheres made from human serum albumin have also been shown to have improved transfection efficiency over naked DNA. Additionally, these vectors have minimized interactions with other intercellular components due to the presence of the albumin protein (Brzoska et al., 2004; Simoes et al., 2004). The advantages to investigating naturally occurring polymers for nanospheres for gene delivery will be discussed further in Chapter 12.

B. Toxicity

The toxicity of cationic polymers is frequently a result of the quantity of polymer required to achieve the optimal +/− charge ratio for the polyplex. In many instances, such as with PEI, the charge ratio that provides optimal efficiency is very near that which induces severe toxicity (Chollet et al., 2004). Polymer length/molecular weight has also been found to influence the toxicity of the polyplexes (Ahn et al., 2004; Fischer et al., 1999; Kramer et al., 2004; Wadhwa et al., 1997). Recent investigations have focused on creating new biodegradable polymers such as poly[α-(4-aminobutyl)-L-glycolic acid] (PAGA) and a network of poly(amino ester) (n-PAE), which are expected to have reduced immunogenicity since they can be easily degraded within the host (Anderson *et al.*, 2003; Lim *et al.*, 2000, 2002). n-PAE has been found to elicit a reduced immune response but equivalent transfection efficiency as compared to PEI (Lim *et al.*, 2002). The development of new biodegradable polymers holds great promise for reducing the toxicity of cationic polymer-based delivery vectors.

IV. TRIGGERED RELEASE

A recurring issue with both lipid and cationic polymer-based non-viral vectors is the release of DNA once the particle is taken into the cell. Many vectors are able to efficiently bind to the target cells; however, the gene expression was lower than expected. A primary example of this is the cationic lipid containing the folate ligand. Particles were small (i.e., less than 50 nm) and association with target cells was observed, yet gene expression was not efficient (Dauty *et al.*, 2002; Zuber *et al.*, 2003). Some polymers bind DNA very tightly, which protects it from nuclease degradation in the serum, but the association is not easily reversible once the vector is inside the cell matrix. In order to enhance the escape of DNA from the vector, new polymers and lipids have been designed that are sensitive to intracellular conditions, such as decreasing pH and denaturing conditions (Asokan and Cho, 2002; Guo and Szoka, 2003). When the engineered vectors are exposed to specific conditions, the particles become unstable and develop defects that result in DNA leaking out of the complex.

A. pH sensitivity

Variations in pH levels have been observed in several cellular pathways as well as within specific cellular compartments (i.e., endosomes) (Drummond *et al.*, 2000; Guo and Szoka, 2003). For over two decades, the pH decrease observed in the endosome has been exploited to enhance DNA release from non-viral vectors (Asokan and Cho, 2002; Yatvin *et al.*, 1980). Destabilization of the vectors largely stems from protonation of neutral or negative components of the complex, which affects the overall structure and molecular interactions of the vector (Guo and Szoka, 2003; Thomas and Tirrell, 2000). Lipids have also been designed to contain an acid-sensitive linker region that is hydrolyzed upon exposure to acidic conditions, creating discontinuity throughout the particle and permitting the leakage of DNA from the complex (Gerasimov *et al.*, 1999; Thompson *et al.*, 1996). Development of pH-sensitive molecules has been shown to increase the efficiency of gene expression by the enhanced release of DNA into the endosomal compartment. Several pH-sensitive molecules are shown in Table 1.1.

Table 1.1. Examples of pH-Sensitive Molecules Used in Nonviral Gene Delivery

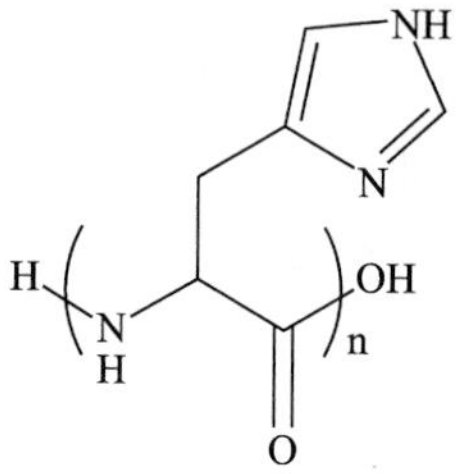

Polyhistidine

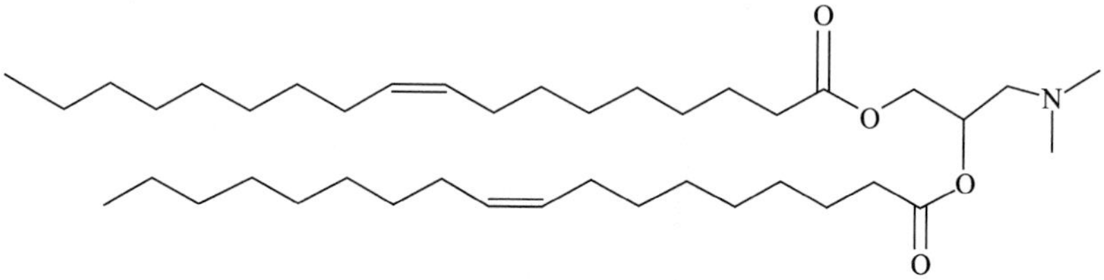

Dioleoyldimethylaminopropane (DODAP)

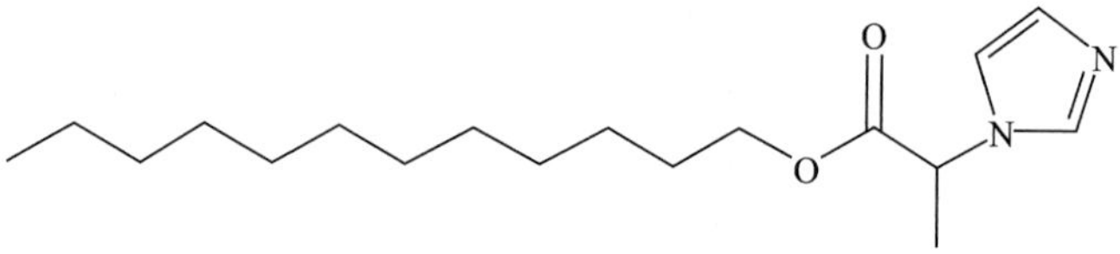

Dodecyl Imidazolyl Propionate

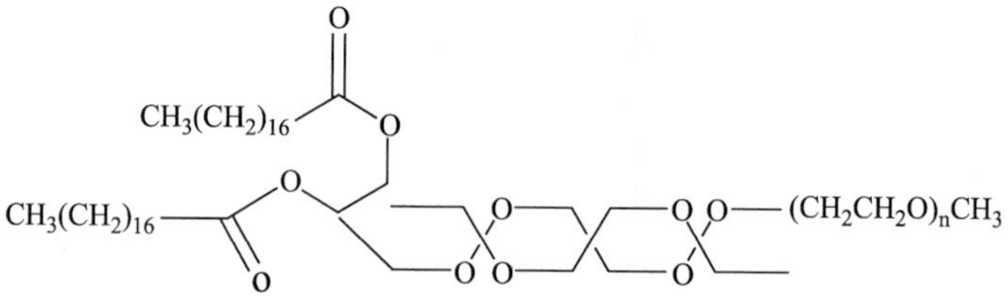

PEG-Diortho Ester-Distearoyl Glycerol Conjugate (POD)

Modified from (Asokan and Cho, 2002; Guo and Szoka, 2003).

B. Redox sensitivity

The cytoplasm of the cell has a reducing environment that is useful in denaturing disulfide bonds. Naturally occurring cationic polymers (i.e., protamines) active in DNA compaction *in vivo* have been shown to be stable at high ionic strengths due to the presence of disulfide linkages. In an environment with free sulfhydryl groups, the linkages are reduced and DNA release was observed (Vilfan *et al.*, 2004). Novel lipids and cationic polymers have cysteine residues incorporated into their core, and frequently the linkage of ligands to the lipids or polymers exists via a reducible moiety (Dauty *et al.*, 2001; Guo and Szoka, 2001, 2003; Kwok *et al.*, 2003). These molecules rely on basic redox chemistry to establish stable disulfide linkages intended to aid in vector efficiency (i.e., attaching target-specific ligand) and stability while also providing a method to increase DNA release from the complex once inside the cell.

V. PHYSICAL DELIVERY METHODS FOR NAKED DNA

Delivery of naked DNA to cells elicits minimal immune response as compared to DNA encapsulated in lipids or cationic polymers. The lack of immunogenicity of naked DNA makes it a good prospect for gene therapy. The limitations with this approach arise in that naked DNA is unprotected against nuclease degradation and the DNA does not have target specificity. Thus, the actual physical delivery of naked DNA must be directed towards the tissues of interest since no target ligands are attached to the DNA. Advances have been made in the development of methodologies to improve targeted delivery of naked DNA. Many of the methods described briefly below will be examined in detail in Part IV. Naked DNA, Oligonucleotides and Physical Methods.

A. High pressure delivery methods

Many methods for the delivery of naked DNA involve the use of high-pressure methods to force the DNA into the desired tissues. Particle bombardment methodology (i.e., gene gun) provides a promising mechanism for gene delivery due to the small quantities of DNA required and the minimal toxicity. To deliver DNA by gene gun, the DNA must first be coated onto gold or tungsten microparticles and then delivered to the cytoplasm of target cells by a pressurized blast from the gene gun (Yang *et al.*, 1990). The success of this method has been severely limited by the small area of tissue to which vector is delivered. A new high-pressure gene gun has recently been developed that uses helium gas to initiate a high-pressure blast, which delivers gold particles suspended in ethanol into target tissue. Delivery with the improved gene gun increased gene delivery to muscle tissue by several orders of magnitude as compared to the conventional gene gun (Dileo et al., 2003b). Penetration of subdermal tissues (i.e., muscle) more than doubled the length of gene expression previously observed.

Hydroporation methods deliver DNA to the target by injecting a large volume of solution with significant force (i.e., a burst of DNA-containing solution). Recent studies by Zhang et al. have shown that by hydrodynamic injection, delivery vectors were able to reach the target tissue (i.e., liver hepatocytes) via tail vein injection by traveling through the hepatic vein (Zhang et al., 2004). The increased pressure from the injection increases membrane permeability by creating defects in the membrane of the hepatocytes, thereby increasing vector uptake. Hydroporation has also recently been explored for gene delivery to the kidneys as well as to muscle tissue (Maruyama et al., 2002; Zhang et al., 2001). Although increased gene delivery and expression have been observed by this delivery method, it remains a very invasive technique that has been found to increase blood pressure and decrease heart rate due to the volume introduced into the system.

Jet injection, like hydroporation, delivers a DNA-containing solution to target tissue via a high-pressure mechanism. The development of a novel low volume 'high speed jet injector' allows for the efficient delivery of small volumes of DNA and is not affected by the length of DNA used (Walther et al., 2001). The modified jet injector can penetrate tissues up to 10 mm deep and has displayed transfection efficiencies similar to those observed with particle bombardment techniques. Gene delivery by jet injection is an appealing mechanism because DNA remained intact throughout the delivery process (i.e., not sheared by the pressure) with no tissue damage or immune response observed.

Another alternative to hydrodynamic injection involves manually massaging liver tissue after intravenous injection of naked DNA (Liu and Huang, 2002a). Mechanical massage of the liver (MML) by applying pressure to the abdomen of the mouse in four short intervals, increased gene uptake into the liver as compared to delivery of naked DNA alone; however, it was approximately an order of magnitude lower than that observed by hydrodynamic injection. Further investigations of the mechanism by which gene expression was enhanced suggest that permeability of liver tissue is increased for several minutes after mechanical massage (Liu et al., 2004a). Delivery of DNA by MML did not have increased toxicity as compared to the control. In vivo studies of mice in hepatic failure revealed that those mice treated with hepatocyte growth factor (HGF) by intravenous injection followed by MML survived, whereas mice treated with empty plasmid or left untreated expired within approximately 20 hr of the diagnosis (Liu and Huang, 2002a). Thus, this less invasive technique may be used to effectively deliver DNA to the liver with reduced toxicity compared to hydrodynamic injection.

B. Electroporation

Electroporation employs the use of an electric field to increase cell permeability and thereby increase the efficiency of DNA uptake into target cells. Naked DNA is delivered by needle into the target tissue, which is then stimulated by the application of an electric field. Gene expression by electroporation has been found to increase by several orders of magnitude over injection of naked DNA alone (Nyode et al., 2004; Wells, 2004). The presence of an applied electric field enhances the permeability of specific tissue near the injection area, allowing for more efficient uptake of DNA into the cells. This method of gene delivery has been used to target a wide variety of tissues but is particularly useful for delivery to more superficial tissues such as skin and muscle. Electroporation of internalized tissues such as kidney and liver has shown gene expression, but major limitations exist due to the invasive nature of the delivery (i.e., surgery to apply electric field to the target tissue).

Serious tissue damage has been associated with electroporation techniques. Modifications to electroporation mechanisms have been designed to optimize gene uptake while reducing the tissue damage introduced by the exposure to high electric currents. This damage may be reduced based on the type of electrodes used as well as the concentration of voltage administered. Meander and caliper electrodes were found to have comparable results for electroporation of the skin, however, the caliper electrode required the pinching of skin between the electrodes (Zhang et al., 2002). Additional comparisons between the needle and plate electrodes for electroporation of internalized tissues determined that the plate electrodes provided more uniform electric fields than the needles but would not be useful in large animals due to the large current that would be required to yield similar results. Liu et al. recently showed that electroporation at lower voltages, delivered by syringe electrodes may have equivalent transfection efficiency as compared to higher voltages, but with significantly less damage to involved tissues (Liu and Huang, 2002b). More recently, short pulses of current, specifically, high followed by low voltages, have provided efficient gene expression with reduced tissue damage (Bureau et al., 2000, 2004).

C. Laser beam gene transduction

Laser beam gene transduction (LBGT) may provide a less damaging and less invasive alternative to gene delivery by electroporation. In this method developed by Ziera et al., the injection of naked DNA into target tissue is followed by the direct application of a pulsed femtosecond laser beam for short intervals (5 sec) (Zeira et al., 2003). The resulting gene expression was comparable to electroporation methods performed under identical conditions. However, histological studies revealed that damage was minimal in tissues that received the laser treatment whereas those treated by electroporation showed significant damage. Thus, this study provides evidence that LBGT may provide a suitable alternative to electroporation for gene delivery to skin tissue.

D. Ultrasound

Ultrasonic gene delivery employs clinical ultrasound techniques to increase the permeability of target tissues and increase the uptake of the vectors. Both high and low intensity ultrasonic velocities have been found to improve the gene expression in a variety of tissues, including muscle, the carotid artery, the heart, and tumors (Wells, 2004). The use of microparticles for encapsulation of naked DNA provides a promising variation of standard ultrasonic techniques. DNA is incorporated into gas-filled particles that generally are coated with lipids, polymers or molecules such as albumin (Unger et al., 2004). These particles may be targeted to specific tissues by the incorporation of peptides or other target molecules to the surface of the microparticle. This technology is attractive because DNA is protected from degradation and is released from the particles due to the ultrasonic energy.

E. Magnetofection

Magnetic nanoparticles coated with polycations (e.g., PEI), which act to stabilize naked DNA attaching to the particle surface, are used to compact DNA for delivery by magnetofection (Huth et al., 2004; Scherer et al., 2002). The particles were directed to a particular location by the introduction of a magnetic field. In vitro studies showed that cells incubated in a magnetic field displayed greater gene expression than those not exposed to the magnet and that the time required for more efficient transfection was reduced from hours to minutes as compared to the controls (Scherer et al., 2002). Magnetofection may also be successful for in vivo transfection of localized areas and has already been shown to work in intestinal tissue. The development of strong magnetic fields should increase the usefulness of this method for internal gene delivery.

F. Photochemical internalization

Photochemical internalization (PCI) relies on the light-sensitive release of DNA contained within endosomal vesicles for more efficient gene delivery. The cells are incubated with photochemical reagents such as $TPPS_2$, which associate with the endosomes and lysosomes. Exposure to light at the appropriate wavelength activates the reagents and leads to the lysis of these intracellular compartments, thereby releasing DNA into the cells. In vitro studies involving the delivery of PEI-encapsulated p53 genes to cells resulted in a 2–3 fold increase in the expression of the p53 gene (Nyode et al., 2004).

VI. PROSPECTS

Many recent advances in non-viral gene therapy have significantly increased the prospect for these methods as successful gene delivery agents. The improved proficiency of cell targeting has increased the potential for cell-specific gene delivery as well as increased uptake as the vectors are attracted to receptors on the cell surface. Additionally, the triggered release technology should enhance the DNA release into the cells, further increasing the gene expression. The increased uptake and release of the DNA help to overcome major barriers limiting non-viral delivery methods. Reducing the toxicity of the vectors via the incorporation of specific cytokine inhibitors would work to further enhance successful gene delivery and expression. By overcoming these barriers, we are increasingly closer to the efficient delivery of genes to target tissues by non-viral gene therapies.

Acknowledgment

The original work from this lab has been supported by NIH grants CA74918, AI48851, DK44935, DK68556, and AR45925.

References

Ahn, C., Chae, S., Bae, Y., and Kim, S. (2004). Synthesis of biodegradable multi-block copolymers of poly(L-lysine) and poly(ethylene glycol) as a non-viral gene carrier. *J. Controlled Release* **97,** 567–574.

Anderson, D., Lynn, D., and Langer, R. (2003). Semi-automated synthesis and screening of a large library of degradable cationic polymers for gene delivery. *Angew. Chem.* **42,** 3153–3158.

Arangoa, M., Duzgunes, N., and Tros de Ilarduya, C. (2003). Increased receptor-mediated gene delivery to the liver by protamine-enhanced-asiolofetuin-lipoplexes. *Gene Thera.* **10,** 5–14.

Asokan, A., and Cho, M. (2002). Exploitation of intracellular pH gradients in the cellular delivery of macromolecules. *J. Pharm. Sci.* **91,** 903–913.

Boussif, O., Lezoualch, F., Zanta, M., Mergney, M., Scherman, D., Demeneix, B., and Behr, J. (1995). A versatile vector for gene and oligonucleotide transfer into cells in culture and *in vivo*: Polyethylenimine. *Proc. Nat. Acad. Sci. USA* **92,** 7297–7301.

Brownlie, A., Uchegbu, I., and Schatzlein, A. (2004). PEI-based vesicle-polymer hybrid gene delivery sustem with improved biocompatibility. *Int. J. Pharm.* **274,** 41–52.

Brzoska, M., Langer, K., Coester, C., Loitsch, S., Wagner, T., and Mallinckrodt, C. (2004). Incorporation of biodegradable nanoparticles into human airway epothelium cells-*in vitro* study of the suitability as a vehicle for drug or gene delivery in pulmonary diseases. *Biochem. Biophys. Res. Commun.* **318,** 562–570.

Bureau, M., Gehl, J., Deleuze, V., Mir, L., and Scherman, D. (2000). Importance of association between permeablization and electrophoretic forces for intramuscular DNA electrotransfer. *Biochim. Biophys. Acta* **1474,** 353–359.

Bureau, M., Naimi, S., Torero Ibad, R., Seguin, J., Georger, C., Arnould, E., Maton, L., Blanche, F., Delaere, P., and Scherman, D. (2004). Intramuscular plasmid DNA electrotransfer: Biodistribution and degradation. *Biochim. Biophys. Acta* **1676,** 138–148.

Chellat, F., Grandjean-Laquerrier, A., Le Naour, R., Fernandes, J., Yahia, L., Guenounou, M., and Laurent-Maquin, D. (2005). Metalloproteinase and cytokine production by THP1 macrophages following exposure to chitosan-DNA nanoparticles. *Biomaterials* **26,** 961–970.

Chollet, P., Favrot, M., Hurbin, A., and Coll, J. (2004). Side-effects of a systemic injection of linear polyethylenimine-DNA complexes. *J. Gene Med.* **4,** 84–91.

Dauty, E., Remy, J., Blessing, T., and Behr, J. (2001). Dimerizable cationic detergents with a low cmc condense plasmid DNA into nanometric particles and transfect cells in culture. *J. Am. Chem. Soc.* **123,** 9227–9234.

Dauty, E., Remy, J., Zuber, G., and Behr, J. (2002). Intracellular delivery of nanometric DNA particles via the folate receptor. *Bioconjugate Chem.* **13,** 831–839.

Dileo, J., Banerjee, R., Whitmore, M., Nayak, J. V., Falo, L. D., and Huang, L. (2003a). Lipid-protamine-DNA-mediated antigen delivery to antigen presenting cells results in enhanced anti-tumor immune responses. *Mol. Ther.* **7,** 640–648.

Dileo, J., Miller, T., Chesnoy, S., and Huang, L. (2003b). Gene transfer to subdermal tissues via a new gene gun design. *Hum. Gene Thera.* **14,** 79–87.

Drummond, D., Zignani, M., and Leroux, J. (2000). Current status of pH- sensitive liposomes in drug delivery. *Prog. Lipid Res.* **39,** 409–460.

Felgner, P., Gadek, T., Holm, M., Roman, R., Chan, H., Wenz, M., Northrop, J., Ringold, G., and Danielsan, M. (1987). Lipofection: A highly efficient, lipid-mediated DNA-transfection procedure. *Proc. Nat. Acad. Sci. USA* **84,** 7413–7417.

Fischer, D., Bieber, T., Li, Y., Elsasser, H., and Kissel, T. (1999). A novel non-viral vector for DNA delivery based on low molecular weight, branched polyethylenimine; Effect of molecular weight on transfection efficiency and cytotoxicity. *Pharm. Res.* **16,** 1273–1279.

Furgeson, D., Chan, W., Yockman, J., and Kim, S. (2003). Modified linear polyethylenimine-cholesterol conjugates for DNA complexation. *Bioconjugate Chem.* **14,** 840–847.

Furgeson, D., Cohen, R., Mahato, R., and Kim, S. (2002). Novel water insoluble lipoparticulates for gene delivery. *Pharm. Res.* **19,** 382–390.

Gerasimov, O., Boomer, J., Qualls, M., and Thompson, D. (1999). Cytosolic drug delivery using pH- and light-sensitive liposomes. *Adv. Drug Delivery Rev.* **38,** 317–338.

Guo, X., and Szoka, F. J. (2001). Steric stabilization of fusogenic liposomes by a low-pH sensitive PEG-diortho ester-lipid conjugate. *Bioconjugate Chem.* **12,** 291–300.

Guo, X., and Szoka, F. J. (2003). Chemical approaches to triggerable lipid vesicles for drug and gene delivery. *Acc. Chem. Res.* **36,** 335–341.

Hong, J., Park, J., Huh, K., Chung, H., Kwon, I., and Jeong, S. (2004). PEGylated polyethylenimine for *in vivo* local gene delivery based on lipiodolized emulsion system. *J. Controlled Release* **99,** 167–176.

Huth, S., Lausier, J., Gersting, S., Rudolph, C., Plank, C., Welsch, U., and Rosenecker, J. (2004). Insights into the mechanisms of magnetofection using PEI-based magnetofectins for gene transfer. *J. Gene Med.* **6,** 923–936.

Kabanov, A., Batrakova, E., and Alakhov, V. (2002). Pluronic block copolymers as novel polymer therapeutics for drug and gene delivery. *J. Controlled Release* **82,** 189–212.

Kichler, A. (2004). Gene transfer with modified polyethylenimines. *J. Gene Med.* **6,** S3–S10.

Kircheis, R., Wightman, L., and Wagner, E. (2001). Design and gene delivery activity of modified polyethylenimines. *Adv. Drug Deli. Rev.* **53,** 341–358.

Koping-Hoggard, M., Varum, K., Issa, M., Danielsen, S., Christensen, B., Stokke, B., and Artursson, P. (2004). Improved chitosan-mediated gene delivery based on easily dissociated chitosan polyplexes of highly defined chitosan oligomers. *Gene Thera.* **11,** 1441–1452.

Kramer, M., Stumbe, J., Grimm, G., Kaufmann, B., Kruger, U., Weber, M., and Haag, R. (2004). Dendritic polyamines: Simple access to new materials with defined treelike structures for application in non-viral gene delivery. *Chembiochem* **5,** 1081–1087.

Kumar, M., Behera, A., Lockey, R., Zhang, J., Bhullar, G., de la Cruz, C., Chen, L., Leong, K., Huang, S., and Mohapatra, S. (2002). Intranasal gene transfer by chitosan- DNA nanospheres protects BALB/c mice against acute respiratory syncytial virus infection. *Hum. Gene Thera.* **13**, 1415–1425.

Kwok, K., Park, Y., Yang, Y., McKenzie, D., Liu, Y., and Rice, K. (2003). *In vivo* gene transfer using sulfhydryl cross-linked PEG-peptide/glycopeptide DNA co-condensates. *J. Pharm. Sci.* **92**, 1174–1185.

Lemieux, P., Guerin, N., Paradis, G., Proulx, R., Chistyakova, L., Kabanov, A., and Alakhov, V. (2000). A combination of poloxamers increases gene expression of plasmid DNA in skeletal muscle. *Gene Thera.* **7**, 986–991.

Li, S., and Huang, L. (1997). *In vivo* gene transfer via intravenous administration of cationic lipid-protamine-DNA (LPD) complexes. *Gene Thera.* **4**, 891–900.

Li, S., Rizzo, M. A., Bhattacharya, S., and Huang, L. (1998). Characterization of catinoi lipid-protamine-DNA (LPD) complexes for intravenous gene delivery. *Gene Therapy* **5**, 930–937.

Li, Z., and Huang, L. (2004). Sustained delivery and expression of plasmid DNA based on biodegradable polyester, poly(D,L-lactide-co-4-hydroxyl-L-proline). *J. Controlled Release* **98**, 437–446.

Lim, Y., Han, S., Kong, H., Lee, Y., Park, J., Jeong, B., and Kim, S. (2000). Biodegradable polyester, poly[a-(4-aminobutyl)-L-glycolic acid], as a non-toxic gene carrier. *Pharm. Res.* **17**, 811–816.

Lim, Y., Kim, S., Suh, H., and Park, J. (2002). Biodegradable, endosome disruptive, and cationic network-type polymer as a highly efficient and nontoxix gene delivery carrier. *Bioconjugate Chem.* **13**, 952–957.

Liu, F., and Huang, L. (2002a). Noninvasive gene delivery to the liver by mechanical massage. *Hepatology* **35**, 1314–1319.

Liu, F., and Huang, L. (2002b). A syringe electrode device for simultaneous injection of DNA and electrotransfer. *Mol. Therapy* **5**, 323–328.

Liu, F., Lei, J., Vollmer, R., and Huang, L. (2004a). Mechanism of liver gene transfer by mechanical massage. *Mol. Therapy* **9**, 452–457.

Liu, F., Shollenberger, L., and Huang, L. (2004b). Non-immunostimulatory non-viral vectors. *FASEB* **18**, 1779–1781.

Mansouri, S., Lavigne, P., Corsi, K., Benderdour, M., Beaumont, E., and Fernandes, J. (2004). Chitosan-DNA nanoparticles as non-viral vectors in gene therapy: Stratagies to improve transfection efficacy. *Eur. J. Pharm. Biopharma.* **57**, 1–8.

Maruyama, H., Higuchi, N., Nishikawa, Y., Hirahara, H., Iino, N., Kameda, S., Kawachi, H., Yaoita, E., Gejyo, F., and Miyazaki, J. (2002). Kidney targeted naked DNA transfer by retrograde renal vein injection in rats. *Human Gene Thera.* **13**, 455–468.

Miyata, K., Kakizawa, Y., Nishiyama, M., Harada, A., Yamasaki, Y., Koyama, H., and Katoaka, K. (2004). Block catiomer polyplexes with regulated densities of charge and disulfide cross-linking directed to enhance gene expresion. *J. Am. Chem. Soc.* **126**, 2355–2361.

Nyode, A., Merlin, J., Leroux, A., Dolivet, G., Erbacher, P., Behr, J., Berg, K., and Guillemin, F. (2004). Enhanced gene transfer and cell death following p53 gene transfer using photochemical internalization of glucosylated PEI-DNA complexes. *J. Gene Med.* **6**, 884–894.

Ogris, M., Brunner, S., Schuller, S., Kircheis, R., and Wagner, E. (1999). PEGylated DNA/transferrin-PEI complexes: Reduced interaction with blood components, extended circulation in blood and potential for systemic gene delivery. *Gene Thera.* **6**, 595–605.

Putnam, D., Zelikin, A., Izumrudov, V., and Langer, R. (2003). Polyhistidine-PEG: DNA nanocomposites for gene delivery. *Biomaterials* **24**, 4425–4433.

Scherer, F., Anton, M., Schillinger, U., Henke, J., Bergemann, C., Kruger, A., Gansbacher, B., and Plank, C. (2002). Magnetofection: Enhancing and targeting gene delivery by magnetic force *in vitro* and *in vivo*. *Gene Thera.* **9**, 102–109.

Simoes, S., Slepushkin, V., Pires, P., Gaspar, R., Pedroso de Lima, M., and Duzgunes, N. (2004). Human serum albumin enhances DNA transfection by lipoplexes and confers resistance to inhibition by serum. *Biochim. Biophys. Acta* **1463,** 459–469.

Tan, Y., Liu, F., Li, Z., Li, S., and Huang, L. (2001). Sequential injection of cationic liposome and plasmid DNA effectively transfects the lung with minimal inflammatory toxicity. *Mol. Thera.* **3,** 673–682.

Thomas, J., and Tirrell, D. (2000). Polymer-induced leakage of cations from dioleoyl phosphatidyl-choline and phosphatidylglycerol liposomes. *J. Controlled Release* **76,** 203–209.

Thompson, D., Gerasimov, O., Wheeler, J., Rui, Y., and Anderson, V. (1996). Triggerable plasma-logen liposomes: Improvements of system efficiency. *Biochim. Biophys. Acta* **1279,** 25–34.

Unger, E., Porter, T., Culp, W., Labell, R., Matsunga, T., and Zutshi, R. (2004). Therapeutic applications of lipid-coated microbubbles. *Adv. Drug Delivery Rev.* **56,** 1291–1314.

Vilfan, I., Conwell, C., and Hud, N. (2004). Formation of native-like mammalian sperm cell chromatin with folded bull protamine. *J. Biol. Chem.* **279,** 20088–20095.

Wadhwa, M., Collard, W., Adami, R., McKenzie, D., and Rice, K. (1997). Peptide-mediated gene delivery: influence of peptide structure on gene expression. *Bioconjugate Chem.* **8,** 81–88.

Walther, W., Stein, U., Fichtner, I., Malcherek, L., Lemm, M., and Schlag, P. (2001). Non-viral *in vivo* gene delivery into tumor using a novel low volume jet-injection technology. *Gene Thera.* **8,** 173–180.

Wells, D. (2004). Gene therapy progress and prospects: electroporation and other physical methods. *Gene Thera.* **11,** 1363–1369.

Whitmore, M., Li, S., Falo, L., Jr., and Huang, L. (2001). Systemic administration of LPD prepared with CpG oligonucleotides inhibits the growth of established pulmonary metastases by stimulat-ing innate and acquired antitumore immune response. *Cancer Immunol. Immunothera.* **50,** 503–514.

Xu, L., Huang, C., Huang, W., Tang, W., Rait, A., Yin, Y., Cruz, I., Xiang, L., Pirollo, K., and Chang, E. (2002). Systemic tumor-targeted gene delivery by anti-transferrin receptor scFv-immunoliposomes. *Mol. Cancer Thera.* **1,** 337–346.

Yang, N., Burkholder, J., Roberts, B., Martinell, B., and McCabe, D. (1990). *In vivo* and *in vitro* gene transfer to mammalian somatic cells by particle bombardment. *Proc. Nat. Acad. Sci. USA* **87,** 9568–9572.

Yatvin, M., Kreutz, W., Horwitz, B., and Shinitzky, M. (1980). pH-sensitive liposomes: Possible clinical implications. *Science* **210,** 1253–1255.

Yu, W., Pirollo, K., Rait, A., Yu, B., Xiang, L., Huang, W., Zhou, Q., Ertem, G., and Chang, E. (2004). A sterically stabilized immunolipoplex for systemic administration of a therapeutic gene. *Gene Ther.* **11,** 1–7.

Zeira, E., Manevitch, A., Khatchatouriants, A., Pappo, O., Hyam, E., Darash-Yahana, M., Tavor, E., Honigman, A., Lewis, A., and Galun, E. (2003). Femto-second infrared laser- an efficient and safe *in vivo* gene delivery system for prolonged expression. *Mol. Thera.* **8,** 342–350.

Zhang, G., Budker, V., Williams, P., Subbotin, V., and Wolff, J. (2001). Efficient expression of naked DNA delivered intraarterialy to limb muscles of nonhuman primates. *Human Gene Thera.* **12,** 427–438.

Zhang, G., Gao, X., Song, Y., Vollmer, R., Stolz, D., Gasiorowski, J., Dean, D., and Liu, D. (2004). Hydroporation as the mechanism of hydrodynamic delivery. *Gene Therapy* **11,** 675–682.

Zhang, L., Nolan, E., Kreitschitz, S., and Rabussay, D. (2002). Enhanced delivery of naked DNA to the skin by non-invasive *in vivo* electroporation. *Biochim. Biophys. Acta* **1572,** 1–9.

Zuber, G., Zammut-Italiano, L., Dauty, E., and Behr, J. (2003). Targeted gene delivery to cancer cells: Directed assembly of nanometric DNA particles coated with folic acid. *Angew. Chem.* **42,** 2666–2699.

2

Barriers to Gene Delivery Using Synthetic Vectors

Martin L. Read,* Ann Logan,* and Leonard W. Seymour[†]
*Molecular Neuroscience Group, Department of Medicine
Wolfson Research Laboratories
University of Birmingham
Birmingham, B15 2TH, United Kingdom
[†]Department of Clinical Pharmacology
University of Oxford
Oxford, OX2 6HE, United Kingdom

ABSTRACT

Progress has been made in the development of different types of nucleic acids such as DNA and siRNA with the potential to form the basis of new treatments for genetic and acquired disorders. The lack of suitable vectors for the

Advances in Genetics, Vol. 53
0065-2660/05 $35.00
DOI: 10.1016/S0065-2660(05)53002-5

delivery of nucleic acids, however, represents a major hurdle to their continued development and therapeutic application. Synthetic vectors based on polycations are promising vectors for gene delivery as they are relatively safe and can be modified by the incorporation of ligands for targeting to specific cell types. However, the levels of gene expression mediated by synthetic vectors are low compared to viral vectors. The aim of this chapter is to give an overview of the main barriers that have been identified as limiting gene transfer using polycation-based synthetic vectors. The chapter is divided into two sections to focus on both extracellular and intracellular barriers. We describe novel strategies that are being used to develop increasingly sophisticated vectors in an attempt to overcome these barriers. For instance, we describe approaches to prolong the plasma circulation of polyplexes by the incorporation onto their surface of hydrophilic polymers such as polyethylene glycol (PEG) and poly[N-(2-hydroxypropyl)methacrylamide] (pHPMA). In addition, strategies to improve transfer of nucleic acids from the outside of the cell to the nucleus are described to overcome barriers such as escape from endocytic vesicles and translocation across the nuclear membrane. Furthermore, we highlight new types of vectors based on reducible polycations that are triggered by the intracellular environment to facilitate efficient cytoplasmic release of nucleic acids. © 2005, Elsevier Inc.

I. INTRODUCTION

There has been rapid progress in the development of nucleic acids based on DNA and RNA with the potential to form the basis of new treatments for genetic and acquired disorders. Indeed, the recent emergence of small interfering RNA (siRNA) to trigger RNA interference has initiated efforts to identify siRNA sequences capable of silencing a vast number of genes in the human genome (Dorsett and Tuschl, 2004). However, a major hurdle to the continued development and successful therapeutic application of nucleic acids is the lack of suitable vectors for their delivery. Synthetic (or non-viral) vectors are ideally suited for this purpose as they are capable of complexing and delivering large quantities of nucleic acids of virtually any size and are generally regarded as being safer than viral vectors. In addition, a large number of studies have demonstrated that the biocompatibility and utility of synthetic vectors can be improved by the incorporation of hydrophilic polymers and targeting ligands (Cotten *et al.*, 1990; Dash *et al.*, 2000).

Two main types of synthetic vectors are currently in development for use in therapeutic applications. Polyplexes are formed by the self-assembly reaction between positively charged polycations such as polyethylenimine (PEI) and negatively-charged DNA, while lipoplexes are formed with lipids such as

DOTAP. A number of clinical studies have been initiated using synthetic vectors for delivery of therapeutic genes. For instance, recent studies have demonstrated that intratumoral administration of DNA/lipid mixtures is safe and can produce clinically significant responses (Hortobagyi *et al.*, 2001; Stopeck *et al.*, 2001; Yoo *et al.*, 2001). However, in general, the transfection efficiency of synthetic vectors is poor compared to viral vectors and will need to be improved significantly to enable successful development in the clinics.

Several methodologies have been used to develop synthetic vectors with improved transfection capabilities. One approach is to screen a panel of polymers to identify properties desirable for gene delivery. In the study by Wolfert *et al.*, for example, several series of cationic polymers were used to evaluate the influence of structural parameters on properties of DNA complexes (Wolfert *et al.*, 1999). They showed, for instance, that cationic polymers with short side chains such as polyvinylamine formed small complexes resistant to destabilization by polyanions but with limited transfection activity. In contrast, cationic polymers with longer side chains such as poly[methacryloyl-Gly-Gly-NH-$(CH_2)_6$-NH_2] showed inefficient complex formation with high positive surface charge that had better transfection activity. A more common approach is to modify existing vectors or develop new types to overcome specific barriers to gene transfer. There are many published examples where this approach has been used. Interaction of vectors with serum proteins, for example, has been identified as being responsible for diminishing their stability resulting in poor blood circulation properties (Dash *et al.*, 1999). However, the attachment of hydrophilic polymers such as PEG or HPMA onto the exterior of polyplexes can diminish the interaction with serum proteins and prolong blood circulation times (Dash *et al.*, 2000; Oupicky *et al.*, 2002a).

The aim of this chapter is to give an overview of the main barriers that have been identified as limiting gene transfer using synthetic vectors based on polycations. This chapter is divided into two sections to focus on both extracellular and intracellular barriers. We also describe some novel strategies that have been used to develop increasingly sophisticated vectors in an attempt to overcome these barriers. In particular, we describe new types of vectors based on reducible polycations that can be triggered by the intracellular environment to facilitate release of nucleic acids.

II. EXTRACELLULAR BARRIERS

There can be many forms of barriers to efficient transgene delivery, and in its widest definition the concept would include issues related to formulation, stability and shelf life; for our purposes here, however, we will focus on the biological challenges that confront a therapeutic nucleic acid from the moment it enters the body until it fulfils its therapeutic purpose. Most of the extracellular barriers are therefore factors that influence biodistribution of the administered vector from the point of delivery to the intended target cells, and they will vary profoundly between routes of delivery. The simplest successful *in vivo* delivery route is probably hydrodynamic delivery to the liver via a superficial vein (Maruyama *et al.*, 2002; Zhang *et al.*, 2004), where there are few if any extracellular barriers to delivery, while the most complex route is probably intravenous injection of agents that are intended to "home" to specific disseminated or widespread targets (Ward *et al.*, 2000). Since this latter involves most of the factors that also influence other delivery routes, it will be discussed in some detail.

Targeted intravenous delivery of genes is desirable for treatment of diseases of the blood and the systemic vasculature, and it is essential for the effective treatment of metastatic cancer where tumor cells may be disseminated in unknown micrometastases around the body. The requirements of a vector suitable for targeted intravenous delivery include the following:

- It should have an adequate circulation in the blood plasma to enable it to reach target cells
- Its physical properties should be compatible with gaining access to the target cells
- It should be able to distinguish target cells from non-target cells in the complex *in vivo* environment

If each of these demands is met the agent should be capable of targeted systemic delivery of therapeutic nucleic acids.

A. Extended plasma circulation

The duration of plasma circulation that is necessary to enable effective systemic targeting is not yet certain. Extrapolation from other areas of medicine may yield clues, for example Doxil (a stealth liposome containing doxorubicin that is designed to accumulate without active targeting within solid tumor nodules) shows an initial plasma half life in humans of 1–3 h (Gabizon *et al.*, 2003). Meanwhile Herceptin, an antibody capable of therapeutic targeting of her2/neu on metastatic breast cancer, has an initial plasma half life in humans of 2–10 days (Tokuda *et al.*, 1999). Accordingly it might be reasonable to expect that successful therapeutic targeting of nucleic acid vectors would require similar circulatory properties, with plasma half lives at least of several hours. Fortunately this may be incorrect, due to the nature of nucleic acid therapeutics; nucleic acids have considerable advantages over other agents in terms of their potency (in principle, delivery of a single gene copy per target cell can be curative), and their potential for promoter-driven selectivity (by careful selection of promoters, aberrant transgene expression within non-target cells may be avoided). It follows that therapeutic delivery of nucleic acids may occur at a much lower proportion of the administered dose, and therefore the plasma kinetics required to enable successful treatment may be shorter than for conventional targeted agents.

Studying plasma kinetics of nucleic acids is normally achieved either by radiolabeling the DNA or RNA, using a range of techniques, Southern blotting or using quantitative (real time) polymerase chain reaction (QPCR). Of these only Southern blotting informs on the integrity of the nucleic acid itself, and data gained using the other techniques may be confounded by the presence of partial degradation products. In the case of radiolabeling, it is also possible that the probe is catabolized and incorporated into a new molecule, hence caution is required particularly for interpreting longer term distribution profiles (Terebesi et al., 1998).

Immediately following intravenous injection, the vector materials are exposed to serum proteins and blood cells. It is a prerequisite for effective systemic delivery that extensive aggregation must not occur, otherwise the aggregated vector is likely to become trapped within the first microvascular bed encountered (Ward et al., 2001). For delivery via superficial veins this is the pulmonary capillary bed, and the size limit above which materials become trapped is estimated at 4 μm (Davis et al., 1985). There are many examples in the literature where vector materials aggregate with components of the blood-stream and are largely eliminated from the circulation within seconds, most notably with vectors formed by self assembly of nucleic acids with simple cationic lipids.

Perhaps surprisingly, polyplexes (formed by self assembly of nucleic acids with polycations) are not usually deposited in the pulmonary capillary bed (Dash et al., 1999; Oupicky et al., 1999a). This is probably because the aggregates formed with components of the serum are smaller. However, these agents do bind serum proteins which may act as opsonins, and they are cleared very rapidly into the liver, probably by scavenging into Kupffer cells or hepatic sinusoidal endothelium (Dash et al., 1999).

There have been fractional improvements in plasma circulation of simple vectors, achieved by manipulation of the formulations. For example, lipopoly-plexes, which are usually formed by initial condensation of the nucleic acid with polycations, followed by addition of cationic lipids, have been reported to show some extension in circulation, although systemic targeting has been limited.

There have also been several attempts to improve plasma circulation by incorporating polyethylene glycol onto the surface of polyplexes, either by simple self-assembly of nucleic acids with hydrophilic-cationic block copoly-mers, or by surface grafting of reactive PEG onto the surface of preformed polyplexes (Toncheva *et al.*, 1998; Wolfert *et al.*, 1996). Unfortunately these agents still appear to be susceptible to binding serum proteins, and there has been little if any improvement in plasma circulation (Oupicky *et al.*, 2000, 1999b).

Nevertheless, it may be a mistake to focus solely on plasma kinetics, for the reason identified above, namely that we do not know what kinetics are required for therapeutic delivery. Wagner and coworkers have shown significant transgene expression within peripheral tumors, despite low levels of plasma circulation (assessed by Southern blotting) (Kircheis *et al.*, 2002; Wolschek *et al.*, 2002). Clearly it may not be essential to gain perfect distribution kinetics prior to observing targeted transduction and potential beneficial effect.

Some groups have taken a more radical approach to achieving extended plasma circulation, for example by crosslinking the surface of polyplex vectors to provide "lateral" stabilization against disruption by serum proteins (Oupicky *et al.*, 2002a). This can be achieved either by crosslinking surface amino groups (Oupicky *et al.*, 2002b), or by using a multivalent polymer to react across the surface. The resulting vectors show far better plasma circulation, with initial clearance half lives as high as 90 mins, and can accumulate to substantial levels within tumors. However, the transfection activity of these vectors is poor, and they are now being designed with bioreversible trigger mechanisms to enable them to become transfection active following arrival at the target site (Oupicky *et al.*, 2002c).

An alternative and successful approach to gaining extended plasma circulation with DNA is based on the use of stabilized plasmid-lipid particles (SPLP) (Cullis *et al.*, 1998; Monck *et al.*, 2000; Tam *et al.*, 2000) which can be further modified by non-covalent linkage of hydrophilic polymers (Wheeler *et al.*, 1999) to provide enhanced circulatory kinetics.

Given this broad range of promising technologies, systemic delivery of genes is becoming gradually more realistic. Indeed, several of these vectors are capable of delivering their nucleic acid payload throughout the bloodstream, and the main restrictions on the success of the strategy have moved downstream.

B. Gaining access to target cells

Possibly the major challenge to successful application of systemic gene delivery now is the requirement for the vectors to leave the bloodstream in order for them to gain access to extravascular targets. Conventional wisdom expects this to occur either through or between vascular endothelial cells, although the intercellular junctions are rather small. The tightest endothelial layers are in the brain, where spaces of just 1–2 nm have been found, and in most vascular beds the most permeable endothelium is in the post capillary venules, where inter-endothelial gaps of 5–10 nm have been observed (reviewed in Seymour,

1992). The most permeable physiological vascular beds are those in the sinuses (notably the liver and spleen) where intercellular spaces of over 100 nm have been reported. Most synthetic vector systems have diameters greater than 50 nm, hence the only obvious targets are either components of the bloodstream, or parenchymal cells of sinusoidal organs such as liver and spleen. However, it may be possible to manipulate the permeability of the vasculature, for example recent reports from Chamberlain's laboratory demonstrate that systemic application of vascular endothelial growth factor (VEGF) dramatically augments the ability of adeno-associated virus (AAV) to gain access to muscle cells following intravenous delivery (Gregorevic et al., 2004). AAV has a diameter of approximately 22 nm, hence, there could be ways forward for systemic delivery of non-viral gene vectors.

The situation in tumors is slightly different, because tumour microvasculature is notoriously disorganized and leaky. One compelling hypothesis was put forward by Hiroshi Maeda (Matsumura and Maeda, 1986), termed the Enhanced Permeability and Retention Effect, which argued that the poor lymphatic drainage of tumors leads to inadequate interstitial fluid convection, leading to hypoxia, and this in turn regulates increases in the permeability of the tumor vasculature. In this way more fluid enters the tumor, bringing oxygen and essential nutrients, and it mainly drains back into the post capillary venules. Macromolecules (and gene delivery vectors) contained within the fluid are often sieved out by the tumor interstitial connective tissue and become trapped within the tumor mass (Maeda et al., 1992; Seymour et al., 1995). Apart from providing a useful means to deliver macromolecular agents into tumors, this also constitutes an important mechanism for the tumor to accumulate proteins that can be subsequently incorporated into clots and form the basis for enlarged tumor growth.

Trapping of agents within tumor extracellular stroma is not obviously conducive to efficient tumor transfection, however, and it remains to be seen whether gene delivery vectors successfully delivered in this way will be able to gain access to the surface of the tumor cells in order to enter them and express the therapeutic transgenes.

C. Discrimination between target and non-target cells

Gene delivery vectors are said to be targeted by transcription if they express therapeutic transgene using target-associated promoter elements, and targeted by transduction if they recognize and transfect selectively the appropriate target cells. In principle either form of targeting may be adequate, but combining them should enable very great selectivity of action. Accordingly several groups have been exploring the possibility of delivering genes via target-associated receptors. Examples of this range from RGD to complex vasculature-specific peptides and antibodies. One strategy that combines the principles of steric and lateral stabilization (described above) with receptor-mediated targeting was exemplified by Fisher et al. (Fisher *et al.*, 2000), who showed receptor-mediated uptake and targeting of polymer-stabilized polyplexes using a range of targeting ligands. This approach has now been taken forward to incorporate biological trigger mechanisms (Oupicky *et al.*, 2002c) (described in detail below) based on intracellular processing. Probably the most elegant *in vivo* tumor-targeting work has come from the group of Ernst Wagner who has incorporated Epidermal Growth Factor (EGF) onto gene delivery vectors, and has shown transduction of remote tumors following intravenous injection (Wolschek *et al.*, 2002). EGF is a particularly promising ligand for tumor-selective targeting since its receptors are upregulated on many tumor types, most notably carcinomas.

III. INTRACELLULAR BARRIERS

A prerequisite for efficient gene transfer is the delivery of nucleic acids from outside the cell to the nucleus. However, this goal is challenging as synthetic vectors need to facilitate a number of distinct steps as depicted schematically in Fig. 2.1, including cell-specific binding, internalisation (e.g., endocytosis), escape from endocytic vesicles, transport through the cytoplasm, translocation across the nuclear membrane and release of DNA for transcription. Viruses have evolved mechanisms to overcome each of these barriers but these functionalities need to be incorporated into the design of synthetic vectors. Another problem is the lack of techniques to study intracellular trafficking of polyplexes to probe and elucidate steps that limit transfection efficiency. To date, the majority of studies have relied on quantitation of gene expression as a measure of a polymer's efficacy. However, the rational design of synthetic vectors will ultimately require a better understanding of the interaction of vectors with cells at each of the intracellular barriers. In the following sections we describe intracellular barriers that limit gene transfer and give examples of strategies currently being employed to overcome them.

A. Endocytic vesicles

Uptake of polyplexes into cells typically occurs by endocytosis where polyplexes are taken up into endocytic vesicles and follow the lysosomal trafficking pathway eventually leading to degradation of the polymer or DNA, or both, by the acidic pH and by various lysosomal enzymes. The ability of polyplexes to avoid this fate is therefore an important requirement to enhance gene transfer upon internalisation (reviewed in Wattiaux *et al.*, 2000). Endosomotropic compounds such as chloroquine are widely used to enhance polyplex-mediated gene transfer *in vitro* by a mechanism that is believed to favor the survival of polyplexes in

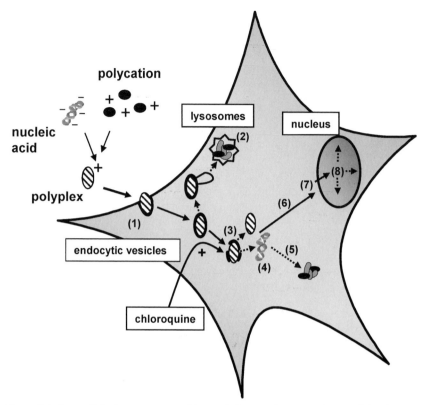

Figure 2.1. Intracellular barriers to gene delivery. Schematic depiction of intracellular obstacles to efficient gene transfer with synthetic gene delivery vectors. Barriers include: (1) cellular uptake; (2) lysosomal degradation; (3) escape from endocytic vesicles; (4) degradation by cytosolic nucleases; (5) vector unpacking; (6) rate of movement across cytoplasm; (7) nuclear entry and (8) targeting to transcription active regions of the nucleus. Chloroquine is thought to favor the survival of synthetic vectors, such as polyplexes, in endocytic vesicles and promote their release into the cytoplasm.

endocytic vesicles and promote their release into the cytoplasm (Forrest and Pack, 2002; Wattiaux *et al.*, 2000). Chloroquine enters cells and accumulates in acidic vesicles where it has been predicted to reach concentrations greater than 50 mM (Erbacher *et al.*, 1996) which is likely to result in osmotic swelling. Previous studies have also suggested that chloroquine protects internalized DNA by inactivating hydrolytic enzymes (Wagner, 1998) and preventing acidification of endocytic vesicles that inhibits transport to later endolysosomal compartments. However, there is still some uncertainty on the precise role of chloroquine in enhancing gene transfer. Results from Erbacher *et al.*, for example, did not show any relationship between neutralization of acidic compartments of cells and the efficiency of transfection (Erbacher *et al.*, 1996). A more recent study by Forrest *et al.* using flow cytometric analysis of pH-sensitive fluorophores conjugated to PLL also indicated that neither the acidic environment nor lysosomal enzymes were responsible for poor gene delivery of PLL-based polyplexes in HEK293 and HepG2 cells and that chloroquine had little to no effect on endocytic trafficking (Forrest and Pack, 2002). Instead, it was concluded that chloroquine most likely acts by enhancing escape from endocytic vesicles, but the mechanism did not appear to involve pH buffering.

Although chloroquine is widely used to enhance gene transfer *in vitro*, it is unlikely that it will be used to assist escape from endocytic vesicles *in vivo* due to concerns over toxicity. Zhang *et al.*, for example, demonstrated in a recent study that acute systemic chloroquine toxicity limited the *in vivo* use of chloroquine to levels substantially below those required for gene delivery (Zhang *et al.*, 2003). The maximum serum levels in rats that could be achieved were 2 μM, whereas chloroquine levels required for optimal gene delivery are greater than 20 μM. Instead, the ability to mediate escape from endocytic vesicles will need to be engineered into cationic polymers to form the basis of efficient synthetic vectors.

1. Escape from endocytic vesicles

A range of polyplexes have been described with the ability to facilitate efficient escape from acidic vesicles by destabilizing membranes and increasing the amount of DNA delivered to the cytoplasm. One of the best known and most widely studied cationic polymers is PEI which is generally regarded as the gold-standard synthetic vector due to its ability to mediate high levels of transfection in a range of cell types (Boussif *et al.*, 1995) and *in vivo* (Aoki *et al.*, 2001; Coll *et al.*, 1999). PEI is available commercially in both a branched and linear form and in acidic endosomes alters the osmolarity leading to their swelling and destabilization which is known as the proton sponge effect (Boussif *et al.*, 1995; Kichler *et al.*, 2001). Itaka *et al.* recently demonstrated the ability of both linear and branched PEI to facilitate rapid escape from endocytic vesicles under confocal microscopy using fluorescein and Cy3 dyes tagged onto plasmid DNA. Most fluorescently-labelled DNA was observed in the cytoplasm, whereas no endosomal escape was observed with PLL/DNA polyplexes (Itaka *et al.*, 2004).

Another common strategy to overcome the endocytic barrier is to conjugate endosomolytic peptides to the synthetic vector that are known to form pores in the lipid membrane. For example, we recently conjugated the membrane-active peptide melittin to PEI in order to enhance cytoplasmic delivery of mRNA (Bettinger *et al.*, 2001). Melittin is a small amphipathic protein from bee venom, *Apis mellifera*, which contains 26 amino acids and a single tryptophan residue. This protein has been shown to integrate into natural

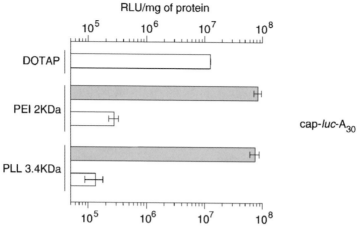

Figure 2.2. Transfection efficiency in B16-F10 cells of short polycation/mRNA polyplexes. B16-F10 cells were transfected with 1 μg of cap-*luc*-A$_{30}$ condensed either by DOTAP at N:P 1.8, PEI 2 kDa at N:P 5 or PLL 3.4 kDa at N:P 7.5 in DMEM without serum in the absence (white bars) or presence (shaded bars) of 100 μM chloroquine. The cells were incubated with the polyplexes for 4 h and the medium discarded and replaced with fresh medium containing 10% FCS. Luciferase activity was measured after 6 h. Results are shown as a mean and standard deviation from triplicate samples. Bettinger *et al.* (2001) © Oxford University Press. Reproduced with permission.

or synthetic membranes and to disrupt a variety of cells and liposomes at micromolar concentrations, which is partly due to its detergent-like sequence. In our study we showed that 25 kDa PEI was ineffective at delivering mRNA encoding for luciferase which was presumably due to the high level of electrostatic interaction between polymer and nucleic acid. However, decreasing the electrostatic interaction between the carrier and mRNA by using shorter polycations gave major increases in expression, with mRNA polyplexes formed using low mw PEI and PLL achieving 5-fold greater levels of luciferase expression than DOTAP/mRNA (Fig. 2.2). As a result of using low mw polycations, however, the resultant polyplexes lost their endosomolytic activity and chloroquine was required to mediate mRNA expression.

Endosomolysis could be restored, however, by conjugating low mw PEI to melittin, and high levels of mRNA expression were demonstrated in the absence of chloroquine. For example, PEI 2 kDa-melittin mediated a high level of transfection of mRNA encoding green fluorescent protein (GFP) in HeLa cells (58.5 ± 2.9%) and post-mitotic cells such as HUVEC (71.6 ± 1.7%), whereas PEI 2 kDa mediated less than 1% GFP-positive cells in both cell types (Fig. 2.3). An overview of the proposed intracellular fate of mRNA delivered by different synthetic vectors in this study is described in Fig. 2.4.

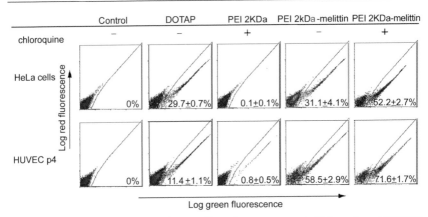

Figure 2.3. Transfection efficiency of PEI 2 KDa-melittin conjugate in HeLa cells and HUVEC. HeLa cells and passage 4 HUVEC were transfected with 1 μg of cap-GFP-A$_{64}$ condensed either by DOTAP at N:P 1.8, PEI 2 kDa at N:P 5 or PEI 2 kDa-melittin at N:P 5 in DMEM with serum. In some cases, cells were treated with 100 μM chloroquine when transfected with PEI 2 kDa and with PEI 2 kDa-melittin. GFP expression was measured after 24 h by flow cytometry analysis. Bettinger *et al.* (2001) © Oxford University Press. Reproduced with permission.

Histidine-rich molecules have also been shown to exhibit membrane destabilization features in an acidic medium resulting in enhanced nucleic acid delivery in the cytosol (Pichon *et al.*, 2001). Indeed, data from several groups has shown that poly-L-histidine mediates an acid-dependent fusion and leakage of negatively charged liposomes (Uster and Deamer, 1985; Wang and Huang, 1984). The ability of histidine-rich vectors to enhance transfection is thought to be related to protonation of the imidazole functionality within the histidine structure. The pKa of the imidazole group in histidine is approximately 6.0 and under slightly acidic conditions the nitrogen heterocyclic ring is protonated. Hence, the imidazole group possesses a buffering capacity in the endosomal pH range, which is likely to facilitate escape from endocytic vesicles. In addition, Gonçalves *et al.* recently showed that histidylated polylysines complexed with DNA had prolonged stability inside early endocytic vesicles that was likely to favor DNA escape into the cytosol (Gonçalves *et al.*, 2002).

B. Transportation to the nucleus

1. Nuclease degradation

Once polyplexes have escaped from endocytic vesicles the next step is to achieve efficient transportation of nucleic acids into the nucleus. There are several factors that can influence this process, including stability of nucleic acids

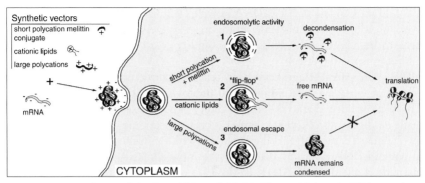

Figure 2.4. Proposed intracellular fate of delivered mRNA. The availability of mRNA to undergo translation in the cytoplasm is dependent on the type of synthetic vector used. 1. When delivered with short polycation containing the endosomolytic peptide melittin (PEI 2 kDa-melittin), polyplexes escape from the endosome and mRNA is efficiently released for translation to occur. In the case of unmodified short polycations (PEI 2 kDa, PLL 3.4 kDa), endosomal escape can be triggered by the buffering agent chloroquine. 2. When delivered with cationic lipids (DOTAP, DOGS), mRNA is released into the cytoplasm following a "flip-flop" process between the cationic lipids and the anionic component of the endosomal membrane. 3. When larger polycations are used, polyplexes escape from the endosome either by release with chloroquine (PLL 54 kDa) or by the "proton sponge" effect (PEI 25 kDa). However, in the cytoplasm the polyplexes do not adequately release mRNA in a form suitable for translation to occur. Bettinger *et al.* (2001) © Oxford University Press. Reproduced with permission.

to nucleases in the cytoplasm, decondensation (or unpacking) of vectors to release nucleic acids and the rate of movement of polyplexes across the cytoplasm to the nucleus. The presence of cytosolic nucleases has been suggested to be a major barrier limiting gene transfer. Early studies demonstrated that plasmid DNA disappeared with an apparent half-life of 50–90 min from the cytoplasm of HeLa and COS cells (Lechardeur *et al.*, 1999). Similarly, another study showed in the cytoplasm of A549 cells, that 2 h after injection, only 50% of plasmid DNA remained intact, as determined by fluorescence *in situ* hybridization (FISH) (Pollard *et al.*, 2001).

Specific inhibitors of nucleases have been used in attempts to improve the gene transfer of plasmid DNA (Niedzinski *et al.*, 2003; Ross *et al.*, 1998; Sperinde *et al.*, 2001). In the study by Niedzinski *et al.*, for example, two endonuclease inhibitors, zinc and aurintricarboxylic acid (ATA), were used to enhance the efficacy of salivary gland transfection with naked plasmid DNA by at least 1000-fold (Niedzinski *et al.*, 2003). In another study by Sperinde *et al.*, a novel peptide DNase II inhibitor, ID2-3, was identified by phage display that was shown to enhance transfection with both lipoplexes and PEI/DNA polyplexes *in vitro* (Sperinde *et al.*, 2001). Indeed, transfection enhancements as high as

270-fold were found with PEI/DNA polyplexes in the presence of ID2-3. However, in order to apply nuclease inhibitors effectively *in vivo,* formulations will need to be devised to increase the possibility of co-localising DNA with the inhibitor at the target cell.

In general, there have been relatively few studies demonstrating the benefit of using nuclease inhibitors to improve gene transfer with polyplexes. This is most likely due to the innate ability of polyplexes to protect DNA from nuclease digestion following complexation with cationic polymers (Dash *et al.,* 1997; Pollard *et al.,* 2001). Pollard et al. showed, for instance, that complexation of plasmid DNA with PEI prevented the intracellular degradation of injected plasmids (Pollard *et al.,* 2001). In the case of synthetic vectors, therefore, an important consideration is the precise timing when unpacking of polyplexes occurs to release the genetic material. If unpacking occurs too early then it is likely the genetic material will be susceptible to nuclease degradation. However, inadequate unpacking will result in lack of access of the genetic material to the transcription/translation machinery.

2. Vector unpacking

Previous studies have demonstrated that the rate of intracellular release of nucleic acids from polyplexes can influence the level of transgene expression (Bettinger *et al.,* 2001; Read *et al.,* 2003; Schaffer *et al.,* 2000). Larger polycations (K_{180}) were shown to remain associated with plasmid DNA in the nucleus of mouse fibroblasts after 48 h, inhibiting gene expression compared with smaller polycations $(K_{19}$ and $K_{36})$ (Schaffer *et al.,* 2000). Itaka *et al.* also recently showed that linear PEI gave rapid and efficient plasmid DNA decondensation in the cytoplasm, which was clearly correlated with earlier gene expression. This was in contrast to branched PEI/DNA polyplexes that remained in the condensed state even after 24 h post transfection (Itaka *et al.,* 2004).

A potential strategy to facilitate intracellular delivery is to use low mw condensing peptides that dissociate readily from DNA. However, the reduced affinity of low mw peptides for DNA means that they can also be easily displaced in physiological conditions. Instead, attempts have been made to overcome this problem by stabilizing low mw polycations bound to DNA with cross-linking agents that can be cleaved or activated by the intracellular environment (Adami and Rice, 1999; Oupicky *et al.,* 2001; Trubetskoy *et al.,* 1999). Adami *et al.,* for example, used glutaraldehyde to cross-link peptide condensates, however, the slow reversal of Schiff bases formed by glutaraldehyde between neighboring peptides meant that increasing the level of cross-linking reduced *in vitro* gene expression (Adami and Rice, 1999). Other cross-linking agents have been used to form caged DNA condensates (Trubetskoy *et al.,* 1999) but these approaches do not produce effective transfection agents.

a. Linear reducible polycations

Recently, we developed a synthetic vector based on a linear reducible polycation (RPC) prepared by oxidative polycondensation of the peptide Cys-Lys$_{10}$-Cys (Read *et al.*, 2003). This design was based on the rationale that, following internalization, the RPC would be cleaved by intracellular reducing conditions to facilitate release of DNA. Initial experiments confirmed that RPC/DNA polyplexes could be destabilized by reduction with increased aggregation of polyplexes (Fig. 2.5) and salt-induced release of DNA (Fig. 2.6). Efficient intracellular release of DNA and gene transfer was also observed in a range of cell types using vectors based on RPC compared to PLL of similar mw (Fig. 2.7). Furthermore, efficient decondensation and release of nucleic acids by RPC-based vectors was demonstrated in LNCaP cells with a 187.3-fold higher level of gene expression observed after 8 h using RPC/DOTAP compared to PLL/DOTAP, which decreased to 3.7 fold after 48 h (Fig. 2.8). In addition, cleavage

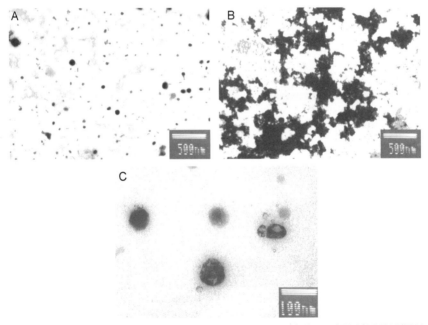

Figure 2.5. Transmission electron micrographs showing the morphology of RPC/DOTAP/DNA complexes. (A) RPC/DOTAP/DNA complexes at N:P 2/2.3 in the absence of DTT, (B) RPC/DOTAP/DNA complexes at N:P 2/2.3 in the presence of 25 mM DTT after 1 h (negative staining, × 15,000 magnification, bar indicates 500 nm) and (C) Higher magnification of RPC/DOTAP/DNA complexes at N:P 2/2.3 (negative staining, × 60,000 magnification, bar indicates 100 nm). Read *et al.* (2003) © John Wiley & Sons Limited. Reproduced with permission.

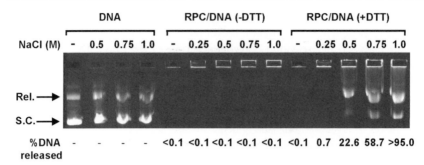

Figure 2.6. Reduction destabilises RPC/DNA polyplexes. RPC/DNA polyplexes were formed at N:P 1 and incubated with 25 mM DTT for 1h. NaCl was added at a range of concentrations (0–1.0 M) and the amount of free DNA assessed by gel electrophoresis. ImageQuant™ software was used to quantify the amount of DNA released (Rel.-relaxed DNA; S.C.-supercoiled DNA). Read *et al.* (2003) © John Wiley & Sons Limited. Reproduced with permission.

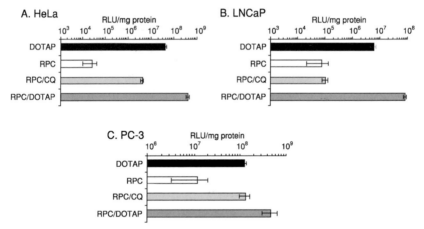

Figure 2.7. Enhanced gene transfer with RPC/DOTAP/DNA complexes. (A) HeLa, (B) LNCaP and (C) PC-3 cells were transfected with 0.5 μg of pCMVLuc condensed by DOTAP at N:P 2.3 (black bars), RPC at N:P 2 (white bars), RPC at N:P 2 in the presence of 100 μM chloroquine (CQ) (light grey bars), or RPC/DOTAP at N:P 2/2.3 (dark grey bars). The cells were incubated with complexes for 4 h in serum-free media and the media was discarded and replaced with fresh media containing 10% FCS. Luciferase activity was measured after 18 h. Results are shown as a mean and standard deviation from triplicate samples. Read *et al.* (2003) © John Wiley & Sons Limited. Reproduced with permission.

LNCaP

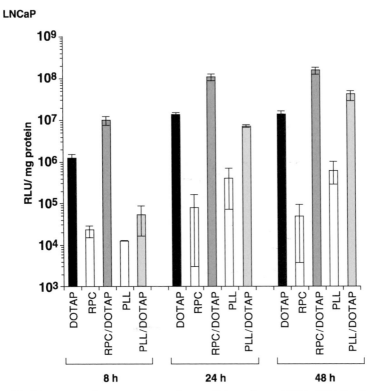

Figure 2.8. Time course of gene expression indicates efficient release of DNA. LNCaP cells were transfected with 0.5 μg of pCMVLuc condensed either by DOTAP at N:P 2.3, RPC at N:P 2, RPC/DOTAP at N:P 2/2.3, PLL at N:P 2, or PLL/DOTAP at N:P 2/2.3. The cells were incubated with the complexes for 4 h and the media discarded and replaced with fresh media containing 10% FCS. Luciferase activity was measured after 8, 24, and 48 h. Results are shown as a mean and standard deviation from triplicate samples. Read *et al.* (2003) © John Wiley & Sons Limited. Reproduced with permission.

of the RPC by the intracellular reducing environment decreased toxicity of the polycation to levels comparable with low mw peptides. A similar strategy has been used by McKenzie *et al.* where low mw peptides containing two to five cysteine groups, such as Cys-Trp-Lys$_8$-Cys-Lys$_8$-Cys, were mixed with DNA prior to spontaneous oxidation to form interpeptide disulfide bonds (McKenzie *et al.*, 2000).

It is envisaged that linear reducible polycations will ideally be suited to form the core of more sophisticated vectors for synthetic gene delivery. Indeed, it was shown that RPC/DNA polyplexes could be laterally stabilized by the addition of the hydrophilic polymer HPMA (Oupicky *et al.*, 2002c). However, a drawback in using this first generation of reducible polycations is that chloroquine (McKenzie *et al.*, 2000; Oupicky *et al.*, 2002c; Read *et al.*, 2003) or the inclusion of lipids such as DOTAP are still required to facilitate escape of RPC/DNA polyplexes from endocytic vesicles and enhance levels of transgene expression.

b. Histidine-rich reducible polycations

In a strategy to overcome this problem we have also recently developed reducible polycations containing both lysine and histidine residues (Read *et al.*, 2004). It was envisaged that the known endosomal buffering capacity of histidine residues would facilitate escape of polyplexes based on these reducible polycations from endocytic vesicles. In particular, two histidine-rich RPCs were prepared by oxidative polycondensation of peptides Cys-His$_3$-Lys$_3$-His$_3$-Cys or Cys-His$_6$-Lys$_3$-His$_6$-Cys containing between 55 (HIS3 RPC) and 70% (HIS6 RPC) histidine content. Initial transfection experiments showed that HIS3 RPC/DNA polyplexes gave modest levels of gene expression in PC-3 cells at levels 15–440 fold lower than those achieved with PEI/DNA. Chloroquine enhanced the transfection activity of HIS3 RPC/DNA polyplexes producing up to a 12.7-fold increase in reporter gene expression. By comparison, HIS6 RPC/DNA polyplexes at (w/w) ratio ≥ 24 gave significant levels of gene expression compared to PEI/DNA that were not enhanced by the addition of chloroquine (Fig. 2.9). The inability of chloroquine to enhance transfection with HIS6 RPC/DNA polyplexes at (w/w) ratios ≥ 24 suggested that transfer of DNA from endocytic vesicles to the cytoplasm had been significantly improved. We then compared the transfection activity of HIS6 RPC and PEI using a GFP expression plasmid and demonstrated that HIS6 RPC mediated significantly greater gene transfer compared to PEI in the majority of cell types tested with up to nearly a 5-fold increase in GFP-positive cells (Fig. 2.10).

Efficient non-viral vectors are also required to deliver RNA molecules for a wide range of therapeutic applications (Bettinger and Read, 2001). However, as previously described, studies so far have demonstrated that unmodified PLL- and PEI-based vectors are generally too stable to release mRNA molecules to enable efficient translation to occur (Bettinger *et al.*, 2001). We have therefore also investigated whether histidine-rich RPCs can be used to deliver RNA molecules such as siRNA. In particular we investigated whether histidine-rich RPCs can be used to deliver siRNA targeted against p75^{NTR} in primary cultures of adult rat dorsal root ganglion (DRG) neurons. Western blotting indicated virtually a complete knockdown in the level of p75^{NTR} following delivery of siRNA (Seq2) targeted to the mRNA of this receptor using HIS6 RPC, whereas there was no decrease in p75^{NTR} using a scrambled version of the siRNA (Scr2). By comparison, PEI and HIS3 RPC were less effective at delivering the Seq2 siRNA with knockdowns of 56% and 25%, respectively, in the levels of p75^{NTR}

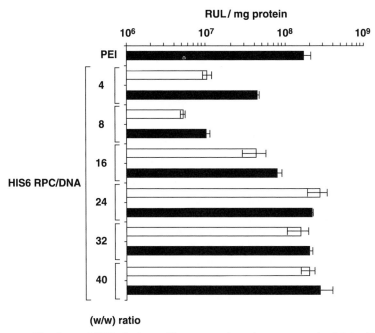

Figure 2.9. Histidine-rich RPCs facilitate chloroquine-independent gene transfer. PC-3 cells were transfected with 0.5 μg pCMVLuc condensed by HIS6 RPC at the indicated (w/w) ratio. For control transfections PEI/DNA polyplexes were formed at N:P 10. Cells were incubated with polyplexes for 4 h in the absence (white bars) or presence (dark gray bars) of 100 μM chloroquine and the media was discarded and replaced with fresh media containing 10% FCS. Luciferase activity was measured after 24 h. Results are shown as a mean and standard deviation from at least three samples.

(Fig. 2.11). These results demonstrated the capability of histidine-rich RPCs to deliver siRNA molecules at levels sufficient to mediate a biological effect.

C. Nuclear barrier

Inefficient transfer of DNA from the cytoplasm to the nucleus has been identified as a major factor restricting transgene expression with only 0.1% of naked DNA or 1% of polyplex DNA reaching the nucleus following microinjection (Pollard et al., 1998). This process is particularly inefficient in post-mitotic and quiescent cells, the majority of cells in vivo, since there is no breakdown of the nuclear envelope in the absence of mitosis (Brunner et al., 2000; Zabner et al., 1995). Cellular and viral proteins enter the nucleus efficiently by means of nuclear localization sequences (NLSs), stretches of amino acids that bind to

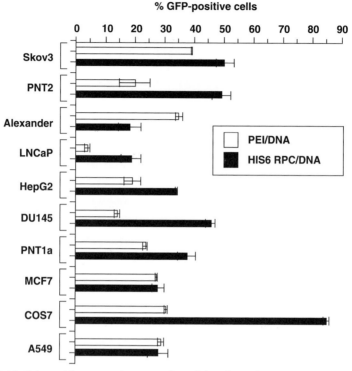

Figure 2.10. Enhanced frequency of gene transfer with histidine-rich RPCs. A range of cell types were transfected as indicated with 0.5 μg pEGFPN1 condensed by PEI at N:P 10 (white bars), or HIS6 RPC at (w/w) ratio of 40 (dark grey bars). GFP expression was measured after 24 h by flow cytometry analysis. Results are shown as a mean and standard deviation from at least three samples.

intracellular transport receptors such as importins and facilitate transfer through the nuclear pore. Early experiments showed that NLSs retain their activity when conjugated as a synthetic peptide to otherwise non-nuclear proteins such as bovine serum albumin and transferrin (Lanford *et al.*, 1986). A widely used strategy to promote nuclear uptake of DNA has therefore been to utilize NLS peptides and exploit intracellular transport mechanisms (Bremner *et al.*, 2001). Results to date have been encouraging and indicated the potential of this approach. For example, coupling of streptavidin-NLS conjugates to biotinylated DNA-enabled linear DNA fragments of between 310–1500 bp to enter the nuclei of digitonin permeabilized cells by an active transport process (Ludtke *et al.*, 1999). Significant increases in gene expression of 10–1000 fold were also observed following ligation of an oligonucleotide-NLS conjugate to one or both

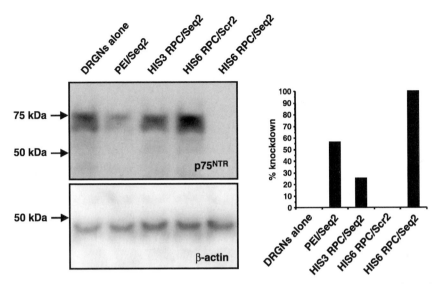

Figure 2.11. Histidine-rich RPCs mediate efficient delivery of siRNA. Western blot analysis of p75^{NTR} (upper panel) and β-actin (lower panel) using DRG cell lysates following transfection with siRNA directed against p75^{NTR} (Seq2) or a scrambled sequence (Scr2) and condensed either by PEI at N:P 10, HIS3 RPC at (w/w) ratio of 66 or HIS6 RPC at (w/w) ratio of 40 as indicated. The relative % knockdowns in p75^{NTR} levels are shown in the bar graph on the right-hand side.

termini of a linear DNA molecule, although no conclusive localization or mechanistic data was given in this study (Zanta et al., 1999).

Although results have been encouraging using linear DNA fragments, the type of chemistry used to conjugate peptides to circular plasmid DNA often results in loss of transcriptional capability. Two studies, for example, used either a photoactivation reaction (Ciolina et al., 1999) or a cyclo-propapyrroloindole group (Sebestyen et al., 1998) to couple NLSs at random locations on the DNA molecule. Sebestyen et al. demonstrated nuclear import in digitonin-permeabilized HeLa cells when greater than 100 NLS peptides were attached to each plasmid, but transgene expression was completely abolished (Sebestyen et al., 1998). More encouragingly, Ciolina et al. found no decrease in expression until modification reached the level of 43 peptides per plasmid, however, despite binding of the conjugates to importin α, no increase in gene expression was observed (Ciolina et al., 1999).

A strategy to prevent loss of gene expression is to attach NLS peptides to DNA at specific locations using peptide nucleic acid (PNA) molecules that have a repeating polyamide backbone rather than a repeating sugar-phosphate backbone and hybridize to DNA in a high-affinity and sequence-specific manner (Nielsen *et al.*, 1991). Branden *et al.* demonstrated the utility of this approach with an 8-fold increase in transfection using PNA-NLS with 25-kDa PEI (Branden *et al.*, 1999). The interaction between DNA and mono PNA is weakened under physiological conditions but this problem has been recently overcome by the use of a homopyrimidine J-base containing bis PNA clamp that forms $(PNA)_2DNA$ triplexes in a highly stable, pH-independent manner (Zelphati *et al.*, 1999). In a recent study, Bremner *et al.* showed a 7-fold increase in gene expression using a NLS peptide/DNA conjugate formed by site-specific linkage of an extended SV40 peptide via a PNA clamp (Bremner *et al.*, 2004).

The exact conditions to successfully utilize the properties of NLS peptides to enhance non-viral gene transfer, however, requires further investigation (Bremner *et al.*, 2004). It is still not clear, for example, what type, length or number of NLS peptides is best suited for this application and NLS sequences that bind to a range of transport receptors, including importin $\beta1$ or transportin 1, have not been thoroughly investigated. Some researchers favor a single NLS and others prefer as great a number of peptides as possible. It has been proposed that a single or cluster of NLS peptide(s) attached to one end of linear DNA is sufficient for nuclear import and that more may actually hinder the process (Ludtke *et al.*, 1999; Zanta *et al.*, 1999). This hindrance could in principle occur through multiple NLSs causing docking of the DNA to several nuclear pore complexes and has been calculated to be physically possible with DNA over 1 kb in size (Maul and Deaven, 1977).

It is also still not clear what delivery vectors should be used with NLS peptide/DNA conjugates to properly exploit the NLS-transport pathways. Previous studies have observed enhanced gene expression with NLS peptide/DNA conjugates using cationic lipids and branched PEI (Zanta *et al.*, 1999). However, the intracellular trafficking and location of DNA release from these types of vectors are thought to differ greatly, with cationic lipids releasing DNA into the cytoplasm and branched PEI/DNA polyplexes remaining largely intact (Godbey *et al.*, 1999; Itaka *et al.*, 2004; Xu and Szoka, 1996). Branched PEI might even be expected to hinder NLS activity by masking the peptide and preventing interaction with transport receptors.

IV. CONCLUSIONS

Progress in the field of gene therapy is hindered by the lack of suitable delivery vectors. Synthetic vectors based on polycations are promising vectors for gene delivery as they are relatively safe and their biocompatibility and utility can be improved by the incorporation of hydrophilic polymers and ligands for targeting to specific cell types. However, the efficiency of gene transfer using synthetic vectors is hindered by both extracellular and intracellular barriers. Many studies have demonstrated that existing vectors can be modified or novel types of vectors developed to overcome specific barriers to transfection. The challenges now lie in integrating many of these different functionalities into a single type of efficient delivery vector. For example, histidine-rich reducible polycations have been designed to incorporate multiple features to overcome several barriers to transfection, including the ability to mediate escape from endocytic vesicles, efficient intracellular release of nucleic acids and a low toxicity profile. Hence, it is envisaged that these types of polycations are ideally suited to form the core of more sophisticated delivery vectors for *in vivo* applications. However, this goal may only be realistic when better techniques to study the interaction of synthetic vectors at each of these different barriers have been developed and quantitation of gene expression is not relied upon predominately as a measure of a vector's efficacy.

Acknowledgments

This work was supported by grants from the Biotechnology and Biological Sciences Research Council (BBSRC) and Cancer Research UK.

References

Adami, R. C., and Rice, K. G. (1999). Metabolic stability of glutaraldehyde cross-linked peptide DNA condensates. *J. Pharm. Sci.* **88,** 739–746.

Aoki, K., Furuhata, S., Hatanaka, K., Maeda, M., Remy, J. S., Behr, J. P., Terada, M., and Yoshida, T. (2001). Polyethylenimine-mediated gene transfer into pancreatic tumor dissemination in the murine peritoneal cavity. *Gene. Ther.* **8,** 508–514.

Bettinger, T., Carlisle, R. C., Read, M. L., Ogris, M., and Seymour, L. W. (2001). Peptide-mediated RNA delivery: A novel approach for enhanced transfection of primary and post-mitotic cells. *Nucleic Acids Res.* **29,** 3882–3891.

Bettinger, T., and Read, M. L. (2001). Recent developments in RNA-based strategies for cancer gene therapy. *Curr. Opin. Mol. Ther.* **3,** 116–124.

Boussif, O., Lezoualc'h, F., Zanta, M. A., Mergny, M. D., Scherman, D., Demeneix, B., and Behr, J. P. (1995). A versatile vector for gene and oligonucleotide transfer into cells in culture and *in vivo*: polyethylenimine. *Proc. Natl. Acad. Sci. USA* **92,** 7297–7301.

Branden, L. J., Mohamed, A. J., and Smith, C. I. (1999). A peptide nucleic acid-nuclear localization signal fusion that mediates nuclear transport of DNA. *Nat. Biotechnol.* **17,** 784–787.

Bremner, K. H., Seymour, L. W., Logan, A., and Read, M. L. (2004). Factors influencing the ability of nuclear localization sequence peptides to enhance non-viral gene delivery. *Bioconjug. Chem.* **15,** 152–161.

Bremner, K. H., Seymour, L. W., and Pouton, C. W. (2001). Harnessing nuclear localization pathways for transgene delivery. *Curr. Opin. Mol. Ther.* **3,** 170–177.

Brunner, S., Sauer, T., Carotta, S., Cotten, M., Saltik, M., and Wagner, E. (2000). Cell cycle dependence of gene transfer by lipoplex, polyplex and recombinant adenovirus. *Gene. Ther.* **7,** 401–407.

Ciolina, C., Byk, G., Blanche, F., Thuillier, V., Scherman, D., and Wils, P. (1999). Coupling of nuclear localization signals to plasmid DNA and specific interaction of the conjugates with importin alpha. *Bioconjug. Chem.* **10**, 49–55.

Coll, J. L., Chollet, P., Brambilla, E., Desplanques, D., Behr, J. P., and Favrot, M. (1999). In vivo delivery to tumors of DNA complexed with linear polyethylenimine. *Hum. Gene. Ther.* **10**, 1659–1666.

Cotten, M., Langle-Rouault, F., Kirlappos, H., Wagner, E., Mechtler, K., Zenke, M., Beug, H., and Birnstiel, M. L. (1990). Transferrin-polycation-mediated introduction of DNA into human leukemic cells: Stimulation by agents that affect the survival of transfected DNA or modulate transferrin receptor levels. *Proc. Natl. Acad. Sci. USA* **87**, 4033–4037.

Cullis, P. R., Chonn, A., and Semple, S. C. (1998). Interactions of liposomes and lipid-based carrier systems with blood proteins: Relation to clearance behaviour in vivo. *Adv. Drug. Deliv. Rev.* **32**, 3–17.

Dash, P. R., Read, M. L., Barrett, L. B., Wolfert, M. A., and Seymour, L. W. (1999). Factors affecting blood clearance and in vivo distribution of polyelectrolyte complexes for gene delivery. *Gene. Ther.* **6**, 643–650.

Dash, P. R., Read, M. L., Fisher, K. D., Howard, K. A., Wolfert, M., Oupicky, D., Subr, V., Strohalm, J., Ulbrich, K., and Seymour, L. W. (2000). Decreased binding to proteins and cells of polymeric gene delivery vectors surface modified with a multivalent hydrophilic polymer and retargeting through attachment of transferrin. *J. Biol. Chem.* **275**, 3793–3802.

Dash, P. R., Toncheva, V., Schacht, E., and Seymour, L. W. (1997). Synthetic polymers for vectorial delivery of DNA: Characterisation of polymer-DNA complexes by photon correlation spectroscopy and stability to nuclease degradation and disruption by polyanions in vitro. *J. Control Release* **48**, 269–276.

Davis, S. S., Hunneyball, I. M., Illum, L., Ratcliffe, J. H., Smith, A., and Wilson, C. G. (1985). Recent advances in the use of microspheres for targeted therapy. *Drugs Exp. Clin. Res.* **11**, 633–640.

Dorsett, Y., and Tuschl, T. (2004). siRNAs: Applications in functional genomics and potential as therapeutics. *Nat. Rev. Drug. Discov.* **3**, 318–329.

Erbacher, P., Roche, A. C., Monsigny, M., and Midoux, P. (1996). Putative role of chloroquine in gene transfer into a human hepatoma cell line by DNA/lactosylated polylysine complexes. *Exp. Cell. Res.* **225**, 186–194.

Fisher, K. D., Ulbrich, K., Subr, V., Ward, C. M., Mautner, V., Blakey, D., and Seymour, L. W. (2000). A versatile system for receptor-mediated gene delivery permits increased entry of DNA into target cells, enhanced delivery to the nucleus and elevated rates of transgene expression. *Gene. Ther.* **7**, 1337–1343.

Forrest, M. L., and Pack, D. W. (2002). On the kinetics of polyplex endocytic trafficking: Implications for gene delivery vector design. *Mol. Ther.* **6**, 57–66.

Gabizon, A., Shmeeda, H., and Barenholz, Y. (2003). Pharmacokinetics of pegylated liposomal Doxorubicin: Review of animal and human studies. *Clin. Pharmacokinet.* **42**, 419–436.

Godbey, W. T., Wu, K. K., and Mikos, A. G. (1999). Tracking the intracellular path of poly(ethylenimine)/DNA complexes for gene delivery. *Proc. Natl. Acad. Sci. USA* **96**, 5177–5181.

Gonçalves, C., Pichon, C., Guerin, B., and Midoux, P. (2002). Intracellular processing and stability of DNA complexed with histidylated polylysine conjugates. *J. Gene. Med.* **4**, 271–281.

Gregorevic, P., Blankinship, M. J., Allen, J. M., Crawford, R. W., Meuse, L., Miller, D. G., Russell, D. W., and Chamberlain, J. S. (2004). Systemic delivery of genes to striated muscles using adeno-associated viral vectors. *Nat. Med.* **10**, 828–834.

Hortobagyi, G. N., Ueno, N. T., Xia, W., Zhang, S., Wolf, J. K., Putnam, J. B., Weiden, P. L., Willey, J. S., Carey, M., Branham, D. L., Payne, J. Y., Tucker, S. D., Bartholomeusz, C., Kilbourn, R. G., De Jager, R. L., Sneige, N., Katz, R. L., Anklesaria, P., Ibrahim, N. K., Murray, J. L., Theriault, R. L., Valero, V., Gershenson, D. M., Bevers, M. W., Huang, L., Lopez-Berestein, G., and Hung, M. C. (2001). Cationic liposome-mediated E1A gene transfer to human breast and ovarian cancer cells and its biologic effects: A phase I clinical trial. *J. Clin. Oncol.* **19**, 3422–3433.

Itaka, K., Harada, A., Yamasaki, Y., Nakamura, K., Kawaguchi, H., and Kataoka, K. (2004). *In situ* single cell observation by fluorescence resonance energy transfer reveals fast intra-cytoplasmic delivery and easy release of plasmid DNA complexed with linear polyethylenimine. *J. Gene. Med.* **6**, 76–84.

Kichler, A., Leborgne, C., Coeytaux, E., and Danos, O. (2001). Polyethylenimine-mediated gene delivery: A mechanistic study. *J. Gene. Med.* **3**, 135–144.

Kircheis, R., Ostermann, E., Wolschek, M. F., Lichtenberger, C., Magin-Lachmann, C., Wightman, L., Kursa, M., and Wagner, E. (2002). Tumor-targeted gene delivery of tumor necrosis factor-alpha induces tumor necrosis and tumor regression without systemic toxicity. *Cancer Gene. Ther.* **9**, 673–680.

Lanford, R. E., Kanda, P., and Kennedy, R. C. (1986). Induction of nuclear transport with a synthetic peptide homologous to the SV40 T antigen transport signal. *Cell* **46**, 575–582.

Lechardeur, D., Sohn, K. J., Haardt, M., Joshi, P. B., Monck, M., Graham, R. W., Beatty, B., Squire, J., O'Brodovich, H., and Lukacs, G. L. (1999). Metabolic instability of plasmid DNA in the cytosol: A potential barrier to gene transfer. *Gene. Ther.* **6**, 482–497.

Ludtke, J. J., Zhang, G., Sebestyen, M. G., and Wolff, J. A. (1999). A nuclear localization signal can enhance both the nuclear transport and expression of 1 kb DNA. *J. Cell Sci.* **112**(Pt. 12), 2033–2041.

Maeda, H., Seymour, L. W., and Miyamoto, Y. (1992). Conjugates of anticancer agents and polymers: Advantages of macromolecular therapeutics *in vivo. Bioconjug. Chem.* **3**, 351–362.

Maruyama, H., Higuchi, N., Nishikawa, Y., Kameda, S., Iino, N., Kazama, J. J., Takahashi, N., Sugawa, M., Hanawa, H., Tada, N., Miyazaki, J., and Gejyo, F. (2002). High-level expression of naked DNA delivered to rat liver via tail vein injection. *J. Gene. Med.* **4**, 333–341.

Matsumura, Y., and Maeda, H. (1986). A new concept for macromolecular therapeutics in cancer chemotherapy: Mechanism of tumoritropic accumulation of proteins and the antitumor agent smancs. *Cancer Res.* **46**, 6387–6392.

Maul, G. G., and Deaven, L. (1977). Quantitative determination of nuclear pore complexes in cycling cells with differing DNA content. *J. Cell Biol.* **73**, 748–760.

McKenzie, D. L., Kwok, K. Y., and Rice, K. G. (2000). A potent new class of reductively activated peptide gene delivery agents. *J. Biol. Chem.* **275**, 9970–9977.

Monck, M. A., Mori, A., Lee, D., Tam, P., Wheeler, J. J., Cullis, P. R., and Scherrer, P. (2000). Stabilized plasmid-lipid particles: Pharmacokinetics and plasmid delivery to distal tumors following intravenous injection. *J. Drug. Target.* **7**, 439–452.

Niedzinski, E. J., Chen, Y. J., Olson, D. C., Parker, E. A., Park, H., Udove, J. A., Scollay, R., McMahon, B. M., and Bennett, M. J. (2003). Enhanced systemic transgene expression after nonviral salivary gland transfection using a novel endonuclease inhibitor/DNA formulation. *Gene. Ther.* **10**, 2133–2138.

Nielsen, P. E., Egholm, M., Berg, R. H., and Buchardt, O. (1991). Sequence-selective recognition of DNA by strand displacement with a thymine-substituted polyamide. *Science* **254**, 1497–1500.

Oupicky, D., Carlisle, R. C., and Seymour, L. W. (2001). Triggered intracellular activation of disulfide crosslinked polyelectrolyte gene delivery complexes with extended systemic circulation *in vivo. Gene. Ther.* **8**, 713–724.

Oupicky, D., Howard, K. A., Konak, C., Dash, P. R., Ulbrich, K., and Seymour, L. W. (2000). Steric stabilization of poly-L-Lysine/DNA complexes by the covalent attachment of semitelechelic poly [N-(2-hydroxypropyl)methacrylamide]. *Bioconjug. Chem.* **11**, 492–501.

Oupicky, D., Konak, C., Dash, P. R., Seymour, L. W., and Ulbrich, K. (1999a). Effect of albumin and polyanion on the structure of DNA complexes with polycation containing hydrophilic nonionic block. *Bioconjug. Chem.* **10**, 764–772.

Oupicky, D., Konak, C., and Ulbrich, K. (1999b). DNA complexes with block and graft copolymers of N-(2-hydroxypropyl)methacrylamide and 2-(trimethylammonio)ethyl methacrylate. *J. Biomater. Sci. Polym. Ed.* **10**, 573–590.

Oupicky, D., Ogris, M., Howard, K. A., Dash, P. R., Ulbrich, K., and Seymour, L. W. (2002a). Importance of lateral and steric stabilization of polyelectrolyte gene delivery vectors for extended systemic circulation. *Mol. Ther.* **5**, 463–472.

Oupicky, D., Ogris, M., and Seymour, L. W. (2002b). Development of long-circulating polyelectrolyte complexes for systemic delivery of genes. *J. Drug. Target.* **10**, 93–98.

Oupicky, D., Parker, A. L., and Seymour, L. W. (2002c). Laterally stabilized complexes of DNA with linear reducible polycations: Strategy for triggered intracellular activation of DNA delivery vectors. *J. Am. Chem. Soc.* **124**, 8–9.

Pichon, C., Goncalves, C., and Midoux, P. (2001). Histidine-rich peptides and polymers for nucleic acids delivery. *Adv. Drug. Deliv. Rev.* **53**, 75–94.

Pollard, H., Remy, J. S., Loussouarn, G., Demolombe, S., Behr, J. P., and Escande, D. (1998). Polyethylenimine but not cationic lipids promotes transgene delivery to the nucleus in mammalian cells. *J. Biol. Chem.* **273**, 7507–7511.

Pollard, H., Toumaniantz, G., Amos, J. L., Avet-Loiseau, H., Guihard, G., Behr, J. P., and Escande, D. (2001). Ca2+-sensitive cytosolic nucleases prevent efficient delivery to the nucleus of injected plasmids. *J. Gene. Med.* **3**, 153–164.

Read, M. L., Bremner, K. H., Oupicky, D., Green, N. K., Searle, P. F., and Seymour, L. W. (2003). Vectors based on reducible polycations facilitate intracellular release of nucleic acids. *J. Gene. Med.* **5**, 232–245.

Read, M. L., Singh, S., Ahmed, Z., Stevenson, M., Briggs, S. S., Oupicky, D., Barrett, L. B., Tirlapur, U., Berry, M., Preece, J. A., Logan, A., and Seymour, L. W. (2005). A versatile reducible polycation-based system for efficient delivery of a broad range of nucleic acid (submitted).

Ross, G. F., Bruno, M. D., Uyeda, M., Suzuki, K., Nagao, K., Whitsett, J. A., and Korfhagen, T. R. (1998). Enhanced reporter gene expression in cells transfected in the presence of DMI-2, an acid nuclease inhibitor. *Gene. Ther.* **5**, 1244–1250.

Schaffer, D. V., Fidelman, N. A., Dan, N., and Lauffenburger, D. A. (2000). Vector unpacking as a potential barrier for receptor-mediated polyplex gene delivery. *Biotechnol. Bioeng.* **67**, 598–606.

Sebestyen, M. G., Ludtke, J. J., Bassik, M. C., Zhang, G., Budker, V., Lukhtanov, E. A., Hagstrom, J. E., and Wolff, J. A. (1998). DNA vector chemistry: The covalent attachment of signal peptides to plasmid DNA. *Nat. Biotechnol.* **16**, 80–85.

Seymour, L. W. (1992). Passive tumor targeting of soluble macromolecules and drug conjugates. *Crit. Rev. Ther. Drug. Carrier. Syst.* **9**, 135–187.

Seymour, L. W., Miyamoto, Y., Maeda, H., Brereton, M., Strohalm, J., Ulbrich, K., and Duncan, R. (1995). Influence of molecular weight on passive tumour accumulation of a soluble macromolecular drug carrier. *Eur. J. Cancer.* **31A**, 766–770.

Sperinde, J. J., Choi, S. J., and Szoka, F. C.Jr. (2001). Phage display selection of a peptide DNase II inhibitor that enhances gene delivery. *J. Gene. Med.* **3**, 101–108.

Stopeck, A. T., Jones, A., Hersh, E. M., Thompson, J. A., Finucane, D. M., Gutheil, J. C., and Gonzalez, R. (2001). Phase II study of direct intralesional gene transfer of allovectin-7, an HLA-

B7/beta2-microglobulin DNA-liposome complex, in patients with metastatic melanoma. *Clin. Cancer. Res.* **7,** 2285–2291.

Tam, P., Monck, M., Lee, D., Ludkovski, O., Leng, E. C., Clow, K., Stark, H., Scherrer, P., Graham, R. W., and Cullis, P. R. (2000). Stabilized plasmid-lipid particles for systemic gene therapy. *Gene. Ther.* **7,** 1867–1874.

Terebesi, J., Kwok, K. Y., and Rice, K. G. (1998). Iodinated plasmid DNA as a tool for studying gene delivery. *Anal. Biochem.* **263,** 120–123.

Tokuda, Y., Watanabe, T., Omuro, Y., Ando, M., Katsumata, N., Okumura, A., Ohta, M., Fujii, H., Sasaki, Y., Niwa, T., and Tajima, T. (1999). Dose escalation and pharmacokinetic study of a humanized anti-HER2 monoclonal antibody in patients with HER2/neu-overexpressing metastatic breast cancer. *Br. J. Cancer.* **81,** 1419–1425.

Toncheva, V., Wolfert, M. A., Dash, P. R., Oupicky, D., Ulbrich, K., Seymour, L. W., and Schacht, E. H. (1998). Novel vectors for gene delivery formed by self-assembly of DNA with poly (L-lysine) grafted with hydrophilic polymers. *Biochim. Biophys. Acta* **1380,** 354–368.

Trubetskoy, V. S., Loomis, A., Slattum, P. M., Hagstrom, J. E., Budker, V. G., and Wolff, J. A. (1999). Caged DNA does not aggregate in high ionic strength solutions. *Bioconjug. Chem.* **10,** 624–628.

Uster, P. S., and Deamer, D. W. (1985). pH-dependent fusion of liposomes using titratable polycations. *Biochemistry* **24,** 1–8.

Wagner, E. (1998). Effects of membrane-active agents in gene delivery. *J. Control. Release* **53,** 155–158.

Wang, C. Y., and Huang, L. (1984). Polyhistidine mediates an acid-dependent fusion of negatively charged liposomes. *Biochemistry* **23,** 4409–4416.

Ward, C. M., Acheson, N., and Seymour, L. W. (2000). Folic acid targeting of protein conjugates into ascites tumour cells from ovarian cancer patients. *J. Drug. Target* **8,** 119–123.

Ward, C. M., Read, M. L., and Seymour, L. W. (2001). Systemic circulation of poly(L-lysine)/DNA vectors is influenced by polycation molecular weight and type of DNA: Differential circulation in mice and rats and the implications for human gene therapy. *Blood* **97,** 2221–2229.

Wattiaux, R., Laurent, N., Wattiaux-De Coninck, S., and Jadot, M. (2000). Endosomes, lysosomes: Their implication in gene transfer. *Adv. Drug. Deliv. Rev.* **41,** 201–208.

Wheeler, J. J., Palmer, L., Ossanlou, M., MacLachlan, I., Graham, R. W., Zhang, Y. P., Hope, M. J., Scherrer, P., and Cullis, P. R. (1999). Stabilized plasmid-lipid particles: Construction and characterization. *Gene. Ther.* **6,** 271–278.

Wolfert, M. A., Dash, P. R., Nazarova, O., Oupicky, D., Seymour, L. W., Smart, S., Strohalm, J., and Ulbrich, K. (1999). Polyelectrolyte vectors for gene delivery: Influence of cationic polymer on biophysical properties of complexes formed with DNA. *Bioconjug. Chem.* **10,** 993–1004.

Wolfert, M. A., Schacht, E. H., Toncheva, V., Ulbrich, K., Nazarova, O., and Seymour, L. W. (1996). Characterization of vectors for gene therapy formed by self-assembly of DNA with synthetic block co-polymers. *Hum. Gene. Ther.* **7,** 2123–2133.

Wolschek, M. F., Thallinger, C., Kursa, M., Rossler, V., Allen, M., Lichtenberger, C., Kircheis, R., Lucas, T., Willheim, M., Reinisch, W., Gangl, A., Wagner, E., and Jansen, B. (2002). Specific systemic non-viral gene delivery to human hepatocellular carcinoma xenografts in SCID mice. *Hepatology* **36,** 1106–1114.

Xu, Y., and Szoka, F. C.Jr. (1996). Mechanism of DNA release from cationic liposome/DNA complexes used in cell transfection. *Biochemistry* **35,** 5616–5623.

Yoo, G. H., Hung, M. C., Lopez-Berestein, G., LaFollette, S., Ensley, J. F., Carey, M., Batson, E., Reynolds, T. C., and Murray, J. L. (2001). Phase I trial of intratumoral liposome E1A gene therapy in patients with recurrent breast and head and neck cancer. *Clin. Cancer Res.* **7,** 1237–1245.

Zabner, J., Fasbender, A. J., Moninger, T., Poellinger, K. A., and Welsh, M. J. (1995). Cellular and molecular barriers to gene transfer by a cationic lipid. *J. Biol. Chem.* **270,** 18997–19007.

Zanta, M. A., Belguise-Valladier, P., and Behr, J. P. (1999). Gene delivery: A single nuclear localization signal peptide is sufficient to carry DNA to the cell nucleus. *Proc. Natl. Acad. Sci. USA* **96,** 91–96.

Zelphati, O., Liang, X., Hobart, P., and Felgner, P. L. (1999). Gene chemistry: Functionally and conformationally intact fluorescent plasmid DNA. *Hum. Gene. Ther.* **10,** 15–24.

Zhang, X., Dong, X., Sawyer, G. J., Collins, L., and Fabre, J. W. (2004). Regional hydrodynamic gene delivery to the rat liver with physiological volumes of DNA solution. *J. Gene. Med.* **6,** 693–703.

Zhang, X., Sawyer, G. J., Dong, X., Qiu, Y., Collins, L., and Fabre, J. W. (2003). The *in vivo* use of chloroquine to promote non-viral gene delivery to the liver via the portal vein and bile duct. *J. Gene. Med.* **5,** 209–218.

3

Pharmacokinetics of Plasmid DNA-Based Non-viral Gene Medicine

Makiya Nishikawa,* Yoshinobu Takakura,* and Mitsuru Hashida†
*Department of Biopharmaceutics and Drug Metabolism
Graduate School of Pharmaceutical Sciences
Kyoto University, Sakyo-ku, Kyoto 606-8501, Japan
†Department of Drug Delivery Research, Graduate School of
Pharmaceutical Sciences, Kyoto University
Sakyo-ku, Kyoto 606-8501, Japan

ABSTRACT

Non-viral gene therapy can be realized by optimization of the pharmacokinetic properties of both the vector and the encoded therapeutic protein. A major obstacle to its successful clinical application is the limited ability of plasmid

Advances in Genetics, Vol. 53
Copyright 2005, Elsevier Inc. All rights reserved.

0065-2660/05 $35.00
DOI: 10.1016/S0065-2660(05)53003-7

DNA, the most convenient gene-coding compound, to distribute within the body after in vivo administration. Under normal conditions, plasmid DNA and its non-viral vector complexes have difficulty in passing through various anatomical and biological barriers. These characteristics greatly limit the number and distribution of cells transduced with the vector, because transgene expression only occurs in cells that are reached by the vector. New approaches to the design of vectors as well as the methods of administration, such as electroporation and a hydrodynamic delivery, have increased the transgene expression in vivo, suggesting that improved distribution of plasmid DNA is possible by these approaches. In this chapter, the basic pharmacokinetic properties of naked plasmid DNA under normal conditions are first reviewed, then the properties of both naked and complexed plasmid DNA are discussed under conditions where significant transgene expression takes place.   © 2005, Elsevier Inc.

I. INTRODUCTION

A vector encoding a therapeutic gene can be a delivery vehicle for a protein, or its precursor form, because the protein translated from the gene encoded in the vector is the active form exhibiting the therapeutic effects after administration of the vector. Therefore, the pharmacokinetic properties of the translated product should directly influence the efficacy of any application of in vivo gene transfer.

If a protein is secreted from the transduced cells into the circulating blood where it exhibits its biological activity, the pharmacokinetic profile of the protein, not the vector, determines the effectiveness of in vivo gene transfer. This can be applied to various proteins including blood coagulation factors, erythropoietin, epidermal growth factor and hepatocyte growth factor. The concentration of these proteins in plasma, or in extracellular fluids, determines the therapeutic efficacy. These characteristics of the minor importance of the cell-type producing the protein offers investigators the option of choosing a suitable target for in vivo gene transfer. In gene therapy protocols for hemophiliacs, not only hepatocytes that produce the coagulation factors in healthy subjects, but also other cells such as fibroblasts or muscle cells have been investigated as target cells producing these factors (Herzog et al., 1999; Kay et al., 2000; Manno et al., 2003; Roth et al., 2001; Snyder et al., 1999). In such cases, the level and duration of transgene expression is closely linked to the efficacy of gene transfer. These parameters governing the expression can be a function of the characteristics of the cells transduced, the vector applied, and the administration method (Nishikawa and Hashida, 2002). For example, muscle cells generally produce very prolonged transgene expression compared with other cells, such as hepatocytes and vascular endothelial cells. In addition, physiological properties, such as blood flow rate to tissues and the structure of blood capillaries, can influence the therapeutic efficacy by altering the rate and extent of delivery of the protein from the transduced cells to the site of action. Therefore, the pharmacokinetic properties of the vector are extremely important in determining the efficacy of *in vivo* gene transfer for secretion proteins, although they have often received little attention in the past.

In contrast to this, intracellular proteins function at an appropriate place within a specific type of cell, therefore, they should be introduced to the appropriate cell by *in vivo* gene transfer. Because the translated protein may not distribute to the outside of the cells transduced, the pharmacokinetic properties of the protein have little effect on therapeutic efficacy. On the other hand, the pharmacokinetic properties of the vector are extremely important; the distribution of the vector to the target cell after administration is essential for transgene expression in the target cell. Genes for metabolic enzymes, structural proteins and transporters need to be delivered to the right target, otherwise, no therapeutic benefits will be obtained although some transgene expression is obtained. Therefore, vectors encoding those intracellular proteins require a well-controlled *in vivo* delivery system.

Thus, the pharmacokinetics—which is the study of the absorption, distribution, metabolism and excretion of vectors for *in vivo* gene transfer—is of significant importance for the development of useful vectors and protocols for *in vivo* gene therapy. Generally speaking, plasmid DNA-based non-viral vectors offer the advantages of safety and versatility over viral vectors (Nishikawa and Huang, 2001). So far, several promising results involving gene transfer using plasmid DNA-based approaches have been reported in preclinical and clinical settings (Morishita *et al.*, 2004). However, these are limited to gene transfer of secretion proteins where the pharmacokinetic behavior of vectors has less significance on the final output. In contrast, few significant improvements have been reported for gene therapy involving intracellular proteins, indicating the difficulties in controlling the pharmacokinetics of vectors, including plasmid DNA.

Gene transfer is expected to occur in cells reached by the vector directly or via the blood circulation. Although conventional pharmaceutical compounds are given by various routes, (e.g., oral, nasal, pulmonary, conjunctival, transdermal and rectal administration), vectors including plasmid DNA are rarely administered by these routes with an expectation of systemic absorption followed by gene transfer at sites in the body other than the administration site. In addition, in contrast to conventional chemical compounds, plasmid DNA is metabolized into fragments and excreted, and both these processes have been shown to contribute little to the overall therapeutic efficacy of plasmid DNA-based approaches. Therefore, distribution is the most important process in the pharmacokinetics of plasmid DNA.

In this chapter, we summarize the tissue distribution properties of plasmid DNA after its injection into the systemic circulation or tissues, in the free or complexed form with targeted or non-targeted non-viral vectors. The contribution of macrophages to the distribution of naked plasmid DNA is discussed in relation to the involvement of scavenger receptor-like mechanisms in the cellular uptake. Then, several options for the improved delivery of plasmid DNA are reviewed with an emphasis on changes in the pharmacokinetics of plasmid DNA.

II. HOW TO TRACE PLASMID DNA *IN VIVO*

Plasmid DNA is chemically stable and easily handled. However, it is unstable after it enters the body. Incubation of plasmid DNA in freshly prepared mouse whole blood results in its degradation with an apparent half-life of about 10 min (Kawabata *et al.*, 1995). Furthermore, faster degradation of plasmid DNA has been detected after intravenous injection into mice. Nucleases mainly contribute to the degradation of plasmid DNA in the body; therefore, the degradation rate is a function of the type and amount of these enzymes in the compartment where the plasmid DNA distributes. For example, naked plasmid DNA injected into skeletal muscle may be considered stable for up to 4 h after intramuscular injection (Satkauskas *et al.*, 2001). As indicated in these previous studies, following administration, plasmid DNA is continuously degraded over time. Breaking down the linkages in plasmid DNA changes its size, a very important factor determining its pharmacokinetic behavior.

Degradation of plasmid DNA *in vivo* makes it very complicated to analyze the tissue distribution of 'biologically active' plasmid DNA. In an actual fact, the boundary between the active and inactive forms of DNA is not clear; degraded, linearized DNA fragments can be biologically active if they possess at least the minimal essential elements for transcription. Furthermore, the transcriptional activity of plasmid DNA depends on its functional form. Adami *et al.* (1998) reported that the open circular and linear forms of plasmid DNA were 90 and 10% as efficient as the supercoiled form. Although monitoring only the 'active' plasmid DNA is important for its therapeutic efficacy, the distribution of total plasmid DNA including the intact and degraded forms is discussed in most publications.

To avoid such a mix-up, the distribution of plasmid DNA in plasma can be examined by using analytical methods that can separate the different structural forms of plasmid DNA. A typical method is the separation of the different forms of plasmid DNA by agarose gel electrophoresis. Houk *et al.* (2001) reported that naked plasmid DNA injected into rats is successively converted from the supercoiled to the open circular form, then the linear form with time. They reported that each form of the DNA had its own unique pharmacokinetic profile of elimination from plasma. Although this method can be used to accurately evaluate the plasma clearance of plasmid DNA, it is time-consuming, and is difficult to use for analyzing the distribution of DNA to tissues and organs. Another method is the polymerase chain reaction (PCR)-based detection of plasmid DNA (Oh et al., 2001). In this case, a selected site of the DNA is amplified by PCR, so the integrity of the fragment is guaranteed. Because of the amplification, the sensitivity of this method is very high but, again, it is also time-consuming, and can only be easily applied to determine plasmid DNA in plasma or other body fluids, and not tissues. Real-time PCR can achieve a more rapid and quantitative determination of plasmid DNA in body fluids.

Using radioisotopes or fluorescent probes is one of the easiest, and most frequently used methods to trace the tissue distribution of plasmid DNA. Several techniques have been developed to this end. These include: ^{32}P-labeling by nick translation (Piatyszek et al., 1988; Rigby et al., 1977), covalent coupling of a fluorescent dye (Neves et al., 2000; Slattum et al., 2003), radioiodination of cytidine (Terebesi et al., 1998), application of a fluorescent peptide nucleic acid clamp (Zelphati et al., 1999), and metabolic labeling using ^{3}H-thymidine 5′-triphosphate (Wasan et al., 1996). As discussed above, an important feature that should be taken into consideration is the changes in the structure of plasmid DNA produced by labeling. Recently, several labeling methods that do not significantly affect the transfection efficacy of plasmid DNA have been proposed (Nishikawa et al., 2003; Slattum et al., 2003; Zelphati et al., 1999). The use of these techniques is suitable for studying the tissue distribution of plasmid DNA. We developed a residualizing radiolabel for plasmid DNA, in which ^{111}In is chelated to diethylenetriaminepentaacetic acid that is covalently conjugated to plasmid DNA through 4-[p-azidosalicylamido]butylamine (Nishikawa et al., 2003). This labeling preserved the overall structure of plasmid DNA and its transcriptional activity was 40–98% that of the original, depending on the number of adducts introduced. Furthermore, compared with its ^{32}P-labeled counterpart, ^{111}In-labeled naked plasmid DNA showed a prolonged retention of radioactivity in the liver, the major organ taking up the DNA from the circulation, after intravenous injection into mice. This slow release of radioactive metabolites makes it easy to estimate the tissue uptake of plasmid DNA as well as its complex. We have shown that the level of transgene expression in the lung following the administration of polyethyleneimine/^{111}In-plasmid DNA complexes correlates well with the amount of radioactivity in the organ. Thus, residualizing radiolabels, which have been developed for proteins (Ali et al., 1988; Deshpande et al., 1990; Thorpe et al., 1993), would be useful for evaluating the tissue distribution of plasmid DNA and developing effective non-viral delivery methods.

III. PHARMACOKINETIC FEATURES OF NAKED PLASMID DNA AFTER INTRAVASCULAR INJECTION

It had been generally considered that naked plasmid DNA is unable to produce any significant transgene products *in vivo* due to its inability to enter cells and to its susceptibility to enzymatic degradation by various extracellular or intracellular nucleases (Barry *et al.*, 1999; Lechardeur *et al.*, 1999; Lew *et al.*, 1995; Pollard *et al.*, 2001). For example, when applied systemically via the vascular system, naked plasmid DNA is pharmacologically inactive and produces very little transgene expression, if any, because it is rapidly scavenged and degraded by the liver nonparenchymal cells, predominantly by Kupffer cells and the liver sinusoidal endothelial cells, as demonstrated in our own series of studies (Hisazumi *et al.*, 2004; Kawabata *et al.*, 1995; Kobayashi *et al.*, 2001).

This was changed by Wolff *et al.* (1990), who showed that a direct injection of naked plasmid DNA into mouse skeletal muscle resulted in significant transgene expression. Furthermore, the group also obtained significant intravascular gene transfer with vector-free naked plasmid DNA by injecting it in a large-volume solution. They reported that a high level of transgene expression could be obtained in mouse liver by injecting naked plasmid DNA in hyperosmotic solution into the portal vein with transient occlusion of the outflow (Budker *et al.*, 1996). More recently, in 1999, Liu *et al.* (1999) and Zhang *et al.* (1999) reported an innovative finding that an astonishingly high level of transgene expression could be obtained in mouse liver and other major organs by a simple intravenous injection of naked plasmid DNA via the tail vein using a large volume of saline introduced at a high velocity.

Thus, naked plasmid DNA can be an effective non-viral vector with the advantage of simplicity, if a suitable method of administration is used. The differences in transgene expression observed when different administration methods are applied correlate with the altered pharmacokinetic features of plasmid DNA. Because plasmid DNA is a huge macromolecule with a high molecular weight of about 2,000 kD or more and a strong anionic charge, its distribution within the body is greatly limited. Therefore, the development of better non-viral vectors or administration methods requires an understanding of the fate of plasmid DNA *in vivo*. First, the tissue distribution characteristics of naked plasmid DNA are discussed in relation to its method of administration: normal intravenous injection, large-volume injection, the so-called hydrodynamics-based procedure (Liu *et al.*, 1999), and the procedure involving local application at a particular site.

A. Normal procedure

When naked ^{32}P-plasmid DNA was injected as a bolus into the tail vein of mice at a dose of 1 mg/kg (20 μg/20 g-body weight mouse), it rapidly disappeared from plasma and 60–70% of the injected dose was recovered in the liver within 5 min of injection (Kawabata et al., 1995). The pharmacokinetic analysis of the tissue distribution based on a clearance concept (Nishikawa et al., 1996) revealed that the apparent hepatic uptake clearance is close to the hepatic plasma flow rate. Under these conditions, naked plasmid DNA resulted in no transgene expression in major organs. This would be attributed to the extensive uptake and degradation of plasmid DNA by the liver nonparenchymal cells, such as Kupffer cells and liver sinusoidal endothelial cells. The hepatic uptake of plasmid DNA was dependent on the concentration and decreased on increasing the dose. In addition, we showed that the hepatic uptake of plasmid DNA is inhibited by calf thymus DNA, polyinosinic acid, dextran sulfate, and heparin, but not by polycytidylic acid and chondroitin sulfate (Kawabata et al., 1995; Kobayashi et al., 2001; Yoshida et al., 1996). These findings indicate that plasmid DNA is taken up by scavenger receptors, which recognize polyanions in a charge- and/or structure-dependent manner (Terpstra et al., 2000). However, we excluded the possibility of class A scavenger receptors (SRA) being involved in the uptake, based on the tissue distribution and uptake experiments using cultured macrophages from SRA-knockout mice (Takakura et al., 1999). Another study using SRA-knockout mice supported the conclusion that SRA are not involved in the uptake of plasmid DNA (Zhu et al., 2001).

The detailed characteristics of the uptake of plasmid DNA were examined using cultured cells. Various cells have the ability to take up plasmid DNA in a concentration- and temperature-dependent manner. The major cells responsible for the uptake of plasmid DNA are macrophages, as indicated in tissue distribution studies of plasmid DNA in which Kupffer cells, liver resident macrophages, make a large contribution to the clearance of plasmid DNA after systemic administration. Under in vitro culture conditions, mouse peritoneal macrophages efficiently took up plasmid DNA (Takagi et al., 1998). Liver sinusoidal endothelial cells, another type of liver nonparenchymal cell, also possess the ability to scavenge a variety of polyanions circulating in the blood (Yamasaki et al., 2002, 2003). A recent study of the tissue distribution of plasmid DNA in rats indicated that liver sinusoidal endothelial cells take up plasmid DNA as effectively as Kupffer cells (Hisazumi et al., 2004). Liver sinusoidal endothelial cells isolated from rats showed a significant uptake and degradation of plasmid DNA in culture. Dendritic cells (DCs), another population of immune cells, are very important as far as both innate and acquired immunity are concerned. Again, the DC cell line, DC2.4, exhibits extensive uptake and degradation of plasmid DNA (Yoshinaga et al., 2002). Primary mouse DCs also exhibited similar processing of plasmid DNA to this cell line. We also observed that bovine brain microvessel endothelial cells, which constitute the blood-brain barrier, are also able to recognize plasmid DNA (Nakamura et al., 1998).

When the uptake of plasmid DNA by any of these types of cells was examined in the presence of neutral or anionic polymers, an almost identical profile of inhibition was observed. The uptake as well as binding were significantly inhibited by excess plasmid DNA, polyinosinic acid, dextran sulfate or heparin, but not by polycytidylic acid and dextran (Hisazumi et al., 2004; Nakamura et al., 1998; Takagi et al., 1998; Yoshinaga et al., 2002). This inhibition pattern resembles closely that of the *in vivo* study described above, supporting the hypothesis that these cells, i.e., macrophages, liver sinusoidal endothelial cells and DCs, make a major contribution to the clearance of plasmid DNA *in vivo*. The exact mechanisms underlying the uptake, including the receptors involved, are yet to be identified. Because of the small number of DCs in the body, their contribution to the distribution of plasmid DNA will be less significant than those of macrophages and liver sinusoidal endothelial cells.

B. Hydrodynamics-based procedure

Hydrodynamics-based gene delivery, involving a large-volume and high-speed intravenous injection of naked plasmid DNA, gives a significantly high level of transgene expression *in vivo* (Liu et al., 1999). Because of the simplicity and the extraordinary efficiency of transgene expression among the current non-viral methods, it has attracted a lot of attention and has been used very frequently as an efficient, simple, and convenient transfection method for laboratory animals. Until recently, however, there was little published information on the pharmacokinetics of the injected plasmid DNA molecules and of the detailed mechanisms underlying the efficient gene transfer.

We investigated the *in vivo* pharmacokinetic characteristics of naked plasmid DNA in mice undergoing the hydrodynamics-based procedure (Kobayashi et al., 2001), in comparison with the normal procedure (up to 200 μl for a 20-g mouse). The time-course profile and the degree of hepatic accumulation of ^{32}P-plasmid DNA were very similar for the hydrodynamics-based and normal procedures: In both cases, plasmid DNA was rapidly eliminated from the blood circulation and taken up mainly by the liver. While the hepatic accumulation profiles of ^{32}P-plasmid DNA were almost identical for the two injection procedures, marked transgene expression could be achieved by the hydrodynamics-based procedure but not by the normal one. The results of our confocal microscopic studies of liver sections indicated widespread intrahepatic distribution of fluorescein-labeled plasmid DNA following the hydrodynamics-based procedure. In addition, polyanions such as poly inosinic acid, dextran sulfate and heparin, which significantly inhibit the hepatic uptake of plasmid DNA injected by the normal procedure, did not affect the hepatic uptake of plasmid DNA given by the hydrodynamics-based procedure. These results indicate that the hepatic uptake process of the plasmid DNA is different from that following normal intravenous injection where receptor-like mechanisms are most likely involved (Emlen et al., 1988; Kawabata et al., 1995; Kobayashi et al., 2001; Yoshida et al., 1996). Under normal conditions, the plasmid DNA molecules, injected slowly via the tail vein with a conventional volume of solution, enter the blood stream and are carried to the heart, and then distributed systemically. Due to the large amount of nucleases in the blood and other compartments such as on the surface of tissues (Emlen et al., 1988), plasmid DNA injected by the normal procedure is likely to be rapidly degraded in the circulation and subsequently in liver cells after being recognized and taken up by liver nonparenchymal cells. On the other hand, part of the plasmid DNA injected by the hydrodynamics-based procedure is directly exposed to the liver cells and some of it is taken up by the cells in intact form before being mixed with blood. This may account for the observation that a high level of transgene expression can be obtained by the hydrodynamics-based procedure, but not by the normal one. In an actual fact, the persistent presence of a significant level of intact plasmid DNA has been demonstrated in the liver (Liu et al., 1999). Also, delayed elimination of plasmid DNA from the plasma pool was observed in the hydrodynamics-based procedure, supporting the reduced degradation of plasmid DNA by nucleases (Kobayashi et al., 2001).

Until recently, little was known about the mechanisms underlying efficient gene transfer by this procedure. Liu et al. (1999) demonstrated that a rapid injection and a large volume of plasmid DNA solution were required to obtain a high level of transgene expression. Following the first reports of the large-volume tail vein injection (Liu et al., 1999; Zhang et al., 1999), Budker et al. (2000) hypothesized that the cellular uptake mechanism of naked pDNA involved an active, receptor-mediated process. In addition, Lecocq et al. (2003) demonstrated in a subcellular distribution study using differential centrifugation methods that ^{35}S-plasmid DNA remained bound to the outside surface of the plasma membrane for at least 1 h after the hydrodynamics-based procedure, supporting the hypothesis that plasmid DNA was internalized slowly via a specific mechanism. On the other hand, we demonstrated in our in vivo pharmacokinetic studies of naked plasmid DNA, involving the normal or hydrodynamics-based procedure, that the hepatic uptake process appeared to be nonspecific (Kobayashi et al., 2001). Liu et al. reported that β-galactosidase and Evans blue were efficiently delivered to hepatocytes by the hydrodynamics-based procedure (Zhang et al., 2004). Notably, in their electron microscopic observations, identifiable membrane defects or pores were detected in the hepatocytes, which were generated by the high-pressure solution. Based on these findings, they concluded that the hydrodynamics-based gene transfer was a nonspecific physical process. We also demonstrated transient hyperpermeability in the hepatic cellular membrane when using the hydrodynamics-based procedure (Kobayashi et al., 2004a). Propidium iodide was effectively incorporated by the liver cells following the hydrodynamics-based procedure and the green fluorescent protein expressed beforehand and accumulating internally in the cytosol was dramatically eliminated from hepatocytes following a large-volume injection of saline. These results suggest a facilitated permeation of propidium iodide and green fluorescent protein through the cell membrane because these molecules are not supposed to cross the plasma membrane of viable cells. Nonspecificity in the cellular uptake process of plasmid DNA was further confirmed by a competitive study. A saturable amount of empty vector did not inhibit transgene expression in the liver following the hydrodynamics-based procedure (Kobayashi et al., 2004a). Furthermore, fluorescein-labeled polystyrene microspheres of 50 nm in diameter were delivered into the intracellular compartment of hepatocytes following the hydrodynamics-based procedure in mice (Kobayashi et al., 2004b).

C. Hydrodynamics at local sites

The principle of the hydrodynamics-based procedure is reproducible and applicable to an organ-restricted gene delivery method, where plasmid DNA is injected *in situ* into tissue-associated vessels, as demonstrated earlier by Wolff's group (Budker et al., 1996, 1998; Zhang et al., 1997, 2001). Zhang et al. (1997) showed that substantial amounts of transgene products were obtained by an injection of naked plasmid DNA-containing solution into the afferent and efferent vessels (i.e., the portal vein, hepatic vein, and bile duct) of the liver in mice, rats, and dogs. The hindlimb muscles were also successfully targeted by an intra-arterial injection of a large volume of naked plasmid DNA into the femoral artery (Budker et al., 1998; Liang et al., 2004; Zhang et al., 2001). The amounts of plasmid DNA in hindleg muscles of mice were much greater after a large-volume injection of naked plasmid DNA than after a normal injection, suggesting that changes take place in the tissue distribution of plasmid DNA (Liang et al., 2004).

IV. PHARMACOKINETIC FEATURES OF PLASMID DNA/NON-VIRAL VECTOR COMPLEX AFTER INTRAVENOUS INJECTION

Plasmid DNA is a huge macromolecule with a strong negative charge, and it has very limited access to the nucleus when administered by the normal procedure as described above. To facilitate the binding to cells, cationic vectors are often used and optimization of the structure and function is a major challenge as far as non-viral vector (carrier) development is concerned.

Formation of a cationic complex of plasmid DNA greatly increases its interaction with the lung endothelial cells upon intravenous injection (Liu *et al.*, 1997; Mahato *et al.*, 1995). Not only the uptake but also the transgene expression is also increased in these cells after intravenous injection of cationic complexes. This is the result of a complicated series of events occurring *in vivo* after intravenous injection of plasmid DNA complex. Serum proteins (Li *et al.*, 1999) and blood cells (Sakurai *et al.*, 2001a,b) have been reported to affect the tissue distribution of intravascularly administered plasmid DNA complex. While plasmid DNA complex is able to avoid first-pass filtration in the lung, only a relatively small DNA complex can pass through the blood vessels and directly interact with the parenchymal cells of each tissue.

A. Lipoplex

An enormous number of cationic lipid/liposome systems have been developed to improve the transfection efficiency of plasmid DNA, some of which have been summarized in a recent publication (Audouy *et al.*, 2002). Cationic liposomes associate with plasmid DNA via an electrostatic interaction, resulting in the formation of a complex called lipoplex (Felgner *et al.*, 1997). The driving force for lipoplex to introduce genes into cells is its electrostatic binding to negatively-charged cellular membranes followed by endocytotic uptake. Some studies have shown that co-lipids, so-called helper lipids, in cationic liposomes are important determinants of the transfection efficiency.

In general, lipoplex is rapidly cleared from the blood circulation after intravenous injection and very little (only a few percent) of the lipoplex remains in the blood 10 min after injection (Mahato *et al.*, 1995, 1998). It immediately distributes to the lung, because it is the first-pass organ after intravenous injection. Then, significant amounts of lipoplex, once captured by the lung, redistribute to the liver. Several reports have suggested that intravenously administered lipoplexes are mainly taken up by liver Kupffer cells (Litzinger *et al.*, 1996; Mahato *et al.*, 1995; McLean *et al.*, 1997). The functional deletion of these cells by gadolinium chloride resulted in a reduced amount of lipoplex taken up by the liver (Sakurai *et al.*, 2002), indicating a significant contribution of Kupffer cells to the pharmacokinetics of lipoplex.

These characteristics of lipoplex, as far as tissue distribution is concerned, do not merely result from the physicochemical properties of lipoplex. A positively charged lipoplex surface would attract various types of blood components. Under *in vitro* conditions, the presence of serum has been shown to strongly affect, in most cases reduce, lipoplex-mediated gene transfer (Li *et al.*, 1999). Yang and Huang (1997) have shown that the inactivation of lipoplex by serum is mediated by negatively-charged serum proteins and inactivation can be overcome by increasing the positive charge of the lipoplex. As shown in these *in vitro* studies, a cationic lipoplex can attract a variety of blood components, which results in alterations in the overall physicochemical properties of the lipoplex. Therefore, such an interaction should alter the tissue distribution of lipoplex following intravenous injection. Li *et al.* (1998) has demonstrated that lipidic vectors become negatively charged after exposure to mouse serum and increase in size. The choice of a helper lipid plays an important role in determining the potential of the lipoplex. In particular, when dioleoylphosphatidylethanolamine (DOPE) is used as a helper lipid, the *in vivo* transfection activity of the lipoplex is much lower than that achieved with a formulation containing cholesterol as a neutral helper lipid. One reason for this is that the cationic lipidic vectors containing DOPE recruit large amounts of serum proteins. Interactions of lipoplex with serum components lead to disintegration of the lipoplex, followed by the release and degradation of plasmid DNA (Li *et al.*, 1999). Release of plasmid DNA from the lipoplex might be due to the interaction of lipoplex with highly negatively charged molecules in serum, as proposed by Xu and Szoka (1996). A DOPE-containing formulation, in spite of its initial efficient accumulation in the lung, is poorly retained there because of its rapid disintegration. This, together with the rapid degradation of plasmid DNA, explains why DOPE-containing formulations do not result in satisfactory transfection of cells *in vivo*.

It is still not known which serum proteins are the major components that are responsible for vector disintegration. Li *et al.* (1998, 1999) have reported that a cationic lipid-protamine-DNA complex was enriched with a protein with a molecular weight corresponding to serum albumin and that further interaction of the complex with serum proteins led to its disintegration, and the release and degradation of plasmid DNA. DOPE-containing formulations disintegrated more rapidly when incubated in serum than cholesterol-containing ones. Barron *et al.* (1998) have shown that a lipoplex interacts with serum complements upon intravenous injection in mice, but this interaction does not affect the transfection efficiency in the lung nor the tissue distribution of the lipoplex. In addition to serum proteins, erythrocytes are also involved in the interaction with lipoplex *in vivo* (Sakurai *et al.*, 2001a,b), an interaction which is also highly dependent on the helper lipid within the lipoplex, DOPE or cholesterol. Lipoplexes without DOPE, in which lipids have a stable and rigid lamellar structure, do not induce fusion between erythrocytes. Although these complexes bind to erythrocytes, a fraction of the erythrocyte-bound complex could still be active as far as transfection is concerned. During its passage through the lung capillaries, some complexes may dissociate from erythrocytes without any loss of lipid components and change in their structure. Then, the complex released may be transferred to the lung endothelial cells and internalized by endocytosis. On the other hand, lipids in the N-[1-(2,3-dioleyloxy)propyl]-n, n, n-trimethylam-monium chloride (DOTMA)/DOPE complex have a highly curved structure with high fluidity, and induce fusion of erythrocytes within a short period. Following fusion, the lipid components of the liposomes are extracted by erythrocytes, resulting in a reduced potential to bind to endothelial cells. These observations also apply to the reduction in transfection activity. Following *in vivo* administration, we also demonstrated that lipoplex binds to blood cells and the interaction between the lipoplex and erythrocytes is important for efficient *in vivo* gene transfer (Sakurai *et al.*, 2001a). DOPE-containing lipoplex forms aggregates with erythrocytes, which will embolize in the capillaries of the lung following intravenous injection. However, erythrocytes bound to cholesterol-containing lipoplex dissociate so that the free lipoplex is able to accumulate in the lung.

To compensate for the lack of specificity of the electrostatic interaction of lipoplex, ligands are introduced into cationic liposomes and sugars are the most extensively investigated ligands so far (Kawakami *et al.*, 2002). The incorporation of sugars into lipoplex formulations increases the amount of the lipoplexes delivered to the cells having the corresponding receptors, such as hepatocytes and liver nonparenchymal cells (Kawakami *et al.*, 2000a,b). However, hepatocyte-targeted delivery of galactosylated lipoplex was achieved only by its intraportal injection, probably due to the nonspecific interaction of the galactosylated lipoplex with serum components as described above.

B. Polyplex

Cationic polymers are another class of non-viral vectors that can be used to increase gene delivery and transfer to target cells *in vivo*. Various types of polymers have been examined with respect to their ability to offer protection from nuclease degradation, deliver to target cells, and increase the transfection efficiency of plasmid DNA. Cationic polymers can condense plasmid DNA more efficiently than cationic liposomes, and this would be beneficial in controlling the tissue distribution of plasmid DNA complexes. These include: poly-L-lysine (PLL), poly-L-ornithine, polyethyleneimine (PEI), chitosan, starburst dendrimer and other novel synthetic polymers. Plasmid DNA/cationic polymer complex, or polyplex, is believed to be taken up by cells via an endocytotic pathway, so its transfection efficiency depends on the release of plasmid DNA into the cytoplasm after cellular uptake. The amount of transgene expression depends on a number of variables including the nature of the polymer used and the amount of plasmid DNA delivered. We demonstrated that the amount of ^{111}In-plasmid DNA delivered to the lung was proportional to the transgene expression in the organ when PEI/plasmid DNA was injected into mice using polyplexes at different N/P ratios (Nishikawa *et al.*, 2003).

As observed with lipoplexes, formation of a cationic complex can increase the interaction of plasmid DNA with various cells. When non-targeted, cationic polyplex is injected intravenously, the interaction with the lung endothelial cells is most marked. Then, transgene expression is also high in these cells. If large aggregates are formed after intravenous injection of polymer/plasmid DNA complex, they will also be trapped by the lung. In some cases, the mixing of plasmid DNA and a cationic carrier in vitro results in very large aggregates with a size close to that of the capillaries (Finsinger et al., 2000; Nishikawa et al., 1998). To prevent aggregation, the surface of the complex can be modified with hydrophilic compounds such as polyethylene glycol. Ogris et al. (1999) have shown that when incubated with plasma, the transferrin-PEI/plasmid DNA complex undergoes aggregation, which leads to reduced delivery to the target. PEGylation appears to be a useful method for prolonging the blood circulation of a polyplex after systemic administration, and PEGylated polyplex has resulted in gene transfer to a tumor without significant toxicity after intravenous injection into tumor-bearing mice (Kircheis et al., 2001). In addition to PEG, other hydrophilic polymers have been shown to extend the circulation time of polyplex (Oupicky et al., 2002; Verbaan et al., 2004). However, it might also shield the cationic charge of the complex that is the driving force for interaction with target cells. To compensate for the loss of affinity for the target cell, the use of ligands is a good way to achieve efficient delivery of polyplex.

The tissue distribution of polyplex is more easily controlled than that of lipoplex, because there is less interaction between the cationic polymers and serum components. Therefore, active targeting to a specific population of cells in the body was attempted as early as 1988 by Wu and Wu. Polymers such as PLL and PEI have been covalently modified with targeting ligand, and these include asialoglycoproteins (Wu and Wu, 1988), carbohydrates (Perales et al., 1994), transferrin (Kircheis et al., 1999), folate (Ward et al., 2002), and antibody (Li et al., 2000). We carried out pharmacokinetic evaluations of the tissue distribution of galactosylated PLL (Gal-PLL)/plasmid DNA complex following intravenous injection in mice (Nishikawa et al., 1998). As mentioned above, naked plasmid DNA is rapidly taken up by the liver. Cell fractionation and confocal imaging of fluorescein-labeled pDNA following intravenous injection into mice showed that plasmid DNA is mainly taken up by sinusoidal cells such as Kupffer cells and liver sinusoidal endothelial cells (Kawabata et al., 1995; Kobayashi et al., 2001). Because the uptake by these cells seems to be mediated by the strong negative charge of plasmid DNA and its clearance is very high, it is very important to mask the negative charge of plasmid DNA in order to control its tissue distribution. After intravenous injection of Gal-PLL/^{32}P-plasmid DNA complex, the hepatic uptake clearance was much greater than that by any other tissue. However, the physicochemical properties of

Gal-PLL used for the complexation markedly affected the pharmacokinetics of the polyplexes. The clearance values demonstrate that polyplexes with a large Gal-PLL (13 or 29 kD for the molecular weight of PLL) have a greater hepatic (target) clearance than those with a small Gal-PLL (1.8 kD), which failed to achieve efficient delivery of plasmid DNA to hepatocytes probably due to complex dissociation before reaching the target. Similar effects of molecular size have also been obtained with galactosylated PEI (Gal-PEI) (Morimoto et al., 2003). Although the transfection potential was highest for PEI with the smallest molecular weight of 1.8 kD, the polyplex composed of Gal-PEI$_{1800}$ and plasmid DNA was the least effective as far as in vivo transgene expression was concerned.

These polyplexes with cationic vectors are internalized by cells via endocytosis resulting in lysosomal degradation. This intracellular pathway greatly limits the efficiency of gene transfer by this approach. In addition to controlling the in vivo pharmacokinetics by using a carrier molecule like Gal-PLL, the control of intracellular sorting of plasmid DNA is a good approach to increasing gene transfer at the target. Wagner et al. (1992) demonstrated increased transgene expression in cultured cells following the addition of fuso-genic peptides, derived from influenza virus hemmaglutinin subunit HA-2, to plasmid DNA complexes. We attached a fusogenic peptide to hepatocyte-targetable polymer and obtained improved transgene expression in the liver, indicating that the peptide also works in whole animals to, at least partially, avoid intracellular degradation (Nishikawa et al., 2000).

V. PHARMACOKINETIC FEATURES AFTER TISSUE INJECTION

Direct tissue injection of plasmid DNA or its complex is used to obtain the expression of a secretion protein that will exhibit its therapeutic activity after entering the blood circulation, or an intracellular protein that will work within the cells in the injected tissue. As stated above, the tissue distribution of 'vector' is only important in the latter case. Because of extensive degradation of naked plasmid DNA entering into the blood circulation, transgene expression after tissue injection is normally highly tissue-specific.

The distribution of plasmid DNA within the injected tissue is a major factor that should be considered as far as the efficacy of in vivo gene transfer after tissue injection is concerned. A needle injection of naked plasmid DNA into various tissues, including skeletal muscle (Wolff et al., 1990), heart (Lin et al., 1990), liver (Hickman et al., 1994), brain (Ono et al., 1990), skin (Raz et al., 1994), thyroid (Sikes et al., 1994), urological organs (Yoo et al., 1999), and tumors (Nomura et al., 1999; Plautz et al., 1993), results in significant transgene expression in the tissue injected. However, cells with transgene expression were normally found only in close proximity to the track of the needle injection

(Hickman *et al.*, 1994; O'Hara *et al.*, 2001). Therefore, dispersion of the injected plasmid DNA within the tissue is an important issue in the delivery of naked plasmid DNA. Compared with conventional, low-molecular weight drugs, plasmid DNA is a huge molecule with a molecular weight of about 2,000 kD or more. This huge size greatly restricts its diffusion within the tissue where plasmid DNA is injected, because the diffusion as well as the absorption of an injected compound into the circulation is governed by its molecular weight (Nara *et al.*, 1992). Furthermore, the complex formation of plasmid DNA with cationic liposomes limits the diffusion within tissues due to the increased size and net charge (Nomura *et al.*, 1997, 1999) Therefore, plasmid DNA locally injected into tissues, such as muscle and skin, may only transfect cells around the injection site, which makes *in vivo* gene transfer tissue-specific. A local injection of lipoplex has been shown to be an effective approach to achieve transgene expression in the lung, brain, tumor and skin (Brigham *et al.*, 1989; Ono *et al.*, 1990; Plautz *et al.*, 1993; Raz *et al.*, 1994).

Some polymers, which do not form condensed complexes with plasmid DNA, have been reported to exhibit an enhancing effect on the transgene expression of naked plasmid DNA in skeletal muscle. Polyvinyl pyrrolidone or polyvinyl alcohol has been used to increase the area and level of transgene expression following intramuscular injection of plasmid DNA (Alila *et al.*, 1997). Use of these polymers resulted in widespread transgene expression within the injected tissues, suggesting altered distribution of plasmid DNA by the presence of such polymers.

VI. CONCLUSION

The key issue for successful *in vivo* gene therapy is to develop a vector and optimize its delivery, so that significant transgene expression *in vivo* can be achieved. Understanding the tissue distribution of plasmid DNA and its complex is a prerequisite for designing an effective approach to improve their efficacy *in vivo*. Pharmacokinetic analysis will give us information about the events occurring in the body following administration and, therefore, it is a very powerful tool for developing a strategy to improve the poor results obtained in gene therapy trials to date.

References

Adami, R. C., Collard, W. T., Gupta, S. A., Kwok, K. Y., Bonadio, J., and Rice, K. G. (1998). Stability of peptide-condensed plasmid DNA formulations. *J. Pharm. Sci.* **87,** 678–683.

Ali, S. A., Eary, J. F., Warren, S. D., Badger, C. C., and Krohn, K. A. (1988). Synthesis and radioiodination of tyramine cellobiose for labeling monoclonal antibodies. *Nucl. Med. Biol.* **15,** 557–561.

Alila, H., Coleman, M., Nitta, H., French, M., Anwer, K., Liu, Q., Meyer, T., Wang, J., Mumper, R., Oubari, D., Long, S., Nordstrom, J., and Rolland, A. (1997). Expression of biologically active human insulin-like growth factor-I following intramuscular injection of a formulated plasmid in rats. *Hum. Gene Ther.* **8,** 1785–1795.

Audouy, S. A., de Leij, L. F., Hoekstra, D., and Molema, G. (2002). In vivo characteristics of cationic liposomes as delivery vectors for gene therapy. *Pharm. Res.* **19,** 1599–1605.

Barron, L. G., Meyer, K. B., and Szoka, F. C., Jr. (1998). Effects of complement depletion on the pharmacokinetics and gene delivery mediated by cationic lipid-DNA complexes. *Hum. Gene Ther.* **9,** 315–323.

Barry, M. E., Pinto-Gonzalez, D., Orson, F. M., McKenzie, G. J., Petry, G. R., and Barry, M. A. (1999). Role of endogenous endonucleases and tissue site in transfection and CpG-mediated immune activation after naked DNA injection. *Hum. Gene Ther.* **10,** 2461–2480.

Brigham, K. L., Meyrick, B., Christman, B., Magnuson, M., King, G., and Berry, L. C., Jr. (1989). In vivo transfection of murine lungs with a functioning prokaryotic gene using a liposome vehicle. *Am. J. Med. Sci.* **298,** 278–281.

Budker, V., Zhang, G., Knechtle, S., and Wolff, J. A. (1996). Naked DNA delivered intraportally expresses efficiently in hepatocytes. *Gene Ther.* **3,** 593–598.

Budker, V., Zhang, G., Danko, I., Williams, P., and Wolff, J. A. (1998). The efficient expression of intravascularly delivered DNA in rat muscle. *Gene Ther.* **5,** 272–276.

Budker, V., Budker, T., Zhang, G., Subbotin, V., Loomis, A., and Wolff, J. A. (2000). Hypothesis: Naked plasmid DNA is taken up by cells in vivo by a receptor-mediated process. *J. Gene Med.* **2,** 76–88.

Deshpande, S. V., Subramanian, R., McCall, M. J., DeNardo, S. J., DeNardo, G. L., and Meares, C. F. (1990). Metabolism of indium chelates attached to monoclonal antibody: Minimal transchelation of indium from benzyl-EDTA chelate in vivo. *J. Nucl. Med.* **31,** 218–224.

Emlen, W., Rifai, A., Magilavy, D., and Mannik, M. (1988). Hepatic binding of DNA is mediated by a receptor on nonparenchymal cells. *Am. J. Pathol.* **133,** 54–60.

Felgner, P. L., Barenholz, Y., Behr, J. P., Cheng, S. H., Cullis, P., Huang, L., Jessee, J. A., Seymour, L., Szoka, F., Thierry, A. R., Wagner, E., and Wu, G. (1997). Nomenclature for synthetic gene delivery systems. *Hum. Gene Ther.* **8,** 511–512.

Finsinger, D., Remy, J. S., Erbacher, P., Koch, C., and Plank, C. (2000). Protective copolymers for non-viral gene vectors: Synthesis, vector characterization and application in gene delivery. *Gene Ther.* **7,** 1183–1192.

Herzog, R. W., Yang, E. Y., Couto, L. B., Hagstrom, J. N., Elwell, D., Fields, P. A., Burton, M., Bellinger, D. A., Read, M. S., Brinkhous, K. M., Podsakoff, G. M., Nichols, T. C., Kurtzman, G. J., and High, K. A. (1999). Long-term correction of canine hemophilia B by gene transfer of blood coagulation factor IX mediated by adeno-associated viral vector. *Nat. Med.* **5,** 56–63.

Hickman, M. A., Malone, R. W., Lehmann-Bruinsma, K., Sih, T. R., Knoell, D., Szoka, F. C., Walzem, R., Carlson, D. M., and Powell, J. S. (1994). Gene expression following direct injection of DNA into liver. *Hum. Gene Ther.* **5,** 1477–1483.

Hisazumi, J., Kobayashi, N., Nishikawa, M., and Takakura, Y. (2004). Significant role of liver sinusoidal endothelial cells in hepatic uptake and degradation of naked plasmid DNA following intravenous injection. *Pharm. Res.* **21,** 1223–1228.

Houk, B. E., Martin, R., Hochhaus, G., and Hughes, J. A. (2001). Pharmacokinetics of plasmid DNA in the rat. *Pharm. Res.* **18,** 67–74.

Kawabata, K., Takakura, Y., and Hashida, M. (1995). The fate of plasmid DNA after intravenous injection in mice: Involvement of scavenger receptors in its hepatic uptake. *Pharm. Res.* **12,** 825–830.

Kawakami, S., Sato, A., Nishikawa, M., Yamashita, F., and Hashida, M. (2000a). Mannose receptor-mediated gene transfer into macrophages using novel mannosylated cationic liposomes. *Gene Ther.* **7,** 292–299.

Kawakami, S., Fumoto, S., Nishikawa, M., Yamashita, F., and Hashida, M. (2000b). *In vivo* gene delivery to the liver using novel galactosylated cationic liposomes. *Pharm. Res.* **17**, 306–313.

Kawakami, S., Yamashita, F., Nishida, K., Nakamura, J., and Hashida, M. (2002). Glycosylated cationic liposomes for cell-selective gene delivery. *Crit. Rev. Ther. Drug Carrier Syst.* **19**, 171–190.

Kay, M. A., Manno, C. S., Ragni, M. V., Larson, P. J., Couto, L. B., McClelland, A., Glader, B., Chew, A. J., Tai, S. J., Herzog, R. W., Arruda, V., Johnson, F., Scallan, C., Skarsgard, E., Flake, A. W., and High, K. A. (2000). Evidence for gene transfer and expression of factor IX in haemophilia B patients treated with an AAV vector. *Nat. Genet.* **24**, 257–261.

Kircheis, R., Schuller, S., Brunner, S., Ogris, M., Heider, K. H., Zauner, W., and Wagner, E. (1999). Polycation-based DNA complexes for tumor-targeted gene delivery *in vivo. J. Gene Med.* **1**, 111–120.

Kircheis, R., Blessing, T., Brunner, S., Wightman, L., and Wagner, E. (2001). Tumor targeting with surface-shielded ligand-polycation DNA complexes. *J. Control. Release* **72**, 165–170.

Kobayashi, N., Kuramoto, T., Yamaoka, K., Hashida, M., and Takakura, Y. (2001). Hepatic uptake and gene expression mechanisms following intravenous administration of plasmid DNA by conventional and hydrodynamics-based procedures. *J. Pharmacol. Exp. Ther.* **297**, 853–860.

Kobayashi, N., Nishikawa, M., Hirata, K., and Takakura, Y. (2004a). Hydrodynamics-based procedure involves transient hyperpermeability in the hepatic cellular membrane: Implication of a nonspecific process in efficient intracellular gene delivery. *J. Gene Med.* **6**, 584–592.

Kobayashi, N., Hirata, K., Chen, S., Kawase, A., Nishikawa, M., and Takakura, Y. (2004b). Hepatic delivery of particulates in the submicron range by a hydrodynamics-based procedure: Implications for particulate gene delivery systems. *J. Gene Med.* **6**, 455–463.

Lechardeur, D., Sohn, K. J., Haardt, M., Joshi, P. B., Monck, M., Graham, R. W., Beatty, B., Squire, J., O'Brodovich, H., and Lukacs, G. L. (1999). Metabolic instability of plasmid DNA in the cytosol: A potential barrier to gene transfer. *Gene Ther.* **6**, 482–497.

Lecocq, M., Andrianaivo, F., Warnier, M. T., Wattiaux-De Coninck, S., Wattiaux, R., and Jadot, M. (2003). Uptake by mouse liver and intracellular fate of plasmid DNA after a rapid tail vein injection of a small or a large volume. *J. Gene Med.* **5**, 142–156.

Lew, D., Parker, S. E., Latimer, T., Abai, A. M., Kuwahara-Rundell, A., Doh, S. G., Yang, Z. Y., Laface, D., Gromkowski, S. H., Nabel, G. J., *et al.* (1995). Cancer gene therapy using plasmid DNA: Pharmacokinetic study of DNA following injection in mice. *Hum. Gene Ther.* **6**, 553–564.

Li, S., Rizzo, M. A., Bhattacharya, S., and Huang, L. (1998). Characterization of cationic lipid-protamine-DNA (LPD) complexes for intravenous gene delivery. *Gene Ther.* **5**, 930–937.

Li, S., Tseng, W. C., Stolz, D. B., Wu, S. P., Watkins, S. C., and Huang, L. (1999). Dynamic changes in the characteristics of cationic lipidic vectors after exposure to mouse serum: Implications for intravenous lipofection. *Gene Ther.* **6**, 585–594.

Li, S., Tan, Y., Viroonchatapan, E., Pitt, B. R., and Huang, L. (2000). Targeted gene delivery to pulmonary endothelium by anti-PECAM antibody. *Am. J. Physiol.* **278**, L504–L511.

Liang, K. W., Nishikawa, M., Liu, F., Sun, B., Ye, Q., and Huang, L. (2004). Restoration of dystrophin expression in mdx mice by intravascular injection of naked DNA containing full-length dystrophin cDNA. *Gene Ther.* **11**, 901–908.

Lin, H., Parmacek, M. S., Morle, G., Bolling, S., and Leiden, J. M. (1990). Expression of recombinant genes in myocardium *in vivo* after direct injection of DNA. *Circulation* **82**, 2217–2221.

Litzinger, D. C., Brown, J. M., Wala, I., Kaufman, S. A., Van, G. Y., Farrell, C. L., and Collins, D. (1996). Fate of cationic liposomes and their complex with oligonucleotide *in vivo. Biochim. Biophys. Acta* **1281**, 139–149.

Liu, Y., Mounkes, L. C., Liggitt, H. D., Brown, C. S., Solodin, I., Heath, T. D., and Debs, R. J. (1997). Factors influencing the efficiency of cationic liposome-mediated intravenous gene delivery. *Nat. Biotechnol.* **15**, 167–173.

Liu, F., Song, Y., and Liu, D. (1999). Hydrodynamics-based transfection in animals by systemic administration of plasmid DNA. *Gene Ther.* **6,** 1258–1266.

Mahato, R. I., Kawabata, K., Nomura, T., Takakura, Y., and Hashida, M. (1995). Physicochemical and pharmacokinetic characteristics of plasmid DNA/cationic liposome complexes. *J. Pharm. Sci.* **84,** 1267–1271.

Mahato, R. I., Anwer, K., Tagliaferri, F., Meaney, C., Leonard, P., Wadhwa, M. S., Logan, M., French, M., and Rolland, A. (1998). Biodistribution and gene expression of lipid/plasmid complexes after systemic administration. *Hum. Gene Ther.* **9,** 2083–2099.

Manno, C. S., Chew, A. J., Hutchison, S., Larson, P. J., Herzog, R. W., Arruda, V. R., Tai, S. J., Ragni, M. V., Thompson, A., Ozelo, M., Couto, L. B., Leonard, D. G., Johnson, F. A., McClelland, A., Scallan, C., Skarsgard, E., Flake, A. W., Kay, M. A., High, K. A., and Glader, B. (2003). AAV-mediated factor IX gene transfer to skeletal muscle in patients with severe hemophilia B. *Blood* **101,** 2963–2972.

McLean, J. W., Fox, E. A., Baluk, P., Bolton, P. B., Haskell, A., Pearlman, R., Thurston, G., Umemoto, E. Y., and McDonald, D. M. (1997). Organ-specific endothelial cell uptake of cationic liposome-DNA complexes in mice. *Am. J. Physiol.* **273,** H387–H404.

Morimoto, K., Nishikawa, M., Kawakami, S., Nakano, T., Hattori, Y., Fumoto, S., Yamashita, F., and Hashida, M. (2003). Molecular weight-dependent gene transfection activity of unmodified and galactosylated polyethyleneimine on hepatoma cells and mouse liver. *Mol. Ther.* **7,** 254–261.

Morishita, R., Aoki, M., Hashiya, N., Makino, H., Yamasaki, K., Azuma, J., Sawa, Y., Matsuda, H., Kaneda, Y., and Ogihara, T. (2004). Safety evaluation of clinical gene therapy using hepatocyte growth factor to treat peripheral arterial disease. *Hypertension* **44,** 203–209.

Nakamura, M., Davila-Zavala, P., Tokuda, H., Takakura, Y., and Hashida, M. (1998). Uptake and gene expression of naked plasmid DNA in cultured brain microvessel endothelial cells. *Biochem. Biophys. Res. Commun.* **245,** 235–239.

Nara, E., Masegi, M., Hatono, T., and Hashida, M. (1992). Pharmacokinetic analysis of drug absorption from muscle based on a physiological diffusion model: Effect of molecular size on absorption. *Pharm. Res.* **9,** 161–168.

Neves, C., Byk, G., Escriou, V., Bussone, F., Scherman, D., and Wils, P. (2000). Novel method for covalent fluorescent labeling of plasmid DNA that maintains structural integrity of the plasmid. *Bioconjug. Chem.* **11,** 51–55.

Nishikawa, M., and Hashida, M. (2002). Non-viral approaches satisfying various requirements for effective *in vivo* gene therapy. *Biol. Pharm. Bull.* **25,** 275–283.

Nishikawa, M., and Huang, L. (2001). Non-viral vectors in the new millennium: Delivery barriers in gene transfer. *Hum. Gene Ther.* **12,** 861–870.

Nishikawa, M., Takakura, Y., and Hashida, M. (1996). Pharmacokinetic evaluation of polymeric carriers. *Adv. Drug Delivery Rev.* **21,** 135–155.

Nishikawa, M., Takemura, S., Takakura, Y., and Hashida, M. (1998). Targeted delivery of plasmid DNA to hepatocytes *in vivo*: optimization of the pharmacokinetics of plasmid DNA/galactosylated poly(L-lysine) complexes by controlling their physicochemical properties. *J. Pharmacol. Exp. Ther.* **287,** 408–415.

Nishikawa, M., Yamauchi, M., Morimoto, K., Ishida, E., Takakura, Y., and Hashida, M. (2000). Hepatocyte-targeted *in vivo* gene expression by intravenous injection of plasmid DNA complexed with synthetic multi-functional gene delivery system. *Gene Ther.* **7,** 548–555.

Nishikawa, M., Nakano, T., Okabe, T., Hamaguchi, N., Yamasaki, Y., Takakura, Y., Yamashita, F., and Hashida, M. (2003). Residualizing indium-111-radiolabel for plasmid DNA and its application to tissue distribution study. *Bioconjug. Chem.* **14,** 955–961.

Nomura, T., Nakajima, S., Kawabata, K., Yamashita, F., Takakura, Y., and Hashida, M. (1997). Intratumoral pharmacokinetics and *in vivo* gene expression of naked plasmid DNA and its cationic liposome complexes after direct gene transfer. *Cancer Res.* **57,** 2681–2686.

Nomura, T., Yasuda, K., Yamada, T., Okamoto, S., Mahato, R. I., Watanabe, Y., Takakura, Y., and Hashida, M. (1999). Gene expression and antitumor effects following direct interferon (IFN)-gamma gene transfer with naked plasmid DNA and DC-chol liposome complexes in mice. *Gene Ther.* **6,** 121–129.

Ogris, M., Brunner, S., Schuller, S., Kircheis, R., and Wagner, E. (1999). PEGylated DNA/transferrin-PEI complexes: Reduced interaction with blood components, extended circulation in blood and potential for systemic gene delivery. *Gene Ther.* **6,** 595–605.

Oh, Y. K., Kim, J. P., Yoon, H., Kim, J. M., Yang, J. S., and Kim, C. K. (2001). Prolonged organ retention and safety of plasmid DNA administered in polyethylenimine complexes. *Gene Ther.* **8,** 1587–1592.

O'Hara, A. J., Howell, J. M., Taplin, R. H., Fletcher, S., Lloyd, F., Kakulas, B., Lochmuller, H., and Karpati, G. (2001). The spread of transgene expression at the site of gene construct injection. *Muscle Nerve* **24,** 488–495.

Ono, T., Fujino, Y., Tsuchiya, T., and Tsuda, M. (1990). Plasmid DNAs directly injected into mouse brain with lipofectin can be incorporated and expressed by brain cells. *Neurosci. Lett.* **117,** 259–263.

Oupicky, D., Ogris, M., Howard, K. A., Dash, P. R., Ulbrich, K., and Seymour, L. W. (2002). Importance of lateral and steric stabilization of polyelectrolyte gene delivery vectors for extended systemic circulation. *Mol. Ther.* **5,** 463–472.

Perales, J. C., Ferkol, T., Beegen, H., Ratnoff, O. D., and Hanson, R. W. (1994). Gene transfer *in vivo*: Sustained expression and regulation of genes introduced into the liver by receptor-targeted uptake. *Proc. Natl. Acad. Sci. USA* **91,** 4086–4090.

Piatyszek, M. A., Jarmolowski, A., and Augustyniak, J. (1988). Iodo-Gen-mediated radioiodination of nucleic acids. *Anal. Biochem.* **172,** 356–359.

Plautz, G. E., Yang, Z. Y., Wu, B. Y., Gao, X., Huang, L., and Nabel, G. J. (1993). Immunotherapy of malignancy by *in vivo* gene transfer into tumors. *Proc. Natl. Acad. Sci. USA* **90,** 4645–4649.

Pollard, H., Toumaniantz, G., Amos, J. L., Avet-Loiseau, H., Guihard, G., Behr, J. P., and Escande, D. (2001). Ca^{2+}-sensitive cytosolic nucleases prevent efficient delivery to the nucleus of injected plasmids. *J. Gene Med.* **3,** 153–164.

Raz, E., Carson, D. A., Parker, S. E., Parr, T. B., Abai, A. M., Aichinger, G., Gromkowski, S. H., Singh, M., Lew, D., Yankauckas, M. A., *et al.* (1994). Intradermal gene immunization: The possible role of DNA uptake in the induction of cellular immunity to viruses. *Proc. Natl. Acad. Sci. USA* **91,** 9519–9523.

Rigby, P. W., Dieckmann, M., Rhodes, C., and Berg, P. (1977). Labeling deoxyribonucleic acid to high specific activity *in vitro* by nick translation with DNA polymerase I. *J. Mol. Biol.* **113,** 237–251.

Roth, D. A., Tawa, N. E., Jr., O'Brien, J. M., Treco, D. A., and Selden, R. F. (2001). Non-viral transfer of the gene encoding coagulation factor VIII in patients with severe hemophilia A. *N Engl. J. Med.* **344,** 1735–1742.

Sakurai, F., Nishioka, T., Saito, H., Baba, T., Okuda, A., Matsumoto, O., Taga, T., Yamashita, F., Takakura, Y., and Hashida, M. (2001a). Interaction between DNA-cationic liposome complexes and erythrocytes is an important factor in systemic gene transfer via the intravenous route in mice: The role of the neutral helper lipid. *Gene Ther.* **8,** 677–686.

Sakurai, F., Nishioka, T., Yamashita, F., Takakura, Y., and Hashida, M. (2001b). Effects of erythrocytes and serum proteins on lung accumulation of lipoplexes containing cholesterol or DOPE as a helper lipid in the single-pass rat lung perfusion system. *Eur. J. Pharm. Biopharm.* **52,** 165–172.

Sakurai, F., Terada, T., Yasuda, K., Yamashita, F., Takakura, Y., and Hashida, M. (2002). The role of tissue macrophages in the induction of proinflammatory cytokine production following intravenous injection of lipoplexes. *Gene Ther.* **9,** 1120–1126.

Satkauskas, S., Bureau, M. F., Mahfoudi, A., and Mir, L. M. (2001). Slow accumulation of plasmid in muscle cells: Supporting evidence for a mechanism of DNA uptake by receptor-mediated endocytosis. *Mol. Ther.* **4,** 317–323.

Sikes, M. L., O'Malley, B. W., Jr., Finegold, M. J., and Ledley, F. D. (1994). *In vivo* gene transfer into rabbit thyroid follicular cells by direct DNA injection. *Hum. Gene Ther.* **5,** 837–844.

Slattum, P. S., Loomis, A. G., Machnik, K. J., Watt, M. A., Duzeski, J. L., Budker, V. G., Wolff, J. A., and Hagstrom, J. E. (2003). Efficient *in vitro* and *in vivo* expression of covalently modified plasmid DNA. *Mol. Ther.* **8,** 255–263.

Snyder, R. O., Miao, C., Meuse, L., Tubb, J., Donahue, B. A., Lin, H. F., Stafford, D. W., Patel, S., Thompson, A. R., Nichols, T., Read, M. S., Bellinger, D. A., Brinkhous, K. M., and Kay, M. A. (1999). Correction of hemophilia B in canine and murine models using recombinant adeno-associated viral vectors. *Nat. Med.* **5,** 64–70.

Takagi, T., Hashiguchi, M., Mahato, R. I., Tokuda, H., Takakura, Y., and Hashida, M. (1998). Involvement of specific mechanism in plasmid DNA uptake by mouse peritoneal macrophages. *Biochem. Biophys. Res. Commun.* **245,** 729–733.

Takakura, Y., Takagi, T., Hashiguchi, M., Nishikawa, M., Yamashita, F., Doi, T., Imanishi, T., Suzuki, H., Kodama, T., and Hashida, M. (1999). Characterization of plasmid DNA binding and uptake by peritoneal macrophages from class A scavenger receptor knockout mice. *Pharm. Res.* **16,** 503–508.

Terebesi, J., Kwok, K. Y., and Rice, K. G. (1998). Iodinated plasmid DNA as a tool for studying gene delivery. *Anal. Biochem.* **263,** 120–123.

Terpstra, V., van Amersfoort, E. S., van Velzen, A. G., Kuiper, J., and van Berkel, T. J. (2000). Hepatic and extrahepatic scavenger receptors: Function in relation to disease. *Arterioscler. Thromb. Vasc. Biol.* **20,** 1860–1872.

Thorpe, S. R., Baynes, J. W., and Chroneos, Z. C. (1993). The design and application of residualizing labels for studies of protein catabolism. *FASEB J.* **7,** 399–405.

Verbaan, F. J., Oussoren, C., Snel, C. J., Crommelin, D. J., Hennink, W. E., and Storm, G. (2004). Steric stabilization of poly(2-(dimethylamino)ethyl methacrylate)-based polyplexes mediates prolonged circulation and tumor targeting in mice. *J. Gene Med.* **6,** 64–75.

Wagner, E., Plank, C., Zatloukal, K., Cotten, M., and Birnstiel, M. L. (1992). Influenza virus hemagglutinin HA-2 N-terminal fusogenic peptides augment gene transfer by transferrin-polylysine-DNA complexes: Toward a synthetic virus-like gene-transfer vehicle. *Proc. Natl. Acad. Sci. USA* **89,** 7934–7938.

Ward, C. M., Pechar, M., Oupicky, D., Ulbrich, K., and Seymour, L. W. (2002). Modification of pLL/DNA complexes with a multivalent hydrophilic polymer permits folate-mediated targeting in vitro and prolonged plasma circulation *in vivo. J. Gene Med.* **4,** 536–547.

Wasan, E. K., Reimer, D. L., and Bally, M. B. (1996). Plasmid DNA is protected against ultrasonic cavitation-induced damage when complexed to cationic liposomes. *J. Pharm. Sci.* **85,** 427–433.

Wolff, J. A., Malone, R. W., Williams, P., Chong, W., Acsadi, G., Jani, A., and Felgner, P. L. (1990). Direct gene transfer into mouse muscle *in vivo. Science* **247,** 1465–1468.

Wu, G. Y., and Wu, C. H. (1988). Receptor-mediated gene delivery and expression *in vivo. J. Biol. Chem.* **263,** 14621–14624.

Xu, Y., and Szoka, F. C., Jr. (1996). Mechanism of DNA release from cationic liposome/DNA complexes used in cell transfection. *Biochemistry* **35,** 5616–5623.

Yamasaki, Y., Sumimoto, K., Nishikawa, M., Yamashita, F., Yamaoka, K., Hashida, M., and Takakura, Y. (2002). Pharmacokinetic analysis of *in vivo* disposition of succinylated proteins targeted to liver nonparenchymal cells via scavenger receptors: Importance of molecular size and negative charge density for *in vivo* recognition by receptors. *J. Pharmacol. Exp. Ther.* **301,** 467–477.

Yamasaki, Y., Hisazumi, J., Yamaoka, K., and Takakura, Y. (2003). Efficient scavenger receptor-mediated hepatic targeting of proteins by introduction of negative charges on the proteins by aconitylation: The influence of charge density and size of the proteins molecules. _Eur. J. Pharm. Sci._ **18,** 305–312.

Yang, J. P., and Huang, L. (1997). Overcoming the inhibitory effect of serum on lipofection by increasing the charge ratio of cationic liposome to DNA. _Gene Ther._ **4,** 950–960.

Yoo, J. J., Soker, S., Lin, L. F., Mehegan, K., Guthrie, P. D., and Atala, A. (1999). Direct _in vivo_ gene transfer to urological organs. _J. Urol._ **162,** 1115–1118.

Yoshida, M., Mahato, R. I., Kawabata, K., Takakura, Y., and Hashida, M. (1996). Disposition characteristics of plasmid DNA in the single-pass rat liver perfusion system. _Pharm. Res._ **13,** 599–603.

Yoshinaga, T., Yasuda, K., Ogawa, Y., and Takakura, Y. (2002). Efficient uptake and rapid degradation of plasmid DNA by murine dendritic cells via a specific mechanism. _Biochem. Biophys. Res. Commun._ **299,** 389–394.

Zelphati, O., Liang, X., Hobart, P., and Felgner, P. L. (1999). Gene chemistry: Functionally and conformationally intact fluorescent plasmid DNA. _Hum. Gene Ther._ **10,** 15–24.

Zhang, G., Vargo, D., Budker, V., Armstrong, N., Knechtle, S., and Wolff, J. A. (1997). Expression of naked plasmid DNA injected into the afferent and efferent vessels of rodent and dog livers. _Hum. Gene Ther._ **8,** 1763–1772.

Zhang, G., Budker, V., and Wolff, J. A. (1999). High levels of foreign gene expression in hepatocytes after tail vein injections of naked plasmid DNA. _Hum. Gene Ther._ **10,** 1735–1737.

Zhang, G., Budker, V., Williams, P., Subbotin, V., and Wolff, J. A. (2001). Efficient expression of naked DNA delivered intraarterially to limb muscles of nonhuman primates. _Hum. Gene Ther._ **12,** 427–438.

Zhang, G., Gao, X., Song, Y. K., Vollmer, R., Stolz, D. B., Gasiorowski, J. Z., Dean, D. A., and Liu, D. (2004). Hydroporation as the mechanism of hydrodynamic delivery. _Gene Ther._ **11,** 675–682.

Zhu, F. G., Reich, C. F., and Pisetsky, D. S. (2001). The role of the macrophage scavenger receptor in immune stimulation by bacterial DNA and synthetic oligonucleotides. _Immunology_ **103,** 226–234.

Section 2

CATIONIC LIPOSOMES

4

What Role Can Chemistry Play in Cationic Liposome-Based Gene Therapy Research Today?

Kostas Kostarelos and Andrew D. Miller

Imperial College Genetic Therapies Centre, Department of Chemistry
Imperial College London
London, SW7 2AY, United Kingdom

ABSTRACT

Gene therapy research is still in trouble owing to a paucity of acceptable vector systems to deliver nucleic acids to patients for therapy. Viral vectors are efficient but may be too dangerous for routine clinical use. Synthetic non-viral vectors

Advances in Genetics, Vol. 53
0065-2660/05 $35.00
DOI: 10.1016/S0065-2660(05)53004-9

are inherently much safer but are currently not efficient enough to be clinically viable. The solution for gene therapy lies with improved synthetic non-viral vectors based upon well-found platform technologies and a thorough under-standing of the barriers to efficient gene delivery and expression (transfection) relevant to clinical applications of interest. Here we introduce and interpret synthetic non-viral vector systems through the **ABCD** nanoparticle structural paradigm that represents, in our view, an appropriate lens through which to view all synthetic, non-viral vector systems applicable to *in vitro* use or *in vivo* applications and gene therapy. Our intention in introducing this paradigm is to shift the focus of organic and physical chemists away from the design of yet another cytofectin, and instead encourage them to appreciate the wider chal-lenges presented by the need to produce tool kits of meaningful chemical components from which to assemble viable, tailor-made nanoparticles for *in vivo* applications and gene therapy, both now and in the future.   © 2005, Elsevier Inc.

I. INTRODUCTION

A. Viral or non-viral?

If we define gene therapy as the delivery of nucleic acids by means of a vector to patients for some therapeutic purpose, then which type of new vector system should be most appropriate to develop, viral or non-viral, synthetic or physical? In our view, synthetic non-viral vector systems represent the only realistic choice for routine *in vivo* applications and gene therapy in the future. Synthetic non-viral vector systems have many potential advantages compared to viral systems, including significantly lower toxicity/immunogenicity and potential for oncogenicity, size-independent delivery of nucleic acids (from oligonucleo-tides to artificial chromosomes), simpler quality control, and substantially easier pharmaceutical and regulatory requirements. Increasing public alarm, particu-larly with viral vectors, may also be strengthening these significant advantages. Ever present in the minds of the public and regulators is the potential for toxic side effects from the use of viral vectors. Therefore, basic clinical confi-dence in non-viral vectors is growing and the various advantages listed above inherent in synthetic non-viral vector systems should ensure substantial clinical uptake once the science and technology of these vector systems can be appro-priately matured for routine clinical use. In our opinion, appropriate synthetic non-viral vector systems for *in vivo* applications and gene therapy should not now be far off in coming, and for reasons of lower toxicity if nothing else, synthetic systems making use of cationic lipids (cytofectins) and liposomes should be paramount.

B. ABCD nanoparticles

Given the numerous permutations of synthetic non-viral vector systems that have been developed, there is a need to find a common language with which to discuss and appreciate these systems in a framework that allows us to relate different, individual systems and hence derive meaningful and realistic structure–activity correlations. Therefore, we would like to introduce the self-assembly **ABCD** nanoparticle concept as an appropriate structural paradigm for synthetic non-viral vector systems used for *in vitro, ex vivo* and/or *in vivo* applications (Fig. 4.1). In **ABCD** nanoparticles, nucleic acids (**A**) are condensed within functional concentric layers of chemical components designed for biological targeting (**D**), biological stability (**C**) and cellular entry/intracellular trafficking (**B**). For the purposes of this review, the **AB** core particle comprises nucleic acids (either DNA or RNA) (**A**) condensed and/or encapsulated by liposomes/micelles (**B**) in a non-covalent manner. Typically, DNA may be in the form of plasmid DNA (pDNA) or oligodeoxynucleotide (ODN); and RNA could be in the form, for example, of messenger RNA (mRNA), oligonucleotide (ON) or small interference RNA (siRNA). According to our new nomenclature, an **AB** core particle is equivalent to the complex formed between simple cationic

Structural paradigm for self-assembly ABCD nanoparticles

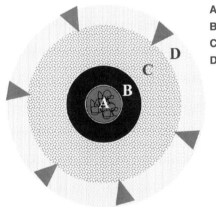

A: nucleic acids (siRNA, mRNA, pDNA)

B: lipid envelope layer

C: stealth/biocompatibility layer

D: biological recognition layer

ABCD nanoparticles constructed
from tool-kits of synthetic chemical
components

AB systems; *in vitro*

Tailor-made delivery solutions
ABC/ABCD systems; *in vivo*

Figure 4.1. **ABCD** nanoparticle concept. Graphic illustration of **ABCD** nanoparticle structure to show how nucleic acids (**A**) are condensed in functional concentric layers of chemical components purpose designed for biological targeting (**D**), biological stability (**C**), and cellular entry (**B**). (See Color Insert.)

liposome/micelle systems and pDNA (lipoplex, LD). In other words, LD parti-cles are one significant type of **AB** core particle. Alternative **AB** core particles could be, for example, formed between simple cationic liposome/micelle sys-tems and siRNA (siRNA lipoplex, LsiR). **AB** core particles could also be ternary LD particles where pDNA is precondensed with an additional cationic agent prior to further complication with cationic liposomes/micelles. For the most part, we would suggest that **AB** core particles are primarily unstable in biological fluids and are destined for primary use *in vitro* and perhaps *ex vivo* but no further.

For use *in vivo*, a biocompatible polymer (**C**-layer) almost certainly needs to be introduced through association or attachment to the surface of each **AB** core particle thereby conferring on **AB** core particles colloidal and structur-al stability in biological fluids. Finally, biological targeting ligands (**D**-layer) are an optional exterior coating designed to maximize the accumulation of nano-particles first in the organ of choice *in vivo* and preferably in target cells of interest within the organ of choice. The role of biological targeting ligands is not obligatory. As we shall see, there are viable **ABC** nanoparticle systems that act by a process of passive targeting, namely tissue accumulation through biophysi-cal means without the need for biological targeting. However, biological target-ing has the potential to reprogram a given nanoparticle system to accumulate into organs of interest by a process of active targeting that seeks to accumu-late nanoparticles in the given organ(s) of interest with more precision, speed and efficiency than passive targeting may allow.

II. AB CORE PARTICLES

A. Cytofectins

Much of our current understanding about synthetic non-viral vector-mediated nucleic acid delivery has been gained by studying the behavior of LD particles *in vitro* and *in vivo*. The general mechanism of delivery appears to be as follows: Nanometric LD particles formed from the combination of cationic liposomes/micelles with nucleic acids in buffered, aqueous solution, enter cells by endocy-tosis triggered through non-specific interactions between complexes and the cell-surface proteoglycans of adherent cells (Fig. 4.2). Once inside, the pH of the endosome compartments drops from pH7 to 5.5 and a proportion of the bound nucleic acids escapes from early-endosomes into the cytosol to perform a therapeutic function there as in the case of RNA [Path B; (Fig. 4.2)], or else traffic from cytosol to the nucleus in order to perform a function there instead, as in the case of DNA [Path C; (Fig. 4.2)]. This process is surprisingly ineffi-cient and every step of the delivery process is problematic (Miller, 1998a,b, 1999).

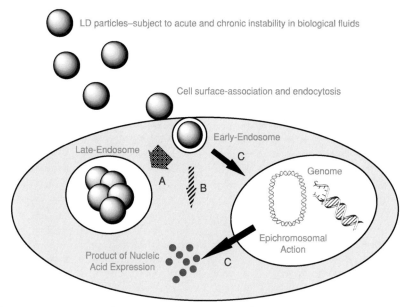

Figure 4.2. Diagram to show process of LD particle cell entry. LD particles that have not succumbed to aggregation and/or serum-inactivation associate with the cell surface and enter usually by endocytosis. The majority in early endosomes become trapped in late endosomes (Path A) and the nucleic acids fail to reach the cytosol. A minority are able to release their bound nucleic acids into the cytosol. Path B is followed by RNA that acts directly in the cytosol. Path C is followed by DNA that enters the nucleus in order to act. The diagram is drawn making the assumption that plasmid DNA (pDNA) has been delivered which is expressed in an epichromosomal manner. Reprinted with permission of Bios. Scientific Publishers from Miller (1999). (See Color Insert.)

As mentioned above, LD particles are generated by the combination of simple cationic liposome/micelle systems with pDNA. Simple cationic liposome/micelle systems are formed from either a single synthetic cationic amphiphile (known as a cytofectin; *cyto-* for cell and *-fectin* for transfection [i.e., gene delivery and expression]) or more commonly from the combination of a cytofectin and a neutral lipid such as dioleoyl L-α-phosphatidylethanolamine (DOPE) **1** or cholesterol (Chol) **2** (Fig. 4.3). There are impressive numbers of cytofectins already described in the literature and available commercially (Ilies and Balaban, 2001; Miller, 1998a,b, 2003; Nicolazzi *et al.*, 2003a) but all have in common a hydrophobic moiety covalently attached to a hydrophilic moiety through a polar linker (Fig. 4.3). While hydrophobic regions are reasonably similar, polar linkers and cationic head groups vary quite substantially. The structures of a number of cytofectins are shown illustrating the structural

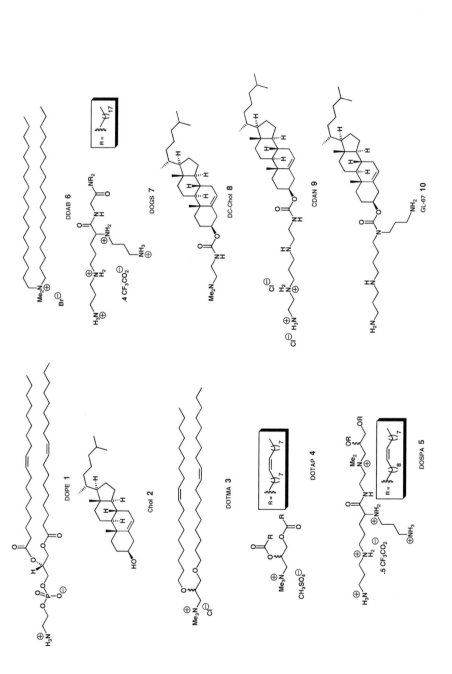

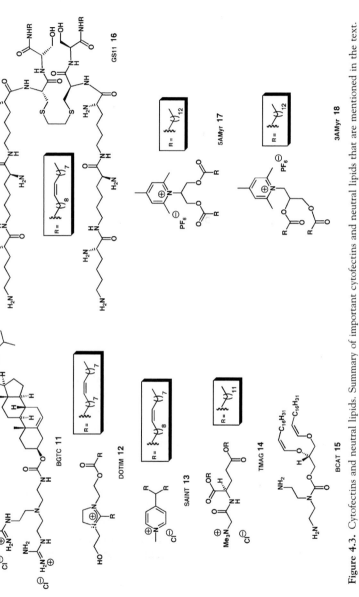

Figure 4.3. Cytofectins and neutral lipids. Summary of important cytofectins and neutral lipids that are mentioned in the text.

diversity that is tolerated without necessarily impairing the efficiency of transfection (Fig. 4.3)! We would like to point out that, in our view, the cytofectin field is now saturated and the creation of further novel structures will make little contribution unless their preparation is associated with the onward generation of **ABC** and **ABCD** nanoparticle systems for *in vivo* applications.

Nevertheless, there are a small number of recent additions to the pantheon of cytofectins that are worth mentioning. N^1-cholesteryloxycarbonyl-3,7-diazanonane-1,9-diamine (CDAN) **9** (Fig. 4.3) synthesis and properties have been described a number of times (Cooper *et al.*, 1998; Stewart *et al.*, 2001; Tagawa *et al.*, 2002). However, CDAN/DOPE cationic liposomes (1:1 m/m, Trojene™) have now been found to exceed competitors and mediate pDNA delivery to cells *in vitro* in minimal growth medium with minimum handling, thereby giving CDAN **9** a new lease on life (Keller *et al.*, 2003b). CDAN/DOPE cationic liposomes (45:55 m/m, siFECTamine®) have also been shown to facilitate the delivery of small interference RNA (siRNA) *in vitro* to cells with even more effect (Spagnou *et al.*, 2004). For this reason, an attractive new solid phase methodology was very recently devised for the synthesis of CDAN **9** in excellent yield that shows a useful way to go for effective synthesis of this increasingly critical cytofectin (Oliver *et al.*, 2004) (Scheme 4.1). Balaban *et al.* (Ilies *et al.*, 2003, 2004) have reported equally attractive, divergent solution-phase syntheses of novel pyridinium amphiphiles that enlarge on the contributions of others in this significant area of cytofectin design (Solodin *et al.*, 1995; van der Woude *et al.*, 1997) that is also growing in importance (Scheme 4.2). Multipurpose Gemini surfactants have also found their way into cytofectins (Kirby *et al.*, 2003; McGregor *et al.*, 2001) (Scheme 4.3).

Finally, one of the most innovative new cytofectins to emerge has been from Thompson *et al.* They developed O-(2R-1,2-di-O-(1′Z,9′Z-octadecadienyl)-glycerol)-N-(bis-2-aminoethyl)-carbamate (BCAT) based on plasmenylcholine synthesis, whose enol ether linkages are primed for acid-catalysed hydrolysis in conditions of acid pH (Boomer and Thompson, 1999; Boomer *et al.*, 2002; Shin and Thompson, 2003) (Scheme 4.4). The notional design objective was to ensure that BCAT should mediate DNA delivery to cells. Thereafter, acid-catalysed BCAT decomposition was expected to take place in acidic endosome compartments leading to enhanced endosmolysis thereby increasing the proportion of bound nucleic acids able to escape from early-endosomes into the cytosol. This approach to the endosomal barrier problem (outlined above) appears to have been less effective than expected owing to the unexpectedly slow rate of enol-ether hydrolysis at pH 5.5. This is unfortunate, but Thompson *et al.* are already at work innovating the next generation of acid-sensitive functional groups.

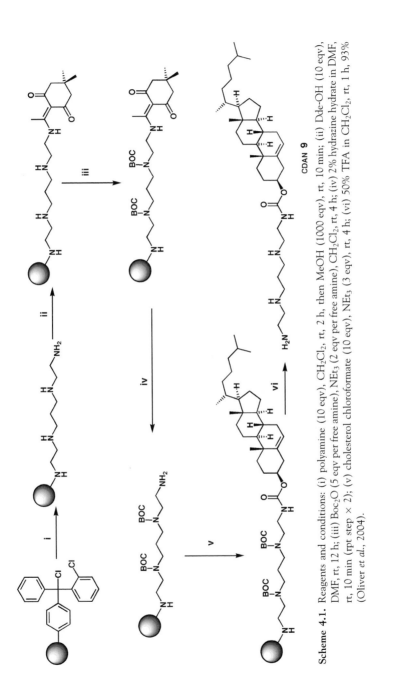

Scheme 4.1. Reagents and conditions: (i) polyamine (10 eqv), CH$_2$Cl$_2$, rt, 2 h, then MeOH (1000 eqv), rt, 10 min; (ii) Dde-OH (10 eqv), DMF, rt, 12 h; (iii) Boc$_2$O (5 eqv per free amine), NEt$_3$ (2 eqv per free amine), CH$_2$Cl$_2$, rt, 4 h; (iv) 2% hydrazine hydrate in DMF, rt, 10 min (rpt step × 2); (v) cholesterol chloroformate (10 eqv), NEt$_3$ (3 eqv), rt, 4 h; (vi) 50% TFA in CH$_2$Cl$_2$, rt, 1 h, 93% (Oliver *et al.*, 2004).

CDAN **9**

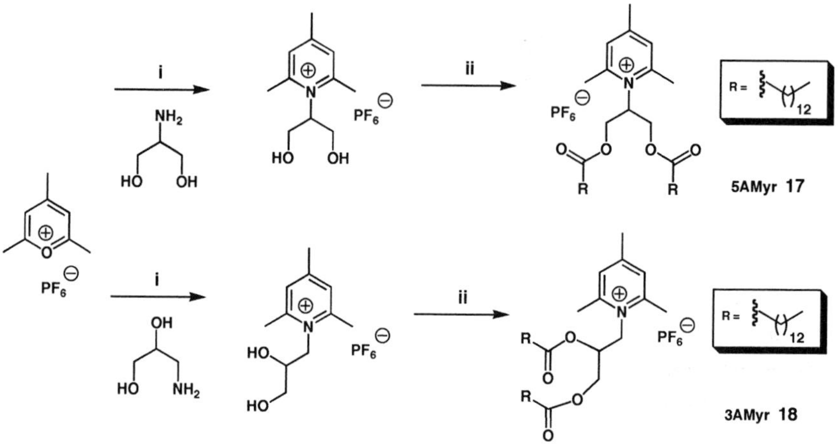

Scheme 4.2. Reagents and conditions: (i) NEt$_3$ (1.2 eqv), EtOH/AcOH, 1–3 h, 50–85%; (ii) NEt$_3$ (2 eqv), myristoyl-Cl (2.2 eqv), AcCN, 3–5 h, reflux (Ilies *et al.*, 2004).

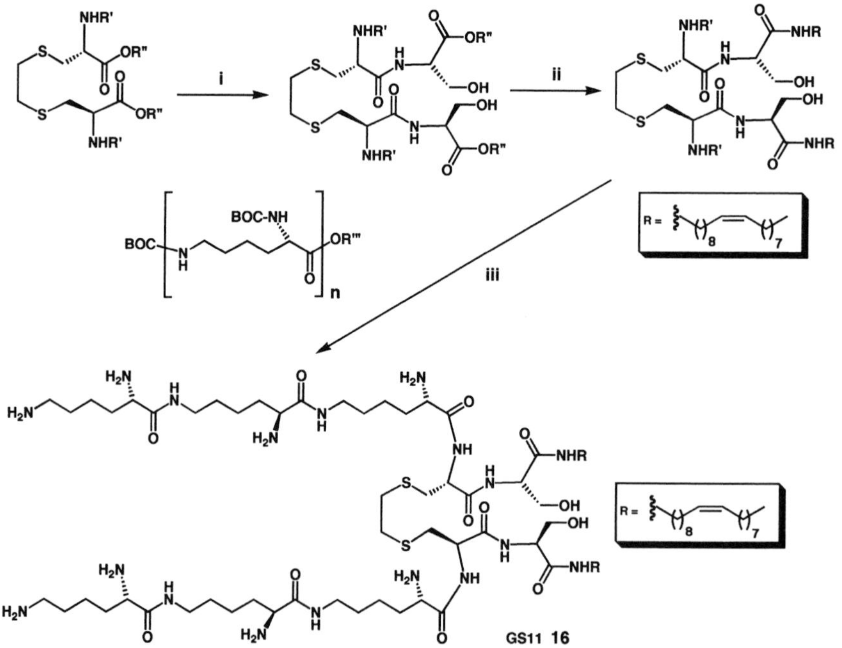

Scheme 4.3. Reagents and conditions: (i) L-Serine (2 eqv), K$_2$CO$_3$ (2 eqv), THF/H$_2$O 1:1 v/v, rt, 72 h; (ii) (a) NHS (2 eqv), DCC (2 eqv), THF, rt, 24 h (b) NEt$_3$ (2 eqv), oleylamine (2 eqv), THF, rt, 48 h; (iii) (a) BOC-protected L-K$_3$ (R''' = *N*-succinimido) (>2 eqv), H$_2$O/NaOH/THF, rt, 48 h (b) MeOH/HCl (McGregor *et al.*, 2001).

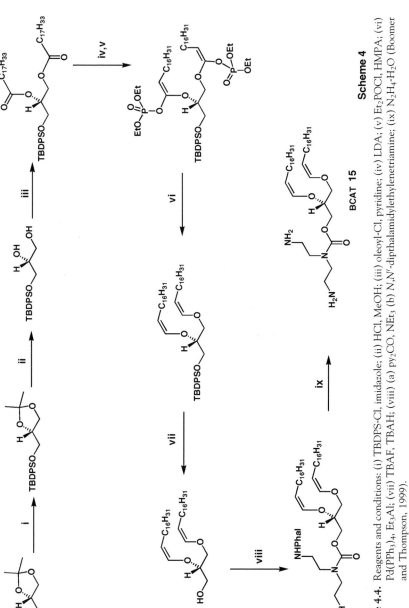

Scheme 4.4. Reagents and conditions: (i) TBDPS-Cl, imidazole; (ii) HCl, MeOH; (iii) oleoyl-Cl, pyridine; (iv) LDA; (v) Et₂POCl, HMPA; (vi) Pd(PPh₃)₄, Et₃Al; (vii) TBAF, TBAH; (viii) (a) py₂CO, NEt₃ (b) *N,N′*-dipthalamidylethylenetriamine; (ix) N₂H₄·H₂O (Boomer and Thompson, 1999).

Scheme 4

B. Characteristics of LD particles

Typically, cytofectin, and neutral lipid components are mixed together in an appropriate mol ratio and then induced or formulated into unilammellar vesicles by any one of a number of methods including reverse phase evaporation (REV), dehydration–rehydration (DRV) and extrusion (Miller, 1998a,b; Tagawa et al., 2002). Alternatively, cytofectins may be assembled into micellar structures after being dispersed in water or aqueous organic solvents (Miller, 1998a,b). Unilammellar vesicles or micelles may then be combined with nucleic acids to form nanometric LD particles. Biophysical structure/activity studies designed to understand the structures of LD particles and their relationships to LD transfection efficiency have been numerous. Unfortunately, the diversity of cytofectin structures, LD particles and biological targets has resulted in considerable inconsistency in the results reported by the research groups concerned. For instance, for LD transfection in vitro the optimal positive/negative charge ratio may be much higher than 1 (Behr et al., 1989; Remy et al., 1995), but more frequently the optimal ratio is reported to be closer to 1 (Alton et al., 1993; Felgner et al., 1994; Fife et al., 1998; McQuillin et al., 1997). Recently we observed that optimal in vitro LD transfection of COS-7 cells and even in vivo LD transfection of Balb/c mice lungs requires LD mixtures with an overall positive/negative charge ratio of <1 (Stewart et al., 2001). This observation has been supported by the results of others (Son et al., 2000).

Similarly diverse views exist concerning the structures of LD mixtures optimal for transfection. In some circumstances, LD mixtures optimal for in vitro transfection appear to be heterogeneous and polydisperse, consisting of a variety of particles and other structures all in dynamic equilibrium (Labat-Moleur et al., 1996; Zabner et al., 1995). These particles and other structures have been variously identified and described by a number of researchers and they include multilammellar lipid/nucleic acid clusters (>100 nm in diameter) (Gershon et al., 1993; Gustafsson et al., 1995; Hui et al., 1996; Lasic et al., 1997; Radler et al., 1997) perhaps with some surface-associated nucleic acids (Eastman et al., 1997), thinly lipid-coated DNA nucleic acid strands (Sternberg et al., 1994) and free nucleic acids (Gustafsson et al., 1995). Such structural observations have led to a substantive debate concerning the relative importance of each of these structural entities for efficient in vitro transfection.

However, LD mixtures optimal for in vitro transfection do not necessarily have to be heterogeneous and polydisperse. Our recent studies using sophisticated cryo-electron microscopy have clearly demonstrated that LD mixtures optimal for in vitro and in vivo transfection may actually consist of properly discrete LD particles (60–250 nm in diameter) exhibiting bilammellar perimeters and striations with a periodicity of 4.2 ± 2 nm, (Fig. 4.4) (Stewart et al., 2001). Small-angle X-ray scattering (SAXS) and other cryo-electron

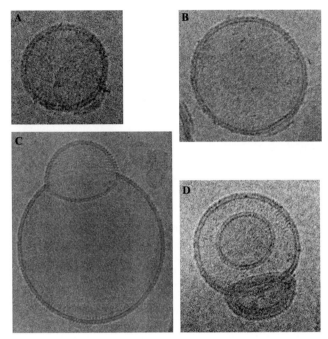

Figure 4.4. Cryo-electron microscopy images of LD particles. These LD particles were formed after the combination of CDAN/DOPE cationic liposomes and pDNA in the [cytofectin]/ [nucleotide] ([cyt]/[nt]) mol ratio of 0.6, optimal for *in vitro* and *in vivo* lung transfection. Final lipid concentration was 0.17 mM. Magnification is 300,000× (1 cm = 30 nm). Reprinted with permission of the Royal Society of Chemistry from Stewart *et al.* (2001).

microscopy studies of LD mixtures have revealed similar periodicities of approx. 6.5 and 3.5 nm that have been shown to result from the encapsulation of DNA molecules in regular periodic arrays within a multilammellar LD assembly (Lasic *et al.*, 1997; Pitard *et al.*, 1999; Radler *et al.*, 1997; Schmutz *et al.*, 1999; Xu *et al.*, 1999) (Fig. 4.5). Therefore, the observed LD particles are most likely composed in a similar way. Hence in this case at least, optimal LD transfection *in vitro* and *in vivo* must be primarily mediated by these discrete, multilammellar LD particles (Stewart *et al.*, 2001). The significance of discrete LD particles for optimal transfection has been supported by the results of at least one other published study comparing LD structure with *in vivo* transfection efficacy (Densmore *et al.*, 1999). Evidence then suggests that the regular multilamellar bilayer structure ($L_{\alpha I}$) of LD particles should undergo a phase change in the endosome forming inverted hexagonal phase structures (H_{II}) that may disrupt endosome membranes and facilitate nucleic acid escape into the cytosol (Koltover *et al.*, 1998) (Fig. 4.5). Lipids like DOPE **1** are well known to prefer H_{II} phases under

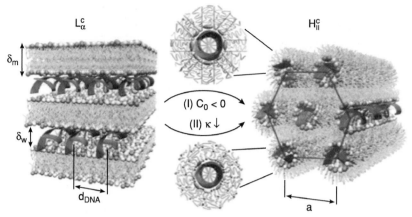

Figure 4.5. LD particle internal structure and dynamics. *Left-hand side:* schematic of the lamellar $L_{\alpha I}$ phase of DNA molecules interacting with cationic bilayers forming a multilayered assembly typical of LD particle composition. DNA double helices are shown as ribbons (blue and purple), head groups of anionic/zwitterionic lipids are shown as white spheres while those of cytofectins are shown as grey spheres. The notation δ_m refers to bilayer thickness, δ_w *to interbilayer separation and* d_{DNA} to DNA interaxial spacing. *Right-hand side:* conversion from lamellar $L_{\alpha I}$ phase to the columnar, inverted hexagonal H_{II} phase thought to be typical of LD particle composition during the transfection process, takes place by two possible routes. The first involves pathway (**I**) typified by negative curvature C_o induced in each cationic monolayer due to the presence of DOPE 1. The second involves pathway (**II**) typified by loss in membrane rigidity κ thereby encouraging phase inversion. Reprinted with permission of AAAS from Koltover *et al.* (1998). (See Color Insert.)

physiological conditions of temperature and pH, and the $L_{\alpha I} \rightarrow H_{II}$ phase transition has been widely implicated as a key facilitator of membrane fusion and membrane disruption events. Hence the inclusion of DOPE 1 in an LD system is likely to facilitate endosome escape of bound nucleic acids through induced $L_{\alpha I} \rightarrow H_{II}$ phase destabilization (Mui *et al.*, 2000).

General biophysical structure/activity studies have generated few proper correlations between LD particles structure, physical attributes and transfection efficiency *in vitro* (Miller, 2003), and in general there have been few attempts to derive unifying biophysical parameters able to account for differences in LD transfection efficiency *in vitro*, yet alone relate these parameters to *in vivo* transfection performance. One exception may be found in the work of Stewart *et al.* (Stewart *et al.*, 2001) wherein the physical properties of a systematic series of cationic liposomes and their corresponding LD mixtures were studied. Liposomes were formulated from DOPE 1 and cholesterol-based polyamine cytofectins such as CDAN 9 (Fig. 4.3). Successful *in vitro* transfection was linked to the ability of cationic liposome systems to (1) provide relatively inefficient neutralization, condensation and encapsulation of nucleic acids into LD particles; and (2) present unprotonated amine functional groups ($pK_a < 8$) at neutral pH with the capacity for substantial endosome buffering, thereby enabling the osmotic shock mechanism to facilitate nucleic acid escape from endosome compartments as their internal pH is reduced from pH 7 to 5.5 (Behr, 1996; Haensler and Szoka, 1993). Critically, both main factors were observed to be under the control of the cytofectin polyamine head group structure. The inclusion of "natural" propylene and butylene spacings between the amine functional groups of head groups appeared to promote efficient neutralization, condensation and encapsulation of nucleic acid. Inclusion of "unnatural" ethylene spacings appeared to promote the reverse effect, but at the same time assisted the perturbation of amine pK_a values from 9–10 to below 7. The appearance of such perturbed pK_a values now also appears to render LD particles metastable and prone to partial aggregation and sedimentation onto cell surfaces *in vitro* (Keller *et al.*, 2003b). Such sedimentation is likely to be beneficial for transfection unless the given LD particles themselves induce cytotoxicity. Consistent with this observation, a correlation has also been established between transfection efficiency and enhanced membrane fluidity in both lipoplex and cellular membranes (Ferrari *et al.*, 2002).

C. Ternary LD particles

Ternary LD particles represent another form of **AB** nanoparticle system. In these systems the cationic character of the cationic liposome/micelle system used to condense and/or encapsulate nucleic acids is supplemented by an additional cationic entity. In the case of pDNA, this cationic entity is frequently used to precondense the pDNA prior to final condensation and encapsulation by cationic liposomes/micelles. This is particularly true of the lipid:protamine:DNA (LPD) and liposome:mu:DNA (LMD) systems prepared using the salmon sperm-derived peptide protamine (Birchall *et al.*, 2000; Chesnoy and Huang, 2000; Dokka *et al.*, 2000; Li and Huang, 1997; Li *et al.*, 1998, 1999, 2000; Mizuarai *et al.*, 2001; Sorgi *et al.*, 1997; Whitmore *et al.*, 1999) and the adenoviral derived peptide μ (mu) respectively (Tagawa *et al.*, 2002). Other cationic entities that have been used include poly-L-lysine (Gao and Huang, 1996; Vitiello *et al.*, 1996), spermidine (Hong *et al.*, 1997), lipopolylysine (Zhou and Huang, 1994), histone proteins (Fritz *et al.*, 1996), chromatin proteins (Namiki *et al.*, 1998), human histone-derived peptides (Schwartz *et al.*, 1999), L-lysine containing synthetic peptides (Vaysse and Arveiler, 2000), not to mention a histidine/lysine (H-K) copolymer (Chen *et al.*, 2000). The formulation process of LMD is illustrated (Fig. 4.6). LPD particles may be prepared in a similar way.

Mature adenovirus consists of an icosahedral, non-enveloped capsid particle (approx. 90 nm) enclosing a core complex that consists of a linear

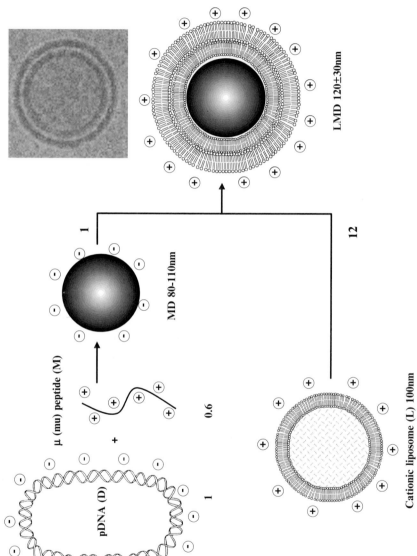

pDNA (D)

μ (mu) peptide (M)

1

0.6

MD 80-110nm

1

12

Cationic liposome (L) 100nm

LMD 120±30nm

dsDNA viral genome (~36 kbp) non-covalently associated with two cationic proteins (proteins V [pV] and VII [pVII]) and the 19-residue mu peptide (Anderson et al., 1989; Hosokawa and Sung, 1976). Mechanistic studies using the mu peptide have revealed how increasing pDNA-peptide interactions lead to progressive base-pair-tilting generating regions of high and low double helical stability, that in turn promote super-coiling followed by pDNA hydrophobic collapse (Preuss et al., 2003; Tecle et al., 2003). In kinetic terms, the process of pDNA condensation and the reverse process of pDNA expansion appear to be equivalent to small single domain protein folding and unfolding, respectively (Tecle et al., 2003). Chaotic behavior is also observed at low peptide/pDNA ratios (0.1–0.3 w/w) that becomes more uniform at higher ratios suggesting that with suboptimal ratios, pDNA is condensing in a multitude of conformations, each representing different stages of hydrophobic collapse in the search for the thermodynamically most stable (i.e., the fully condensed pDNA molecule). This represents yet another analogy with protein folding. At higher ratios, peptide/pDNA complexes formed appear to be increasingly irreversible consistent with the formation of kinetically and/or thermodynamically stable, condensed pDNA molecules (Tecle et al., 2003). Such stable states could create problems for the successful transcription of DNA post delivery to cells, yet another barrier to successful delivery of DNA to cells that is yet to be understood!

Both LPD and LMD systems are able to form discrete, essentially monodisperse particles. DOTAP/Chol-based LPD systems were even more effective and were found to formulate into discrete particles (approx. 135 ± 42 nm) (Li et al., 1998). DC-Chol/DOPE-based LMD systems were found to formulate into discrete mono-disperse particles (approx. 120 ± 30 nm) (Tagawa et al., 2002) (Fig. 4.6). LMD particles can be formulated reproducibly that are amenable to long-term storage at $-80\,^{\circ}\mathrm{C}$ and stable up to a pDNA concentration of 5 mg/ml (nucleotide concentration 15 mM), a concentration appropriate for facile use in vivo (Tagawa et al., 2002). Using LD systems, nucleotide concentrations >4 mM are difficult to achieve owing to ready LD particle aggregation above this concentration threshold (Cooper et al., 1998; Lee et al., 1996; Stewart et al., 2001). Moreover, LMD transfections appear to be significantly more time and dose efficient in vitro than LD transfections. LMD transfection times as short as 10 min and DNA doses as low as 0.001 μg/well

---

Figure 4.6. LMD formulation. Schematic illustration of LMD particle formation. Initially, pDNA (D) is introduced under vortex mixing to mu peptide (M) in the ratio of 0.6:1 w/w forming MD particles. These are then added under vortex mixing to cationic liposomes (L) in a ratio of 12:1 w/w with respect to pDNA, resulting in the formation of bilammellar LMD particles. *Inset:* cryo-electron microscopy image of LMD particle prepared with DC-Chol/DOPE cationic liposomes and pDNA (1 cm = 55 nm) (Tagawa et al., 2002).

result in significant gene expression. Furthermore, LMD transfections will also take place in the presence of biological fluids (*e.g.*, up to 100% serum), conditions typically intractable to LD transfections, suggesting that LMD formulations exhibit an additional element of stability. In consequence, LMD transfection of murine lung *in vivo* was up to six-fold more dose efficient than transfection with GL-67:DOPE:DMPE-PEG^{5000} (1:2:0.05 m/m/m) (one of the best synthetic non-viral vector systems reported to date for lung transfection). LMD has been called an artificial virus-like nanoparticle (VNP) on the basis that cryo-electron microscopy shows LMD particles to consist of a mu: DNA (MD) particle encapsulated within a cationic bilammellar liposome (Fig. 4.6).

However, this additional element of LMD stability is unlikely to be adequate for general *in vivo* applications and gene therapy (Perouzel *et al.*, 2003). Indeed, corresponding LPD particles are readily modified by serum causing gradual vector disintegration, release of DNA and probable reticulo-endothelial system (RES) scavenging (Li *et al.*, 1998, 1999). Released DNA is also noted to be susceptible to extracellular nuclease digestion. Furthermore, LPD particles have been found to promote a systemic, Th1-like innate immune response in mice, much more appropriate for a DNA vaccine than for gene therapy (Whitmore *et al.*, 1999). However, the general impression given is that LPD-like LMD systems could have a role to play clinically for the passive delivery of genes to lung but are not appropriate for targeted gene delivery to other tissues (Li *et al.*, 1999).

Studies carried out by confocal microscopy on dividing tracheal epithelial cells suggest that endocytosis is not a significant barrier to LMD transfection. However, the nuclear envelope remains a highly significant barrier. LMD particles were found to enter cells rapidly (mins), and disintegrate almost immediately leaving mu peptide free to migrate to the nuclear zone (within 15 min) and pDNA to enter after a further 15–30 min. There is every possibility that both cytofectin and perhaps even mu peptide are exercising fusogenic behavior with respect to early endosome membranes (Drummond *et al.*, 2000; Kamata *et al.*, 1994; Kichler *et al.*, 1997). However, LMD does not appear to facilitate pDNA entry into the nucleus of growth arrested (aphidicolin-treated) cells suggesting that the nuclear pore complex remains a significant barrier to LMD transfection even though mu peptide has been shown to possess strong nuclear localization sequence (NLS) characteristics (Keller *et al.*, 2003a). The obvious solution is to ensure that mu peptide and pDNA remain in association for long enough within non-dividing (quiescent) cells for the DNA to utilize the NLS characteristics of the mu peptide to cross the nuclear membrane (Keller *et al.*, 2003a; Preuss *et al.*, 2003). Evidence from DNA trafficking and expression studies using NLS peptides covalently or non-covalently associated with the pDNA appear to support this suggestion amply (Vaysse *et al.*, 2004; Zanta *et al.*,

1999), assisted by the presence of such elements as the SV40 enhancer in pDNA structure (Young *et al.*, 2003).

Very recently, a new ternary LD system known as the Multifunctional envelope-type nano-device (MEND) system was described (Kogure *et al.*, 2004). The formulation process compares in an interesting way to the LMD and LPD systems involving a cationic DNA/polycation complex interacting with an anionic fusogenic lipid film prior to sonication into large but discrete particles (402 ± 73 nm) whose charge can be modified by the post-insertion of stearyl octa-arginine (STR-R8) to give transfection competent particles (Fig. 4.7)— Without a doubt, an imaginative, alternative way to arrive at condensed discrete particles. In the cases of LMD, LPD, and perhaps MEND particles, these represent systems that can be formulated in a reproducible and scalable manner, that are resistant to aggregation in low ionic strength media, are amenable to long-term storage and give properly reproducible transfection outcomes. Therefore, these are ideal platforms upon which to build viable lipid-based synthetic, non-viral vector systems for DNA delivery *in vivo* by a process of modular upgrading through systematic chemical adaptation with appropriate tool-kits of known or newly designed chemical components.

III. ABD PARTICLES

A. Synthetic ABD particles

Some fascinating examples of **ABD** particles have emerged in recent years notable for some *in vivo* viability although somewhat irregular in formulation. For instance, peptides consisting of an oligo-L-lysine moiety linked to a peptide moiety specific for cell surface integrin proteins have been combined with LD systems (Colin *et al.*, 1998, 2000; Cooper *et al.*, 1999; Jenkins *et al.*, 2000). Credible enhancements of at least an order of magnitude in *in vitro* transfection have been observed over and above the results of binary LD transfection owing to the involvement of integrin-mediated cell uptake (Colin *et al.*, 1998, 2000; Cooper *et al.*, 1999). Furthermore, enhancements to *in vivo* transfection have been reported as well, but the mechanism of these so-called lipid: integrin-targeting peptide:DNA (LID) systems does not actually appear to be integrin-receptor dependent in this case (Jenkins *et al.*, 2000).

Modular adaptation of LMD particles has arguably resulted in alternative **ABD** systems whose behavior has given more clarity. A glyco-LMD variant was prepared by a *post-modification* strategy in which neoglycolipid micelles were combined with pre-formulated LMD particles (**AB** system) in order to encourage insertion of neoglycolipid molecules into the outer leaflet membranes of LMD particles using their hydrophobic lipid moieties. The syntheses of neoglycolipids

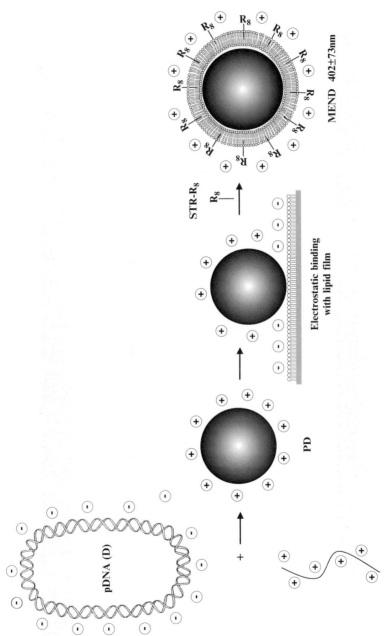

pDNA (D)

Polycation (P)

PD

Electrostatic binding
with lipid film

STR-R₈

MEND 402±73nm

is shown given the particular use of an aminoxy functional group to couple reducing sugars to the lipid moiety without the requirement for any protecting groups illustrating the high chemoselectivity of the coupling reaction (Scheme 4.5). This coupling reaction takes place in aqueous as well as organic solvents, ideal given the range of reducing sugars coupled (Perouzel *et al.*, 2003). The resulting glyo-LMD particles (**ABD** nanoparticles) were stable in high-salt medium (but not 100% serum) and mediated enhanced non-specific transfection of cells *in vitro* (Perouzel *et al.*, 2003). The aminoxy coupling reaction shown in this scheme is turning out to have widespread implications for the construction of viable **ABC** and **ABCD** systems (IC-Vec Ltd) for *in vivo* applications.

A peptido-LMD variant was also prepared very recently by a *pre-modification* strategy in which lipopeptides of two classes were formulated into cationic liposomes prior to LMD formulation. LMD formulations were prepared using both CDAN **9** and DC-Chol **8** cytofectins. The synthesis of one lipopeptide is shown, notable for the application of a novel variation of the Fukuyama-Mitsunobu reaction (Scheme 4.6) that now appears to have general applications in the synthesis of complex secondary amines (Waterhouse *et al.*, 2005). The peptide sequence used (tenascin peptide sequence: PLAEIDGIELA) was previously shown to target $\alpha_9\beta_1$-integrin proteins predominant on upper airway epithelial cells in mammals. When peptido-LMD systems were prepared using CDAN **9** cytofectin, we observed no evidence of receptor-mediated enhancement of transfection. Instead, even with as little as 0.05 mol% of lipopeptide present in each peptido-LMD particle, transfection was at least 10-fold more effective than found for corresponding LMD systems without peptide present, irrespective of whether the cells under investigation expressed $\alpha_9\beta_1$-integrin proteins or not! Such non-specific peptide enhancement is interesting but not necessarily desirable. When peptido-LMD systems were prepared using DC-Chol **8** cytofectin, thereby reducing the overall positive charge of each particle, a modest element of specific enhancement was observed (2-fold) over a general background enhancement that was otherwise non-specific (Waterhouse *et al.*, 2005).

Non-specific enhancements appear to be a hazard (or perhaps even an advantage under some circumstances) of using **ABD** systems. There has been some apparent success in using neoglycolipids as **D**-layer targeting agents inserted

---

Figure 4.7. MEND formulation. Schematic illustration of MEND particle formation. Initially, cationic PD particles are formed from pDNA (D) and cationic polymer (P) (usually pLL). These associate electrostatically with a negatively charged mono-layer lipid film and are then encouraged to form particles by a process of hydration and sonication. Final post-modification with STR-R$_8$ results in the formulation of cationic MEND particles (see text for references).

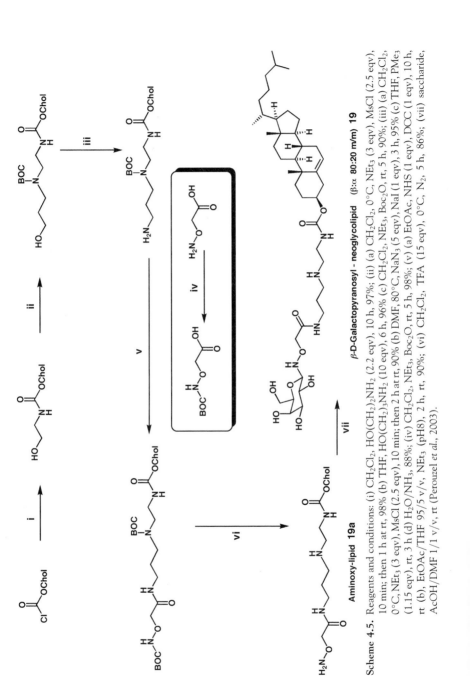

Scheme 4.5. Reagents and conditions: (i) CH$_2$Cl$_2$, HO(CH$_2$)$_2$NH$_2$ (2.2 eqv), 10 h, 97%; (ii) (a) CH$_2$Cl$_2$, 0°C, NEt$_3$ (3 eqv), MsCl (2.5 eqv), 10 min; then 1 h at rt, 98% (b) THF, HO(CH$_2$)$_3$NH$_2$ (10 eqv), 6 h, 96% (c) CH$_2$Cl$_2$, NEt$_3$, Boc$_2$O, rt, 5 h, 90%; (iii) (a) CH$_2$Cl$_2$, 0°C, NEt$_3$ (3 eqv), MsCl (2.5 eqv), 10 min; then 2 h at rt, 90% (b) DMF, 80°C, NaN$_3$ (5 eqv), NaI (1 eqv), 3 h, 95% (c) THF, PMe$_3$ (1.15 eqv), rt, 3 h (d) H$_2$O/NH$_3$, 88%; (iv) CH$_2$Cl$_2$, NEt$_3$, Boc$_2$O, rt, 5 h, 98%; (v) (a) EtOAc, NHS (1 eqv), DCC (1 eqv), 10 h, rt (b), EtOAc/THF 95/5 v/v, NEt$_3$ (pH8), 2 h, rt, 90%; (vi) CH$_2$Cl$_2$, TFA (15 eqv), 0°C, N$_2$, 5 h, 86%; (vii) saccharide, AcOH/DMF 1/1 v/v, rt (Perouzel *et al.*, 2003).

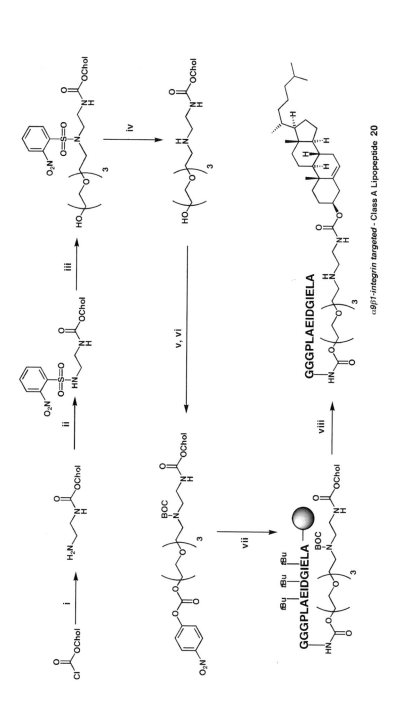

α9β1-integrin targeted - Class A Lipopeptide 20

into LD complexes. For instance, Behr *et al.* reported *in vitro* galactose-receptor mediated uptake into hepatoma cells of LD complexes formulated with a triantennary galactolipid (Remy *et al.*, 1995). By contrast, Kawakami *et al.* have suggested liver targeting *in vivo* using LD complexes formulated with a mannosyl neoglycolipid as targeting agent. However, non-specific mannosyl-induced LD stabilization leading to longer circulation times seems to be sufficient explanation to account for these results too (Kawakami *et al.*, 2000, 2001). An additional example of non-specific enhancements in **ABD** systems has been provided from experiments with the Transferrin (Tf) protein. Tf has been used quite frequently for **D**-layer biological targeting on the basis that transferrin receptors (TfR) are found routinely at the surface of vascular endothelial cells associated with tumors or the blood brain barrier and are rapidly internalized upon binding transferrin (da Cruz *et al.*, 2001; de Ilarduya and Duzgunes, 2000; Joo and Kim, 2002; Simoes *et al.*, 1999; Tan *et al.*, 2001; Yanagihara *et al.*, 2000). However, whilst *in vitro* and *ex vivo* transfections with a Tf-LD **ABD** system is enhanced relative to binary LD transfection, the mechanism is quite clearly TfR independent, the protein instead acting primarily to promote endosome disruption and subsequent escape of complexed DNA into the cytosol (da Cruz *et al.*, 2001; Simoes *et al.*, 1999), and even entry into the nucleus (Joshee *et al.*, 2002). In addition, Tf is an acidic protein, negatively charged at neutral pH. Accordingly, the association of Tf with binary LD systems seems to reduce the overall positive charge and provides simultaneously a combined stereo-electronic barrier to biological fluid components allowing *in vitro* transfection to take place, even in 60% serum. In the latter context, human serum albumin (HSA) has been deliberately combined with binary LD systems in order to create negatively charged, sterically protected complexes appropriate for *in vitro* transfection in the presence of up to 30% serum and even for lung or spleen transfection *in vivo* (Simoes *et al.*, 2000).

Chang *et al.* have suggested that a simple reformulation of Tf-LD **ABD** systems is sufficient to give a highly compact particle with a relatively uniform size (50–90 nm) comprising a dense Tf-DNA core enveloped by a membrane coated with Tf molecules spiking the surface. This system appears to render enhanced stability, improved *in vivo* gene transfer efficiency, and long-term

---

Scheme 4.6. Reagents and conditions: (i) $H_2NCH_2CH_2NH_2$ (200 eqv), 2 days, 65%; (ii) (a) 2-NsCl (1.3 eqv), NEt_3 (1.5 eqv), CH_2Cl_2, 14 h, 87% (b) BnBr (1.1 eqv), Ag_2O (1.5 eqv), 20 h, generating 55%; (iii) tetraethyleneglycol (TEG) (1.3 eqv), DTBAD (1.5 eqv) slow addition over 1 h in CH_2Cl_2, PPh_2py (1.5 eqv), CH_2Cl_2, 3 h, 71%; (iv) Na (10 eqv), $C_{10}H_8$ (10 eqv), $-30°C$, 45 min, 74%; (v) Boc_2O (1 eqv), NEt_3 (1.1 eqv), CH_2Cl_2, 10 h, 84%; (vi) NEt_3 (2 eqv), DMAP (2 eqv), *p*-nitrophenyl chloroformate (3 eqv), CH_2Cl_2, 10 h, 92%; (vii) fully protected peptide (Fmoc deprotected on *N*-terminus) (0.5 eqv), NEt_3 (2.5 eqv), DMF, 18 h; (viii) 95% v/v TFA/H_2O, 90 min, 10% over two steps (Waterhouse *et al.*, 2005).

efficacy for systemic p53 gene therapy of human prostate cancer when used in combination with conventional radiotherapy (Xu *et al.*, 2002a). Others have reported the need to introduce protamine, giving the equivalent of a Tf-LPD **ABD** system that is able to transfect cells in a number of tissues post intravenous (i.v.) injection (de Ilarduya *et al.*, 2002); otherwise alternative Tf-LD **ABD** systems have been administered directly (intra-tumoral injection) to subcutaneous mouse xenograft models of human prostate cancer (Seki *et al.*, 2002), or else by intra-arterial (i.a.) administration into hepatocellular carcinoma (Seol *et al.*, 2000). Such data is certainly consistent with the possibility of some TfR-mediated uptake of Tf-LD particles by some tissues (Voinea *et al.*, 2002).

Chang *et al.* have also developed alternatives to the Tf-LD **ABD** systems. For instance, using an anti-TfR antibody variable region fragment (anti-TfR scFv), they have produced an anti-TfR scFv-LD system with the scFv covalently attached to a number of cytofectins that appears to show some promise for systemic p53 gene therapy in a number of human tumor models including human breast cancer metastasis (Xu *et al.*, 2001, 2002b). This anti-TfR scFv-LD system has been further improved by the inclusion of a cationic peptide (HoKC) to precondense pDNA during complex formation (i.e., in a similar way to LMD and LPD systems) (Yu *et al.*, 2004b). However, the final *coup-de-grâce* has been to demonstrate that anti-TfR scFv-containing **ABD** systems are in fact inferior to a complete **ABCD** system in which a post-coating strategy was employed taking preformed LD particles (**AB** core particles) that were sequentially conjugated with PEG polymer (**C**-layer) and then anti-TfR scFv (**D**-layer) (Yu *et al.*, 2004a). However, there may yet be a future for simple monoclonal antibody (MAb)-LD **ABD** systems with or without covalent coupling of the antibody to cytofectins (Mohr *et al.*, 1999; Tan *et al.*, 2003); and even for **ABD** systems involving lectins (Yanagihara and Cheng, 1999)!

B. Semi-synthetic ABD particles

Difficulties experienced in working with fully synthetic **ABD** particles have also resulted in the development of some semi-synthetic virosome systems. The term virosome was originally coined in reference to combinations of liposomes and various virus glycoproteins but is now more generally used to refer to various types of viral/non-viral hybrid vector systems. Of these, the HVJ-liposome system is instructive. This semi-synthetic system is prepared from a combination of UV-irradiated virions of the Hemagglutinating Virus of Japan (HVJ; Sendai virus) and liposomes in which are encapsulated nucleic acids complexed with the High Mobility Group 1 (HMG-1) protein (Kaneda, 1999; Yonemitsu *et al.*, 1998). The HMG-1 protein is there to assist nuclear access and localization of delivered nucleic acids as well as promoting gene stabilization within the nuclear envelope (Kaneda, 1999; Kaneda *et al.*, 1989). Conventional HVJ-liposomes are negatively charged (Aoki *et al.*, 1997; Kaneda, 1999; Morishita *et al.*, 1994; Yonemitsu *et al.*, 1998), however an HVJ-cationic liposome system has recently been developed, based on the cytofectin DC-Chol **8**, that appears to transfect various mammalian cell types *in vitro* 100–800-fold more effectively than conventional HVJ liposomes (Saeki *et al.*, 1997). In addition, HVJ-cationic liposomes prepared with the cytofectin N-(α-trimethylammonioacetyl)-didodecyl-D-glutamate chloride (TMAG) **14** (Fig. 4.3), have also proved able to mediate delivery of nucleic acids to tracheal and bronchiolar epithelial cells *in vivo* with reasonable efficiency (Yonemitsu *et al.*, 1997). One major reason for the success of HVJ-liposome systems is the presence of the hemagglutinin-neuraminidase (H_N) and fusion (F) glycoproteins in the liposome bilayer (Fig. 4.8). These are fusogenic proteins that allow HVJ-liposomes to interact with cell surface sialic residues, fuse with the cell membrane and then release encapsulated nucleic acids directly into the cytosol, bypassing endocytosis altogether (Yonemitsu *et al.*, 1998). For this reason, HVJ liposomes have also been called fusogenic liposomes (Nakanishi *et al.*, 1999). The clear success of HVJ-cationic liposomes has resulted in the development of a number of other cationic virosome systems including systems prepared with the influenza membrane fusion protein hemagglutinin that were used to deliver genes to cells *in vitro* (Schoen *et al.*, 1999), cationic lipid-reconstituted influenza-virus envelopes used to deliver an ODN to cells *in vitro* (Waelti and Gluck, 1998), and LD complexes prepared from DOTMA/DOPE cationic liposomes and pDNA doped with the partially purified G glycoprotein of the vesicular stomatitis virus envelope (VsV-G) (Abe *et al.*, 1998). These semi-synthetic **ABD** nanoparticle systems are sure to remain of great interest going forward.

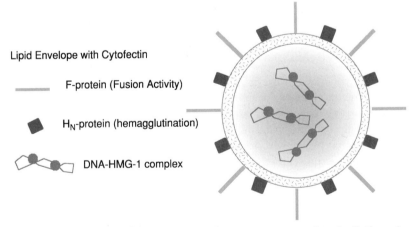

Figure 4.8. HVJ-cationic liposome system. Cytofectins are incorporated in the lipid envelope. Reprinted with permission of Bios. Scientific Publishers from Miller (1999). (See Color Insert.)

IV. ABC PARTICLES

Particles of this type appear to be altogether more promising for *in vivo* applications and gene therapy owing to passive targeting of particles stabilized with the aid of **C**-layer stealth/biological stabilization molecules. Passive targeting is the process by which stabilized nanoparticles accumulate with time into organs of choice by means of long-term circulation in biological fluids, without the requirement for **D**-layer active targeting agents. By far the most popular **C**-layer molecule is polyethyleneglycol (PEG). PEG provides a steric barrier to interaction with biological fluid components and prevents uptake of liposomal vesicles by cells of the RES (Lasic and Papahadjopoulos, 1995; Miller, 1998b). Safinya *et al.* have recently demonstrated that only PEG with a molecular weight of 2000 Da and above gives adequate stealth characteristics (Martin-Herranz *et al.*, 2004). Hong *et al.* reported one of the first attempts to generate an **ABC** complex (Hong *et al.*, 1997). In this instance, DDAB/Chol cationic liposome-based LD particles were stabilized for storage by inclusion of N-{ω-methoxypoly (oxyethylene)-α-oxycarbonyl}-DSPE (PEG-PE) and partially stabilized in the circulation *in vivo*. However, >1 mol% of PEG-PE proved sufficient to reduce lung *in vivo* transfection efficacy to a fraction of the transfection level mediated by DDAB/Chol cationic liposomes. This alone is indicative of a necessary compromise between the requirement to include PEG-PE for stabilization purposes countered by the requirement to keep levels modest in order to prevent "steric blocking" of LD transfection.

There are essentially three different ways in which **ABC** nanoparticles may be formulated with an exterior PEG **C**-layer (Fig. 4.9). These are:

1. *Pre-modification:* where a PEG-lipid is formulated into cationic liposomes prior to the addition of nucleic acids (Hong *et al.*, 1997).
2. *Post-modification:* where PEG-lipids in the form of micelles are combined with preformulated **AB** nanoparticle systems in the expectation that free and micellar PEG-lipids will transfer from free solution or micellar state and insert their hydrophobic lipid moieties into the outer leaflet membranes of **AB** nanoparticles (Perouzel *et al.*, 2003).
3. *Post-coupling:* where PEG-polymers are equipped with reactive functional groups that bioconjugate in aqueous conditions with complementary functional groups presented on the outside surface of the **AB** nanoparticle (Keller *et al.*, 2003a; Perouzel *et al.*, 2003).

Free amino functional groups on the surface of LMD particles can be modified easily by post-coupling with a PEG-succinide ester, giving a simple LMD-based **ABC** nanoparticle (Keller *et al.*, 2003a). Perhaps surprisingly, these particles were observed to enter cells with ease even though prevailing opinion would

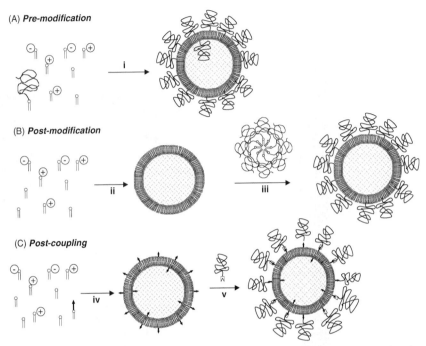

Figure 4.9. C-layer stealth molecule incorporation strategies. Top: *Pre-modification* implies that PEG lipids are incorporated into cationic liposomes directly (step i) prior to the addition of any given nucleic acid; Middle: *Post-modification* strategy implies that simple cationic liposome (or LD, LsiR, LMD or LPD-like) systems are prepared in advance (step ii) and then PEG lipid micelles are incubated (step iii) with the liposome (or LD, LsiR, LMD or LPD-like) particles to encourage micelle breakdown and insertion of PEG lipids via their hydrophobic moieties into the outer leaflet membrane of whichever type of particle is being prepared; Bottom: *Post-coupling* strategy implies that simple cationic liposome (or LD, LsiR, LMD or LPD-like) systems are formulated with a coupling-lipid that enters lipid membranes (step iv). This coupling-lipid comprises a polar, functional group (black arrow) with very high chemoselectivity for certain complementary functional groups introduced into the termini (checked complementary-arrow shape) of modified PEG-molecules, allowing for highly efficient coupling in the subsequent conjugation step (step v). Post-coupling is flexible enough to allow for the introduction of other biological compatibility/stealth molecules and/or biological targeting molecules (see text for references).

have suggested that the PEG stealth layer might provide a steric barrier to cellular uptake. Instead, cellular uptake was found to be rapid (mins) and substantial, but particles appeared to be entrapped in endosome compartments post cell entry and no measurable transfection was observed (Keller *et al.*, 2003a; Perouzel *et al.*, 2003). PEG has undeniable biocompatibility/stealth properties, but clearly facilitates the cellular uptake of attached nanoparticles in spite of these effects. Unfortunately, PEG also appears to completely block subsequent endosome escape. The obvious solution appears to be some form of triggered release of attached PEG once nanoparticles become trapped in endosome compartments revealing naked LMD particles (**AB** core nanoparticles) that are able to effect endosmolysis and onwards transfection (Keller *et al.*, 2003a). Other polymers such as pluronic acid and oligosaccharides also promote DNA uptake into cells (Chen *et al.*, 2004; Perouzel *et al.*, 2003). This appears fundamental. Any given **ABC** (or even **ABCD**) systems delivering DNA is only likely to be properly clinically viable once triggerable, meaning that they should be completely stable and non-reactive in extracellular fluids but unstable once recognized and/or internalized by target cells in the organ of choice. This paradox goes to the heart of the matter. **ABC** (or even **ABCD**) nanoparticles that are not triggerable cannot be effective particles for DNA delivery by their very nature.

In the absence of triggered release, time-dependent release of PEG-lipids has turned out to be a reliable if not entirely effective alternative. According to this approach, PEG-lipids with hydrophobic moieties of variable chain-length will have variable affinities for the outer leaflet membrane of the **AB** core particle into which they are inserted. The shorter the alkyl chain the lower will be the affinity and the lower will be the PEG-lipid residence time in the membrane (usually referred to in terms of half-life $t_{1/2}$). Ideally, PEG-lipids should be retained as far as the organ of choice *in vivo* and even up until the target cells, before dissociation and the release of naked **AB** core nanoparticles to enter cells. This characterizes something of the rationale leading to the stabilized plasmid-lipid particle (SPLP) systems, the most developed of **ABC** nanoparticle systems to date. The first generation SPLP system contained DOPE 1 (84 mol%), low levels (6 mol%) of cationic lipid dioleyldimethylammonium chloride (DODAC) and quite high levels of a PEG-Ceramide with an arachidoyl acyl group (PEG-CerC$_{20}$) (10 mol%) (Wheeler *et al.*, 1999). The surface tenacity of PEG-CerC$_{20}$ ($t_{1/2} > 13$ days) proved such an intractable steric barrier to transfection *in vitro* that PEG-CerC$_{20}$ was replaced by PEG-CerC$_8$ ($t_{1/2} < 1.2$ min) with an octanoyl acyl group. Entrapment of pDNA was then accomplished by a detergent dialysis procedure (55–70% efficient), giving second-generation DOPE:DODAC:PEG-CerC$_8$ SPLP particles containing DODAC (24–30 mol%) and PEG-CerC$_8$ (15 mol%) (diameter approx. 100 ± 40 nm) (Zhang *et al.*, 1999), that were able to effect transfection of cells *in vitro* and regional delivery *in vivo*. The formulation procedure is illustrated diagrammatically (Fig. 4.10). One of the most important aspects about SPLP particles is their very structural integrity (no changes in size or DNA encapsulation at 4°C for 5 months).

SPLP particles were designed for passive targeting. That is to say, particles were designed for long term circulation *in vivo* enabling the gradual

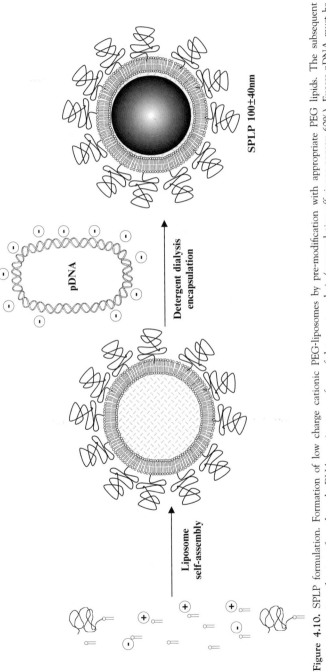

Figure 4.10. SPLP formulation. Formation of low charge cationic PEG-liposomes by pre-modification with appropriate PEG lipids. The subsequent introduction of condensed pDNA requires some form of detergent analysis (encapsulation efficiency approx 60%). Excess pDNA must be removed afterwards by column chromatography in order to yield stabilized plasmid-lipid particles (SPLP) ready for use (see text for references).

Liposome self-assembly

pDNA

Detergent dialysis encapsulation

SPLP 100±40nm

partition of particles into interstitial spaces in diseased tissue (such as tumor) by extravasation from blood via the enlarged sinusoidal gaps that typically exist between the endothelial cells which line the vasculature in diseased and inflamed organs. Given the severity of the extracellular *in vivo* environment, only PEG-CerC$_{20}$ SPLP particles were found to give modest, detectable transfection in animals post intravenous administration (Tam *et al.*, 2000). Other systems, such as the PEG-CerC$_8$ SPLP particles, were not sufficiently robust to effect passive targeting even though their transfection properties were, in principle, superior. In other words, in seeking to find the balance between extracellular stability and the instability necessary to promote transfection, the requirement for extracellular stability was too overwhelming with the result that SPLP particles were developed with appropriate extracellular stability but insufficient local instability for effective transfection once at the target cells. Improvements continue to be made in order to increase DNA encapsulation efficiency and to improve transfection efficiency (Lee *et al.*, 2003; Palmer *et al.*, 2003, 2004; Saravolac *et al.*, 2000). SPLP systems are now under evaluation in Phase I clinical trials (Protiva, unpublished data), the first **ABC** nanoparticle to be so evaluated. Data is eagerly awaited. Undoubtedly, SPLP systems represent a key synthetic, non-viral platform technology, and one that others are aiming to increasingly emulate and improve upon (Hayes *et al.*, 2004).

The absence of triggered release is still perceived to be the main limitation of SPLP systems. Accordingly, Szoka *et al.* have assembled their own SPLP system taking advantage of a designed ortho ester PEG-lipid known as polyethyleneglycol-diorthoester-distearoylglycerol conjugate lipid (POD) **21**. POD **21** is one of the best bespoke pH-triggered PEG-lipids to date and the structure and synthesis are shown (Guo and Szoka, 2001; Guo *et al.*, 2004) (Scheme 4.7). The key design feature of such a pH-triggered PEG-lipid is that the triggerable linker should be completely stable at pH 7 and sufficiently destabilized at pH 5.5 to completely and irreversibly dissociate, at the very least, within 1 hour. Such a requirement is severe but is essential to ensure quantitative release of nucleic acid from endosomes! Biophysical release studies were performed with POD- loaded liposomes suggesting that $t_{1/2}$ at pH 5.5 was approx. 10 min (Guo *et al.*, 2003). Hence, proof of concept studies were carried out with a POD-SPLP formulation (DOPE/DOTAP/POD 68:12:20 m/m/m) into which pDNA was encapsulated (40–45% efficiency) by detergent dialysis (as above) giving 60 nm particles. These were found to mediate transfection *in vitro* much less effectively than simple DOTAP/DOPE cationic liposomes mixed with pDNA. However, POD-SPLP systems were up to three orders of magnitude more effective at transfection than equivalent pH-insensitive nanoparticle systems formulated with PEG-DSG rather than POD **21** (Choi *et al.*, 2003). Both POD-SPLP particles and particles of the equivalent pH-insensitive systems were found to enter cells in line with data obtained with LMD-based

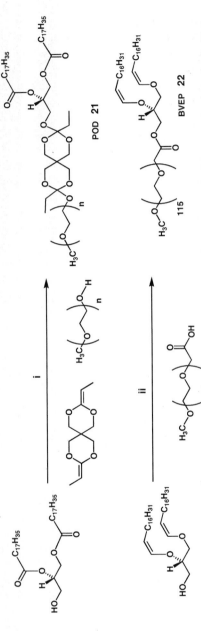

Scheme 4.7. Reagents and conditions: (i) pTSA, THF, 40°C; (ii) DCC, DMAP, methoxy-poly(ethyleneglycol)-carboxymethyl (MPEGA) (Boomer and Thompson, 1999; Guo and Szoka, 2001).

ABC nanoparticles (Keller *et al.*, 2003a). Therefore, the clear implication is that the enhanced transfection efficiency of the POD-SPLP system was the result of the triggered release of PEG in the endosome leading to considerably enhanced endosmolysis and pDNA escape to the nucleus. Proof of concept studies *in vivo* are now awaited, so too are alternative next generation triggerable ABC nanoparticle systems for DNA delivery.

Other ABC nanoparticles have been described mostly with regard to triggered or time dependent release properties and provide useful supporting studies. For instance, ABC nanoparticle systems have recently been constructed using short- and long-chain SAINT-PEG lipids constructed from pyridinium cytofectins that have variable length residence times in the same way that PEG-ceramide lipids do (Rejman *et al.*, 2004). Alternatively, Thompson *et al.* have adapted their chemical routes to cytofectin BCAT and diplasmenylcholine in order to prepare a novel acid labile PEG-lipid (R)-1,2-di-O-(1'Z, 9'Z-octadeca-dienyl)-glyceryl-3-(ω-methoxy-poly(ethylene) glycolate 5000) (BVEP) **22** (Boomer and Thompson, 1999; Boomer *et al.*, 2003; Thompson *et al.*, 2004) (Scheme 4.7). Elegant though the idea is, biophysical release studies were performed with BVEP-loaded liposomes suggesting that $t_{1/2}$ at pH 4.5 was approx. 4 h (Boomer *et al.*, 2003). In this respect, the vinyl ether functional group is probably just too stable to provide a triggered release that is rapid and effective enough. Otherwise, two other studies are worthy of note. PEG-lipids appear to stabilize LD particles prepared from bis(guanidinium)-*tren*-cholesterol (BGTC)/DOPE liposomes. Subsequent ABC nanoparticles will transfect murine lung *in vivo* even without triggered release suggesting that topical lung administration may be exceptional (Pitard *et al.*, 2001). Finally, Scherman *et al.* have been demonstrating how the judicious introduction of anionic PEG moieties can tune the systemic circulation lifetimes of ABC nanoparticles (Nicolazzi *et al.*, 2003b, 2004).

V. ABCD PARTICLES

For most *in vivo* applications and gene therapies, true **ABCD** nanoparticle systems are perhaps the best proposition. The number of these is growing in spite of the obvious technical problems surrounding reproducible and scalable formulation of an **AB** core particle alongside controlled and reliable association of **C** and **D** layer molecules. This remains an ongoing problem. Pardridge *et al.* have reported really impressive results using an **ABCD** system comprising an LD core particle prepared from cationic liposomes with minimal cytofectin. This LD core is doped with PEG-PE variants, one for stabilization and one for the covalent attachment of an anti-TfR monoclonal antibody (OX26) specific for TfR that is enriched at the blood brain barrier (BBB) and also in peripheral organs such as

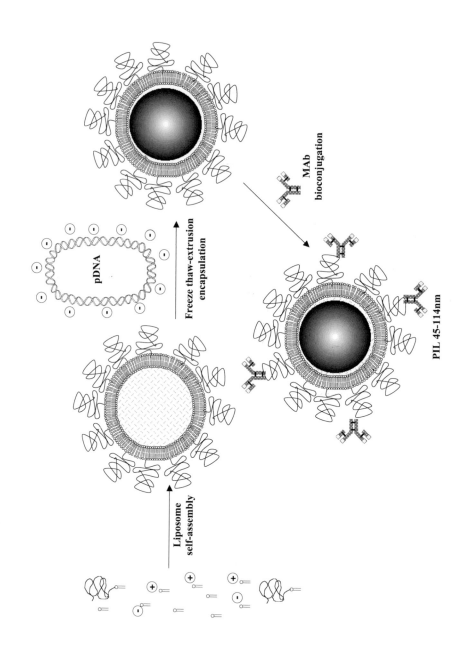

pDNA

Freeze thaw-extrusion
encapsulation

MAb
bioconjugation

Liposome self-assembly

PIL, 45-114nm

liver and spleen (Shi and Pardridge, 2000; Shi *et al.*, 2001). Their main **ABCD** system is comprised of POPC/DDAB/DSPE-PEG^{2000} (19.2:0.2:0.6 m/m/m) liposomes where the DSPE-PEG^{2000} is distributed DSPE-PEG^{2000}: DSPE-PEG^{2000}-Maleimide (95:5 m/m). Particles were prepared by initial mixing of all of the lipids together in chloroform solution followed by solvent evaporation, rehydration, sonication, pDNA addition, concluding with multiple freeze-thaw cycles and extrusion. This arduous process (20% efficient) yielded **ABC** nanoparticles that were coupled to OX26 antibody overnight, yielding complete **ABCD** particles (35–50 OX26 MAb/particle; 45–114 nm) after final Sepharose Cl4B gel filtration to remove excess unreacted antibody (Shi *et al.*, 2001) (Fig. 4.11).

These **ABCD** nanoparticles are almost completely neutral in charge and do not possess any triggered release system. Therefore, the impressive biological data showing transfection in liver, spleen and brain must be a consequence of active targeting mechanisms involving TfR interactions. From these beginnings, Pardridge *et al.* have further developed the formulation protocols for these pegylated immunoliposome (PIL) systems and demonstrated that PILs have impressively low levels of associated systemic toxicity (Zhang *et al.*, 2003b). Additional applications of PILs include the delivery of pDNA-directed epidermal growth factor receptor (EGFR) antisense mRNA using particles "armed" with both an anti-TfR MAb and an anti-insulin receptor (INSR) MAb, the first to promote crossing of the BBB and the second to promote transport of pDNA to the nucleus across plasma and nuclear membranes in the target brain tumor (Zhang *et al.*, 2002a,b). Dual targeting was deemed essential to ensure targeting across both the tumor cell membrane and microvasculature barrier to deep within cancer cells. Furthermore, TfR MAb-targeted PILs were used to mediate pDNA-directed short hairpin interference RNA (shRNAi) downregulation of transgenic luciferase activity *in vivo* in rat cranial brain tumors by up to 90% (Zhang *et al.*, 2003a).

Others are also moving towards success with zathe active targeting of tumors after i.v. injection of alternative **ABCD** systems prepared using MAbs or folate as the targeting ligand (Rejman *et al.*, 2004; Hofland *et al.*, 2002; Reddy and Low, 2000). One of the most interesting variations on this theme has been the creation of folate receptor (FR)-targeted DDAB/CHEMS/f-PEG-PE liposomes comprising folate (f) conjugated PEG-PE (f-PEG-PE), cholesteryl

---

Figure 4.11. PIL formulation. Formation of low charge cationic PEG-liposomes by pre-modification with appropriate PEG lipids. The subsequent introduction of condensed pDNA requires extensive freeze-thaw and extrusion steps (encapsulation efficiency approx 20%). Excess pDNA must be removed afterwards. Final monoclonal antibody (MAb) coupling then takes place to yield pegylated immunoliposomes (PIL) particles ready for use (see text for references).

hemisuccinate (CHEMS) and the cytofectin dimethyldioctadecylammonium bromide (DDAB) **6**, that are combined with poly-L-lysine (pLL) condensed pDNA to give **ABCD** nanoparticles competent to mediate FR-specific delivery of pDNA to cells *in vitro* (Shi *et al.*, 2002). However, whether or not this system will operate well *in vivo* remains to be seen.

VI. THE FUTURE ROLE OF CHEMISTRY—B, C, AND D LAYER INNOVATIONS

Previously, we noted that platform technologies like LMD and SPLP systems should be the only meaningful way forward for cationic liposome/micelle-based systems for *in vivo* applications and gene therapy. Our reasons for stating this were that these systems represent well-characterized transfection vehicles constructed from tool-kits of well-defined chemical components, that can be formulated in a reproducible and scalable manner, giving rise to reproducible transfection outcomes. We would now like to take the opportunity to update our comments in light of the self-assembly **ABCD** nanoparticle concept presented in this review. Hence, in our revised view the most appropriate way forward for cationic liposome/micelle-based synthetic non-viral vector systems, is now the creation of **ABC** and **ABCD** nanoparticle systems that have been self-assembled in a modular and sequential fashion from tool-kits of well-defined chemical components. These systems must formulate with nucleic acids of choice in a reproducible and scalable manner giving discrete and well-defined particles (with a narrow polydispersity centered on 100 nm diameter or less). Furthermore, these particles should be able to give reproducible transfection outcomes (or siFection outcomes, as appropriate) *in vivo* with minimal toxicity. The combination of all these features should ensure regulatory confidence in the therapeutic applications of such **ABC** and **ABCD** nanoparticle systems. Furthermore, the very self-assembly and modular build characteristics of such nanoparticle systems should ensure that nanoparticles can be tailor-made for individual nucleic acid delivery requirements by using a range of different tool-kits of well-defined chemical components. It is in the synthesis and integration of these tool-kits of chemical components that the future opportunity for chemistry now lies.

As stated above, in our opinion, the cytofectin (**B**-layer) field is saturated. There are numerous examples of cytofectins in the literature and the field of non-viral gene therapy is unlikely to benefit from further additions. However, there is real benefit in the synthesis of new lipids to facilitate the attachment and function of the biological compatibility **C**-layer. For instance, lipids containing the aminoxy functional group shown above (Scheme 4.5) can be prepared for formulation into complexes with nucleic acids and then used to post-couple PEG-dialdehydes in aqueous medium. The result of this aqueous post-coupling procedure appears to be the highly efficient, reliable and non-disruptive introduction of a biocompatible/stealth **C**-layer resulting in very high quality **ABC/ABCD** nanoparticles (IC-Vec Ltd, unpublished results). Mindful of this technology, IC-Vec Ltd has gone on to develop siFECTplus™ nanoparticles for the functional delivery of siRNA to cells in organs *in vivo* (IC-Vec Ltd, unpublished results). Publication of relevant data and a description of these siFECTplus™ nanoparticles are expected next year. Post coupling through aminoxy functional groups is clearly potentially effective, but what of other functional groups? Alternatively, requirements for **B** and **C** layer innovation could combine in the quest for alternatives to the pH-triggerable PEG-lipids of Szoka or Thompson *et al.* (Scheme 4.7). The recent review of Guo and Szoka (Guo and Szoka, 2003) was compiled not only to illustrate pH triggering but also redox potential, temperature and even enzymatic triggering processes. There is plenty of room for chemical innovation here!

Then there is the question of PEG itself. This remains the mainstay for most *in vivo* applications involving viable **ABC** and **ABCD** nanoparticles. However, this large and unwieldy biocompatibility/stealth molecule has already been shown to be refractory for transfection with pDNA (Keller *et al.*, 2003a; Perouzel *et al.*, 2003). Triggered release of PEG from the **AB** core once nanoparticles have entered cells, seems imperative in order for effective pDNA transfection to take place (Choi *et al.*, 2003), although the presence of attached PEG may in fact be much less a problem for siFection (delivery of siRNA) (IC-Vec Ltd, unpublished data). Nevertheless, efforts should be put into finding alternative hydrophilic polymers that can mimic the biocompatibility and stealth properties of PEG without the refractory characteristics and lack of biodegradability. Some alternative hydrophilic polymers have already been described by Seymour *et al.*, including poly-[N-(2- hydroxypropyl)methacrylamide] (pHMPA) (Dash *et al.*, 2000). Once again, there should be a host of other possibilities once the enthusiasm of polymer chemists can be engaged on this problem. Once prepared, each prospective surrogate of PEG will have to be rigorously evaluated for biocompatibility and stealth properties coupled with low toxicity and adequate biodegradability. However, this process should be encouraged at the earliest opportunity.

Finally, there is a real requirement for more and better bioconjugation methodologies for the coupling of biological targeting moieties to core **AB** or **ABC** nanoparticles. Thus far, bioconjugation methodologies have been few and rather inefficient, including the aqueous coupling between free thiol groups and maleimide functional groups, or free amino groups and succinimide-activated esters. Moreover, there is usually little effort to characterize and confirm the results of most bioconjugation reactions that have been described in the literature, and correspondingly little real effort to separate bioconjugation products from reactants! This is woeful and also needs addressing at the earliest opportunity. Aminoxy-aldehyde aqueous functional group coupling and robust high performance liquid chromatography (HPLC) analyses have recently been developed (IC-Vec Ltd, unpublished results) and appear to represent an effective means of **D**-layer bioconjugation. However, once again, much more chemical diversity is required for reproducible and scalable aqueous coupling of peptides, proteins and/or oligosaccharide targeting moieties to core **AB** or **ABC** nanoparticles.

Curiously, one of the most important barriers to effective nucleic acid delivery is that of formulation, a problem for physical chemistry! LD particles formed from cationic liposome/micelles and nucleic acids have been found typically difficult to formulate in a reproducible and scalable manner. Furthermore, they are susceptible to aggregation (in low ionic strength medium), are difficult to store long-term and do not, as a consequence, mediate reproducible nucleic acid delivery even *in vitro* and *ex vivo*. This formulation barrier cannot be underestimated. Any synthetic non-viral vector system that is intended for *in vivo* applications or gene therapy must be amenable to reproducible and scalable formulation with the nucleic acid of choice rendering particles that are both discrete, nanometric in dimension (≤ 120 nm in diameter), and essentially monodisperse in character. Furthermore, the capacity for long-term storage, preferably without the requirement for refrigeration, is indispensable as well. Should any of these characteristics be ignored or overlooked in the development of new synthetic, non-viral vector systems, then these systems are unlikely to satisfy increasingly stringent regulatory requirements for gene therapy clinical trials and cannot be expected to be of much use in other *in vivo* applications either. Recent research, our own included, has amply demonstrated that any attempts at systematic improvements of synthetic non-viral vector systems are destined to be fruitless unless the most fundamental problems associated with achieving reproducible and scalable formulations, resistance to aggregation, long-term storage and properly reproducible transfection outcomes are convincingly solved prior to future attempts at systematic improvements.

VII. MAIN ABBREVIATIONS

DOPE	dioleoyl-L-α-phosphatidylethanolamine
Chol	cholesterol
DOTMA	N-[1-(2,3-dioleyloxy)propyl]-N,N,N-trimethyl ammonium chloride
DOTAP	1,2-dioleoyloxy-3-(trimethylammonio)propane
DOSPA	2,3-dioleyloxy-N-[2-(sperminecarboxamido)ethyl]-N,N-dimethyl-1-propanaminium trifluoroacetate

DDAB	dimethyldioctadecylammonium bromide
DOGS	dioctadecylamidoglycylspermine
DC-Chol	3β-[N-(N',N'-dimethylaminoethane)carbamoyl]cholesterol
CDAN	N^1-cholesteryloxycarbonyl-3,7-diazanonane-1,9-diamine
BGTC	bis-guanidinium-tren-cholesterol
DOTIM	1-[2-(oleoyloxy)ethyl]-2-oleyl-3-(2-hydroxyethyl)imidazolinium chloride
SAINT	Synthetic Amphiphiles INTerdisciplinary
TMAG	N-(α-trimethylammonioacetyl)-didodecyl-D-glutamate chloride
BCAT	O-(2R-1,2-di-O-(1'Z,9'Z-octadecadienyl)-glycerol)-N-(bis-2-aminoethyl)carbamate
GS11	Gemini Surfactant 11
5AMyr	1-(1,3-dimyristoyloxypropane-2-yl)-2,4,6-trimethylpyridinium hexafluorophosphate
3AMyr	1-(2,3-dimyristoyloxypropyl)-2,4,6-trimethylpyridinium hexafluorophosphate
DODAC	dioleyldimethylammonium chloride
PEG-PE	N-[ω-methoxypoly(oxyethylene)-α-oxycarbonyl]-DSPE
PEG	polyethylene glycol
DSPE	distearoyl-L-α- phosphatidylethanolamine
PEG-CerC$_{20}$	PEG-Ceramide bioconjugate with an arachidoyl acyl group
PEG-CerC$_8$	PEG-Ceramide bioconjugate with an octanoyl acyl group
POD	polyethyleneglycol-diorthoester-distearoylglycerol conjugate
DSG	distearoyl glycerol
BVEP	(R)-1,2-di-O-(1'Z,9'Z-octadecadienyl)-glyceryl-3-(ω-methoxy-poly(ethylene) glycolate 5000)
pHMPA	poly-[N-(2-hydroxypropyl)methacrylamide]
CHEMS	cholesteryl hemisuccinate
pLL	poly-L-lysine
LD	lipoplex (cationic liposome/micelle-nucleic acid complex)
LPD	liposome:polycation:DNA (lipid:protamine:DNA)
LMD	liposome:mu:DNA
mu	μ peptide (of adenovirus)
MEND	multifunctional envelope-type nano device
STR-R$_8$	stearyl octaarginine
SPLP	stabilized plasmid-lipid particles
HVJ	hemagglutinating virus of Japan (Sendai virus)
PIL	pegylated immunoliposome
ODN	oligodeoxynucleotide
ON	oligonucleotide
pDNA	plasmid DNA
siRNA	small interference RNA

RES	reticulo-endothelial system
Tf	transferrin
TfR	transferrin receptor
EGFR	epidermal growth factor receptor
INSR	insulin receptor
f	folate
FR	folate receptor

References

Abe, A., Miyanohara, A., and Friedmann, T. (1998). Enhanced gene transfer with fusogenic liposomes containing vesicular stomatitis virus G glycoprotein. *J. Virol.* **72**, 6159–6163.

Alton, E. W. F. W., Middleton, P. G., Caplen, N. J., Smith, S. N., Steel, D. M., Munkonge, F. M., Jeffery, P. K., Geddes, D. M., Hart, S. L., Williamson, R., Fasold, K. I., Miller, A. D., Dickinson, P., Stevenson, B. J., McLachlan, G., Dorin, J. R., and Porteous, D. J. (1993). Non-invasive liposome-mediated gene delivery can correct the ion transport defect in cystic fibrosis mutant mice. *Nat. Genet.* **5**, 135–142.

Anderson, C. W., Young, M. E., and Flint, S. J. (1989). Characterization of the adenovirus 2 virion protein, mu. *Virology* **172**, 506–512.

Aoki, M., Morishita, R., Higaki, J., Moriguchi, A., Kida, I., Hayashi, S., Matsushita, H., Kaneda, Y., and Ogihara, T. (1997). *In vivo* transfer efficiency of antisense oligonucleotides into the myocardium using HVJ-liposome method. *Biochem. Biophys. Res. Commun.* **231**, 540–545.

Behr, J. P. (1996). The proton sponge, a means to enter cells viruses never thought of. *M/S-Med. Sci.* **12**, 56–58.

Behr, J. P., Demeneix, B., Loeffler, J. P., and Perez-Mutul, J. (1989). Efficient gene transfer into mammalian primary endocrine cells with lipopolyamine-coated DNA. *Proc. Natl. Acad. Sci. USA* **86**, 6982–6986.

Birchall, J. C., Kellaway, I. W., and Gumbleton, M. (2000). Physical stability and *in-vitro* gene expression efficiency of nebulised lipid-peptide-DNA complexes. *Int. J. Pharm.* **197**, 221–231.

Boomer, J. A., Inerowicz, H. D., Zhang, Z. Y., Bergstrand, N., Edwards, K., Kim, J. M., and Thompson, D. H. (2003). Acid-triggered release from sterically stabilized fusogenic liposomes via a hydrolytic DePEGylation strategy. *Langmuir* **19**, 6408–6415.

Boomer, J. A., and Thompson, D. H. (1999). Synthesis of acid-labile diplasmenyl lipids for drug and gene delivery applications. *Chem. Phys. Lipids* **99**, 145–153.

Boomer, J. A., Thompson, D. H., and Sullivan, S. M. (2002). Formation of plasmid-based transfection complexes with an acid-labile cationic lipid: Characterization of *in vitro* and *in vivo* gene transfer. *Pharm. Res.* **19**, 1292–1301.

Chen, H. L., Hu, Q. H., and Liang, W. Q. (2004). Effect of Pluronic on cellular uptake of cationic liposomes-mediated antisense oligonucleotides. *Pharmazie* **59**, 131–133.

Chen, Q. R., Zhang, L., Stass, S. A., and Mixson, A. J. (2000). Co-polymer of histidine and lysine markedly enhances transfection efficiency of liposomes. *Gene Ther.* **7**, 1698–1705.

Chesnoy, S., and Huang, L. (2000). Structure and function of lipid-DNA complexes for gene delivery. *Annu. Rev. Biophys. Biomol. Struct.* **29**, 27–47.

Choi, J. S., MacKay, J. A., and Szoka, F. C., Jr. (2003). Low-pH-sensitive PEG-stabilized plasmid-lipid nanoparticles: Preparation and characterization. *Bioconjug. Chem.* **14**, 420–429.

Colin, M., Harbottle, R. P., Knight, A., Kornprobst, M., Cooper, R. G., Miller, A. D., Trugnan, G., Capeau, J., Coutelle, C., and Brahimi-Horn, M. C. (1998). Liposomes enhance delivery and expression of an RGD-oligolysine gene transfer vector in human tracheal cells. *Gene Ther.* **5,** 1488–1498.

Colin, M., Maurice, M., Trugnan, G., Kornprobst, M., Harbottle, R. P., Knight, A., Cooper, R. G., Miller, A. D., Capeau, J., Coutelle, C., and Brahimi-Horn, M. C. (2000). Cell delivery, intracellular trafficking and expression of an integrin-mediated gene transfer vector in tracheal epithelial cells. *Gene Ther.* **7,** 139–152.

Cooper, R. G., Etheridge, C. J., Stewart, L., Marshall, J., Rudginsky, S., Cheng, S. H., and Miller, A. D. (1998). Polyamine analogues of 3β-[N-(N',N'-dimethylaminoethane)carbamoyl]-cholesterol (DC- Chol) as agents for gene delivery. *Chem. Eur. J.* **4,** 137–152.

Cooper, R. G., Harbottle, R. P., Schneider, H., Coutelle, C., and Miller, A. D. (1999). Peptide mini-vectors for gene delivery. *Angew. Chem. Int. Ed.* **38,** 1949–1952.

da Cruz, M. T. G., Simoes, S., Pires, P. P. C., Nir, S., and de Lima, M. C. P. (2001). Kinetic analysis of the initial steps involved in lipoplex-cell interactions: Effect of various factors that influence transfection activity. *Biochim. Biophys. Acta* **1510,** 136–151.

Dash, P. R., Read, M. L., Fisher, K. D., Howard, K. A., Wolfert, M., Oupicky, D., Subr, V., Strohalm, J., Ulbrich, K., and Seymour, L. W. (2000). Decreased binding to proteins and cells of polymeric gene delivery vectors surface modified with a multivalent hydrophilic polymer and retargeting through attachment of transferrin. *J. Biol. Chem.* **275,** 3793–3802.

de Ilarduya, C. T., and Duzgunes, N. (2000). Efficient gene transfer by transferrin lipoplexes in the presence of serum. *Biochim. Biophys. Acta* **1463,** 333–342.

de Ilarduya, T. C., Arangoa, M. A., Moreno-Aliaga, M. J., and Duzgunes, N. (2002). Enhanced gene delivery *in vitro* and *in vivo* by improved transferrin-lipoplexes. *Biochim. Biophys. Acta* **1561,** 209–221.

Densmore, C. L., Giddings, T. H., Waldrep, J. C., Kinsey, B. M., and Knight, V. (1999). Gene transfer by guanidinium-cholesterol: Dioleoylphosphatidyl-ethanolamine liposome-DNA complexes in aerosol. *J. Gene Med.* **1,** 251–264.

Dokka, S., Toledo, D., Shi, X., Ye, J., and Rojanasakul, Y. (2000). High-efficiency gene transfection of macrophages by lipoplexes. *Int. J. Pharm.* **206,** 97–104.

Drummond, D. C., Zignani, M., and Leroux, J. (2000). Current status of pH-sensitive liposomes in drug delivery. *Prog. Lipid Res.* **39,** 409–460.

Eastman, S. J., Siegel, C., Tousignant, J., Smith, A. E., Cheng, S. H., and Scheule, R. K. (1997). Biophysical characterization of cationic lipid: DNA complexes. *Biochim. Biophys. Acta* **1325,** 41–62.

Felgner, J. H., Kumar, R., Sridhar, C. N., Wheeler, C. J., Tsai, Y. J., Border, R., Ramsey, P., Martin, M., and Felgner, P. L. (1994). Enhanced gene delivery and mechanism studies with a novel series of cationic lipid formulations. *J. Biol. Chem.* **269,** 2550–2561.

Ferrari, M. E., Rusalov, D., Enas, J., and Wheeler, C. J. (2002). Synergy between cationic lipid and co-lipid determines the macroscopic structure and transfection activity of lipoplexes. *Nucleic Acids Res.* **30,** 1808–1816.

Fife, K., Bower, M., Cooper, R. G., Stewart, L., Etheridge, C. J., Coombes, R. C., Buluwela, L., and Miller, A. D. (1998). Endothelial cell transfection with cationic liposomes and herpes simplex-thymidine kinase mediated killing. *Gene Ther.* **5,** 614–620.

Fritz, J. D., Herweijer, H., Zhang, G., and Wolff, J. A. (1996). Gene transfer into mammalian cells using histone-condensed plasmid DNA. *Hum. Gene Ther.* **7,** 1395–1404.

Gao, X., and Huang, L. (1996). Potentiation of cationic liposome-mediated gene delivery by polycations. *Biochemistry* **35,** 1027–1036; 9286.

Gershon, H., Ghirlando, R., Guttman, S. B., and Minsky, A. (1993). Mode of formation and structural features of DNA-cationic liposome complexes used for transfection. *Biochemistry* **32,** 7143–7151.

Guo, X., Huang, Z., and Szoka, F. C. (2004). Improved preparation of PEG-diortho ester-diacyl glycerol conjugates. *Methods Enzymol.* **387**, 147–152.

Guo, X., MacKay, J. A., and Szoka, F. C., Jr. (2003). Mechanism of pH-triggered collapse of phosphatidylethanolamine liposomes stabilized by an ortho ester polyethyleneglycol lipid. *Biophys. J.* **84**, 1784–1795.

Guo, X., and Szoka, F. C., Jr. (2001). Steric stabilization of fusogenic liposomes by a low-pH sensitive PEG-diortho ester-lipid conjugate. *Bioconjug. Chem.* **12**, 291–300.

Guo, X., and Szoka, F. C., Jr. (2003). Chemical approaches to triggerable lipid vesicles for drug and gene delivery. *Acc. Chem. Res.* **36**, 335–341.

Gustafsson, J., Arvidson, G., Karlsson, G., and Almgren, M. (1995). Complexes between cationic liposomes and DNA visualized by cryo-TEM. *Biochim. Biophys. Acta* **1235**, 305–312.

Haensler, J., and Szoka, F. C., Jr. (1993). Polyamidoamine cascade polymers mediate efficient transfection of cells in culture. *Bioconjug. Chem.* **4**, 372–379.

Hayes, M. E., Drummond, D. C., Kirpotin, D. B., Park, J. W., Marks, J. D., Benz, C. C., and Hong, K. (2004). Assembly of nucleic acid-lipid nanoparticles from aqueous-organic monophase. *Biophys. J.* **86**, 36A.

Hofland, H. E., Masson, C., Iginla, S., Osetinsky, I., Reddy, J. A., Leamon, C. P., Scherman, D., Bessodes, M., and Wils, P. (2002). Folate-targeted gene transfer *in vivo*. *Mol. Ther.* **5**, 739–744.

Hong, K., Zheng, W., Baker, A., and Papahadjopoulos, D. (1997). Stabilization of cationic liposome-plasmid DNA complexes by polyamines and poly(ethylene glycol)-phospholipid conjugates for efficient *in vivo* gene delivery. *FEBS Lett.* **400**, 233–237.

Hosokawa, K., and Sung, M. T. (1976). Isolation and characterization of an extremely basic protein from adenovirus type 5. *J. Virol.* **17**, 924–934.

Hui, S. W., Langner, M., Zhao, Y. L., Ross, P., Hurley, E., and Chan, K. (1996). The role of helper lipids in cationic liposome-mediated gene transfer. *Biophys. J.* **71**, 590–599.

Ilies, M. A., and Balaban, A. T. (2001). Recent developments in cationic lipid-mediated gene delivery and gene therapy. *Expert Opin. Ther. Patents* **11**, 1729–1752.

Ilies, M. A., Seitz, W. A., Caproiu, M. T., Wentz, M., Garfield, R. E., and Balaban, A. T. (2003). Pyridinium-based cationic lipids as gene-transfer agents. *Eur. J. Org. Chem.* 2645–2655.

Ilies, M. A., Seitz, W. A., Ghiviriga, I., Johnson, B. H., Miller, A., Thompson, E. B., and Balaban, A. T. (2004). Pyridinium cationic lipids in gene therapy: A structure-activity correlation study. *J. Med. Chem.* **47**, 3744–3754.

Jenkins, R. G., Herrick, S. E., Meng, Q. H., Kinnon, C., Laurent, G. J., McAnulty, R. J., and Hart, S. L. (2000). An integrin-targeted non-viral vector for pulmonary gene therapy. *Gene Ther.* **7**, 393–400.

Joo, S. Y., and Kim, J. S. (2002). Enhancement of gene transfer to cervical cancer cells using transferrin-conjugated liposome. *Drug Dev. Ind. Pharm.* **28**, 1023–1031.

Joshee, N., Bastola, D. R., and Cheng, P. W. (2002). Transferrin-facilitated lipofection gene delivery strategy: Characterization of the transfection complexes and intracellular trafficking. *Hum. Gene Ther.* **13**, 1991–2004.

Kamata, H., Yagisawa, H., Takahashi, S., and Hirata, H. (1994). Amphiphilic peptides enhance the efficiency of liposome-mediated DNA transfection. *Nucleic Acids Res.* **22**, 536–537.

Kaneda, Y. (1999). Development of a novel fusogenic viral liposome system (HVJ-liposomes) and its applications to the treatment of acquired diseases. *Mol. Membr. Biol.* **16**, 119–122.

Kaneda, Y., Iwai, K., and Uchida, T. (1989). Increased expression of DNA cointroduced with nuclear protein in adult rat liver. *Science* **243**, 375–378.

Kawakami, S., Sato, A., Nishikawa, M., Yamashita, F., and Hashida, M. (2000). Mannose receptor-mediated gene transfer into macrophages using novel mannosylated cationic liposomes. *Gene Ther.* **7**, 292–299.

Kawakami, S., Sato, A., Yamada, M., Yamashita, F., and Hashida, M. (2001). The effect of lipid composition on receptor-mediated *in vivo* gene transfection using mannosylated cationic liposomes in mice. *Stp Pharma Sci.* **11**, 117–120.

Keller, M., Harbottle, R. P., Perouzel, E., Colin, M., Shah, I., Rahim, A., Vaysse, L., Bergau, A., Moritz, S., Brahimi-Horn, C., Coutelle, C., and Miller, A. D. (2003a). Nuclear localisation sequence templated non-viral gene delivery vectors: Investigation of intracellular trafficking events of LMD and LD vector systems. *ChemBioChem.* **4**, 286–298.

Keller, M., Jorgensen, M. R., Perouzel, E., and Miller, A. D. (2003b). Thermodynamic aspects and biological profile of CDAN/DOPE and DC-Chol/DOPE lipoplexes. *Biochemistry* **42**, 6067–6077.

Kichler, A., Mechtler, K., Behr, J. P., and Wagner, E. (1997). Influence of membrane-active peptides on lipospermine/DNA complex mediated gene transfer. *Bioconjug. Chem.* **8**, 213–221.

Kirby, A. J., Camilleri, P., Engberts, J. B. F. N., Feiters, M. C., Nolte, R. J. M., Soderman, O., Bergsma, M., Bell, P. C., Fielden, M. L., Rodriguez, C. L. G., Guedat, P., Kremer, A., McGregor, C., Perrin, C., Ronsin, G., and van Eijk, M. C. P. (2003). Gemini surfactants: New synthetic vectors for gene transfection. *Angew. Chem. Int. Ed.* **42**, 1448–1457.

Kogure, K., Moriguchi, R., Sasaki, K., Ueno, M., Futaki, S., and Harashima, H. (2004). Development of a non-viral multifunctional envelope-type nano device by a novel lipid film hydration method. *J. Control. Rel.* **98**, 317–323.

Koltover, I., Salditt, T., Radler, J. O., and Safinya, C. R. (1998). An inverted hexagonal phase of cationic liposome-DNA complexes related to DNA release and delivery. *Science* **281**, 78–81.

Labat-Moleur, F., Steffan, A. M., Brisson, C., Perron, H., Feugeas, O., Furstenberger, P., Oberling, F., Brambilla, E., and Behr, J. P. (1996). An electron microscopy study into the mechanism of gene transfer with lipopolyamines. *Gene Ther.* **3**, 1010–1017.

Lasic, D. D., and Papahadjopoulos, D. (1995). Liposomes revisited. *Science* **267**, 1275–1276.

Lasic, D. D., Strey, H., Stuart, M. C. A., Podgornik, R., and Frederik, P. M. (1997). The structure of DNA-liposome complexes. *J. Am. Chem. Soc.* **119**, 832–833.

Lee, A. C. H., Evans, J. C., Shaw, J. R., Swann, L. E., McClintock, K., and MacLachlan, I. (2003). A single-dose intravenous pharmacokinetics and biodistribution study of the Stable Plasmid-Lipid Particle Pro-1 in healthy and tumor-bearing A/J mice. *Mol. Ther.* **7**, 549.

Lee, E. R., Marshall, J., Siegel, C. S., Jiang, C., Yew, N. S., Nichols, M. R., Nietupski, J. B., Ziegler, R. J., Lane, M. B., Wang, K. X., Wan, N. C., Scheule, R. K., Harris, D. J., Smith, A. E., and Cheng, S. H. (1996). Detailed analysis of structures and formulations of cationic lipids for efficient gene transfer to the lung. *Hum. Gene Ther.* **7**, 1701–1717.

Li, B., Li, S., Tan, Y., Stolz, D. B., Watkins, S. C., Block, L. H., and Huang, L. (2000). Lyophilization of cationic lipid-protamine-DNA (LPD) complexes. *J. Pharm. Sci.* **89**, 355–364.

Li, S., and Huang, L. (1997). *In vivo* gene transfer via intravenous administration of cationic lipid-protamine-DNA (LPD) complexes. *Gene Ther.* **4**, 891–900.

Li, S., Rizzo, M. A., Bhattacharya, S., and Huang, L. (1998). Characterization of cationic lipid-protamine-DNA (LPD) complexes for intravenous gene delivery. *Gene Ther.* **5**, 930–937.

Li, S., Tseng, W. C., Stolz, D. B., Wu, S. P., Watkins, S. C., and Huang, L. (1999). Dynamic changes in the characteristics of cationic lipidic vectors after exposure to mouse serum: Implications for intravenous lipofection. *Gene Ther.* **6**, 585–594.

Martin-Herranz, A., Ahmad, A., Evans, H. M., Ewert, K., Schulze, U., and Safinya, C. R. (2004). Surface functionalized cationic lipid-DNA complexes for gene delivery: PEGylated lamellar complexes exhibit distinct DNA-DNA interaction regimes. *Biophys. J.* **86**, 1160–1168.

McGregor, C., Perrin, C., Monck, M., Camilleri, P., and Kirby, A. J. (2001). Rational approaches to the design of cationic gemini surfactants for gene delivery. *J. Am. Chem. Soc.* **123**, 6215–6220.

McQuillin, A., Murray, K. D., Etheridge, C. J., Stewart, L., Cooper, R. G., Brett, P. M., Miller, A. D., and Gurling, H. M. (1997). Optimization of liposome mediated transfection of a neuronal cell line. *Neuroreport* **8**, 1481–1484.

Miller, A. D. (1998a). Cationic liposome systems in gene therapy. *Curr. Res. Mol. Ther.* **1**, 494–503.

Miller, A. D. (1998b). Cationic liposomes for gene therapy. *Angew. Chem. Int. Ed.* **37**, 1768–1785.

Miller, A. D. (1999). Non-viral delivery systems for gene therapy. *In* "Understanding Gene Therapy" (N. R. Lemoine, ed.). Bios Scientific Publishers, Oxford.

Miller, A. D. (2003). The problem with cationic liposome/micelle-based non-viral vector systems for gene therapy. *Curr. Med. Chem.* **10**, 1195–1211.

Mizuarai, S., Ono, K., You, J., Kamihira, M., and Iijima, S. (2001). Protamine-modified DDAB lipid vesicles promote gene transfer in the presence of serum. *J. Biochemistry* **129**, 125–132.

Mohr, L., Schauer, J. I., Boutin, R. H., Moradpour, D., and Wands, J. R. (1999). Targeted gene transfer to hepatocellular carcinoma cells *in vitro* using a novel monoclonal antibody-based gene delivery system. *Hepatology* **29**, 82–89.

Morishita, R., Gibbons, G. H., Kaneda, Y., Ogihara, T., and Dzau, V. J. (1994). Pharmacokinetics of antisense oligodeoxyribonucleotides (cyclin B1 and CDC 2 kinase) in the vessel wall *in vivo*: Enhanced therapeutic utility for restenosis by HVJ-liposome delivery. *Gene* **149**, 13–19.

Mui, B., Ahkong, Q. F., Chow, L., and Hope, M. J. (2000). Membrane perturbation and the mechanism of lipid-mediated transfer of DNA into cells. *Biochim. Biophys. Acta* **1467**, 281–292.

Nakanishi, M., Mizuguchi, H., Ashihara, K., Senda, T., Eguchi, A., Watabe, A., Nakanishi, T., Kondo, M., Nakagawa, T., Masago, A., Okabe, J., Ueda, S., Mayumi, T., and Hayakawa, T. (1999). Gene delivery systems using the Sendai virus. *Mol. Membr. Biol.* **16**, 123–127.

Namiki, Y., Takahashi, T., and Ohno, T. (1998). Gene transduction for disseminated intraperitoneal tumor using cationic liposomes containing non-histone chromatin proteins: Cationic liposomal gene therapy of carcinomatosa. *Gene Ther.* **5**, 240–246.

Nicolazzi, C., Garinot, M., Mignet, N., Scherman, D., and Bessodes, M. (2003a). Cationic lipids for transfection. *Curr. Med. Chem.* **10**, 1263–1277.

Nicolazzi, C., Mignet, N., de la Figuera, N., Cadet, M., Ibad, R. T., Seguin, J., Scherman, D., and Bessodes, M. (2003b). Anionic polyethyleneglycol lipids added to cationic lipoplexes increase their plasmatic circulation time. *J. Control. Rel.* **88**, 429–443.

Nicolazzi, C., Mignet, N., de la Figuera, N., Cadet, M., Ibad, R., Seguin, J., Scherman, D., and Bessodes, M. (2004). Anionic polyethyleneglycol lipids added to cationic lipoplexes increase their plasmatic circulation time. *Mol. Ther.* **9**, 475.

Oliver, M., Jorgensen, M. R., and Miller, A. D. (2004). The facile solid-phase synthesis of cholesterol-based polyamine lipids. *Tetrahedron Lett.* **45**, 3105–3108.

Palmer, L., Jeffs, L. B., Chan, K. P. Y., McClintock, K., Giesbrecht, C., Heyes, J., Lee, A. C. H., and MacLachlan, I. (2004). Systemic delivery and tumor gene expression using polycation-precondensed DNA prepared as stable plasmid lipid particles. *Mol. Ther.* **9**, 679.

Palmer, L. R., Chen, T., Lam, A. M., Fenske, D. B., Wong, K. F., MacLachlan, I., and Cullis, P. R. (2003). Transfection properties of stabilized plasmid-lipid particles containing cationic PEG lipids. *Biochim. Biophys. Acta* **1611**, 204–216.

Perouzel, E., Jorgensen, M. R., Keller, M., and Miller, A. D. (2003). Synthesis and formulation of neoglycolipids for the functionalization of liposomes and lipoplexes. *Bioconjug. Chem.* **14**, 884–898.

Pitard, B., Oudrhiri, N., Lambert, O., Vivien, E., Masson, C., Wetzer, B., Hauchecorne, M., Scherman, D., Rigaud, J. L., Vigneron, J. P., Lehn, J. M., and Lehn, P. (2001). Sterically stabilized BGTC-based lipoplexes: Structural features and gene transfection into the mouse airways *in vivo*. *J. Gene Med.* **3**, 478–487.

Pitard, B., Oudrhiri, N., Vigneron, J. P., Hauchecorne, M., Aguerre, O., Toury, R., Airiau, M., Ramasawmy, R., Scherman, D., Crouzet, J., Lehn, J. M., and Lehn, P. (1999). Structural characteristics of supramolecular assemblies formed by guanidinium-cholesterol reagents for gene transfection. *Proc. Natl. Acad. Sci. USA* **96**, 2621–2626.

Preuss, M., Tecle, M., Shah, I., Matthews, D. A., and Miller, A. D. (2003). Comparison between the interactions of adenovirus-derived peptides with plasmid DNA and their role in gene delivery mediated by liposome-peptide-DNA virus-like nanoparticles. *Org. Biomol. Chem.* **1**, 2430–2438.

Radler, J. O., Koltover, I., Salditt, T., and Safinya, C. R. (1997). Structure of DNA-cationic liposome complexes: DNA intercalation in multilamellar membranes in distinct interhelical packing regimes. *Science* **275**, 810–814.

Reddy, J. A., and Low, P. S. (2000). Enhanced folate receptor mediated gene therapy using a novel pH-sensitive lipid formulation. *J. Control. Rel.* **64**, 27–37.

Rejman, J., Wagenaar, A., Engberts, J. B., and Hoekstra, D. (2004). Characterization and transfection properties of lipoplexes stabilized with novel exchangeable polyethylene glycol-lipid conjugates. *Biochim. Biophys. Acta* **1660**, 41–52.

Remy, J. S., Kichler, A., Mordvinov, V., Schuber, F., and Behr, J. P. (1995). Targeted gene transfer into hepatoma cells with lipopolyamine-condensed DNA particles presenting galactose ligands: A stage toward artificial viruses. *Proc. Natl. Acad. Sci. USA* **92**, 1744–1748.

Saeki, Y., Matsumoto, N., Nakano, Y., Mori, M., Awai, K., and Kaneda, Y. (1997). Development and characterization of cationic liposomes conjugated with HVJ (Sendai virus): Reciprocal effect of cationic lipid for *in vitro* and *in vivo* gene transfer. *Hum. Gene Ther.* **8**, 2133–2141.

Saravolac, E. G., Ludkovski, O., Skirrow, R., Ossanlou, M., Zhang, Y. P., Giesbrecht, C., Thompson, J., Thomas, S., Stark, H., Cullis, P. R., and Scherrer, P. (2000). Encapsulation of plasmid DNA in stabilized plasmid-lipid particles composed of different cationic lipid concentration for optimal transfection activity. *J. Drug Target.* **7**, 423–437.

Schmutz, M., Durand, D., Debin, A., Palvadeau, Y., Etienne, A., and Thierry, A. R. (1999). DNA packing in stable lipid complexes designed for gene transfer imitates DNA compaction in bacteriophage. *Proc. Natl. Acad. Sci. USA* **96**, 12293–12298.

Schoen, P., Chonn, A., Cullis, P. R., Wilschut, J., and Scherrer, P. (1999). Gene transfer mediated by fusion protein hemagglutinin reconstituted in cationic lipid vesicles. *Gene Ther.* **6**, 823–832.

Schwartz, B., Ivanov, M. A., Pitard, B., Escriou, V., Rangara, R., Byk, G., Wils, P., Crouzet, J., and Scherman, D. (1999). Synthetic DNA-compacting peptides derived from human sequence enhance cationic lipid-mediated gene transfer *in vitro* and *in vivo*. *Gene Ther.* **6**, 282–292.

Seki, M., Iwakawa, J., Cheng, H., and Cheng, P. W. (2002). p53 and PTEN/MMAC1/TEP1 gene therapy of human prostate PC-3 carcinoma xenograft, using transferrin-facilitated lipofection gene delivery strategy. *Hum. Gene Ther.* **13**, 761–773.

Seol, J. G., Heo, D. S., Kim, H. K., Yoon, J. H., Choi, B. I., Lee, H. S., Kim, N. K., and Kim, C. Y. (2000). Selective gene expression in hepatic tumor with trans-arterial delivery of DNA-liposome/transferrin complex. *In Vivo* **14**, 513–517.

Shi, G., Guo, W., Stephenson, S. M., and Lee, R. J. (2002). Efficient intracellular drug and gene delivery using folate receptor-targeted pH-sensitive liposomes composed of cationic/anionic lipid combinations. *J. Control. Rel.* **80**, 309–319.

Shi, N., Boado, R. J., and Pardridge, W. M. (2001). Receptor-mediated gene targeting to tissues *in vivo* following intravenous administration of pegylated immunoliposomes. *Pharm. Res.* **18**, 1091–1095.

Shi, N. Y., and Pardridge, W. M. (2000). Noninvasive gene targeting to the brain. *Proc. Natl. Acad. Sci. USA* **97**, 7567–7572.

Shin, J., and Thompson, D. H. (2003). Direct synthesis of plasmenylcholine from allyl-substituted glycerols. *J. Org. Chem.* **68**, 6760–6766.

Simoes, S., Slepushkin, V., Pires, P., Gaspar, R., de Lima, M. C. P., and Duzgunes, N. (2000). Human serum albumin enhances DNA transfection by lipoplexes and confers resistance to inhibition by serum. *Biochim. Biophys. Acta* **1463**, 459–469.

Simoes, S., Slepushkin, V., Pires, P., Gaspar, R., de Lima, M. P., and Duzgunes, N. (1999). Mechanisms of gene transfer mediated by lipoplexes associated with targeting ligands or pH-sensitive peptides. *Gene Ther.* **6,** 1798–1807.

Solodin, I., Brown, C. S., Bruno, M. S., Chow, C. Y., Jang, E. H., Debs, R. J., and Heath, T. D. (1995). A novel series of amphiphilic imidazolinium compounds for *in vitro* and *in vivo* gene delivery. *Biochemistry* **34,** 13537–13544.

Son, K. K., Patel, D. H., Tkach, D., and Park, A. (2000). Cationic liposome and plasmid DNA complexes formed in serum-free medium under optimum transfection condition are negatively charged. *Biochim. Biophys. Acta* **1466,** 11–15.

Sorgi, F. L., Bhattacharya, S., and Huang, L. (1997). Protamine sulfate enhances lipid-mediated gene transfer. *Gene Ther.* **4,** 961–968.

Spagnou, S., Miller, A. D., and Keller, M. (2004). Lipidic Carriers of siRNA: Differences in the formulation, cellular uptake, and delivery with plasmid DNA. *Biochemistry* **43,** 13348–13356.

Sternberg, B., Sorgi, F. L., and Huang, L. (1994). New structures in complex formation between DNA and cationic liposomes visualized by freeze-fracture electron microscopy. *FEBS Lett.* **356,** 361–366.

Stewart, L., Manvell, M., Hillery, E., Etheridge, C. J., Cooper, R. G., Stark, H., van-Heel, M., Preuss, M., Alton, E. W. F. W., and Miller, A. D. (2001). Physico-chemical analysis of cationic liposome-DNA complexes (lipoplexes) with respect to *in vitro* and *in vivo* gene delivery efficiency. *J. Chem. Soc. Perkin Trans.* **2,** 624–632.

Tagawa, T., Manvell, M., Brown, N., Keller, M., Perouzel, E., Murray, K. D., Harbottle, R. P., Tecle, M., Booy, F., Brahimi-Horn, M. C., Coutelle, C., Lemoine, N. R., Alton, E. W., and Miller, A. D. (2002). Characterisation of LMD virus-like nanoparticles self-assembled from cationic liposomes, adenovirus core peptide mu and plasmid DNA. *Gene Ther.* **9,** 564–576.

Tam, P., Monck, M., Lee, D., Ludkovski, O., Leng, E. C., Clow, K., Stark, H., Scherrer, P., Graham, R. W., and Cullis, P. R. (2000). Stabilized plasmid-lipid particles for systemic gene therapy. *Gene Ther.* **7,** 1867–1874.

Tan, P. H., King, W. J., Chen, D., Awad, H. M., Mackett, M., Lechler, R. I., Larkin, D. F. P., and George, A. J. T. (2001). Transferrin receptor-mediated gene transfer to the corneal endothelium. *Transplantation* **71,** 552–560.

Tan, P. H., Manunta, M., Ardjomand, N., Xue, S. A., Larkin, D. F., Haskard, D. O., Taylor, K. M., and George, A. J. (2003). Antibody targeted gene transfer to endothelium. *J. Gene Med.* **5,** 311–323.

Tecle, M., Preuss, M., and Miller, A. D. (2003). Kinetic study of DNA condensation by cationic peptides used in non-viral gene therapy: Analogy of DNA condensation to protein folding. *Biochemistry* **42,** 10343–10347.

Thompson, D. H., Shin, J., Boomer, J., and Kim, J. M. (2004). Preparation of plasmenylcholine lipids and plasmenyl-type liposome dispersions. *Methods Enzymol.* **387,** 153–168.

van der Woude, I., Wagenaar, A., Meekel, A. A., ter Beest, M. B., Ruiters, M. H., Engberts, J. B., and Hoekstra, D. (1997). Novel pyridinium surfactants for efficient, nontoxic *in vitro* gene delivery. *Proc. Natl. Acad. Sci. USA* **94,** 1160–1165.

Vaysse, L., and Arveiler, B. (2000). Transfection using synthetic peptides: Comparison of three DNA-compacting peptides and effect of centrifugation. *Biochim. Biophys. Acta* **1474,** 244–250.

Vaysse, L., Harbottle, R., Bigger, B., Bergau, A., Tolmachov, O., and Coutelle, C. (2004). Development of a self-assembling nuclear targeting vector system based on the tetracycline repressor protein. *J. Biol. Chem.* **279,** 5555–5564.

Vitiello, L., Chonn, A., Wasserman, J. D., Duff, C., and Worton, R. G. (1996). Condensation of plasmid DNA with polylysine improves liposome-mediated gene transfer into established and primary muscle cells. *Gene Ther.* **3,** 396–404.

Voinea, M., Dragomir, E., Manduteanu, I., and Simionescu, M. (2002). Binding and uptake of transferrin-bound liposomes targeted to transferrin receptors of endothelial cells. *Vascul. Pharmacol.* **39**, 13–20.

Waelti, E. R., and Gluck, R. (1998). Delivery to cancer cells of antisense L-myc oligonucleotides incorporated in fusogenic, cationic-lipid-reconstituted influenza-virus envelopes (cationic virosomes). *Int. J. Cancer* **77**, 728–733.

Waterhouse, J. E., Harbottle, R. P., Keller, M., Kostarelos, K., Coutelle, C., Jorgensen, M. R., and Miller, A. D. (2005). Synthesis and application of integrin targeting lipopeptides in targeted gene delivery. *Chem. Biochem.*, in press.

Wheeler, J. J., Palmer, L., Ossanlou, M., MacLachlan, I., Graham, R. W., Zhang, Y. P., Hope, M. J., Scherrer, P., and Cullis, P. R. (1999). Stabilized plasmid-lipid particles: Construction and characterization. *Gene Ther.* **6**, 271–281.

Whitmore, M., Li, S., and Huang, L. (1999). LPD lipopolyplex initiates a potent cytokine response and inhibits tumor growth. *Gene Ther.* **6**, 1867–1875.

Xu, L., Frederik, P., Pirollo, K. F., Tang, W. H., Rait, A., Xiang, L. M., Huang, W., Cruz, I., Yin, Y., and Chang, E. H. (2002a). Self-assembly of a virus-mimicking nanostructure system for efficient tumor-targeted gene delivery. *Hum. Gene Ther.* **13**, 469–481.

Xu, L., Huang, C. C., Huang, W., Tang, W. H., Rait, A., Yin, Y. Z., Cruz, I., Xiang, L. M., Pirollo, K. F., and Chang, E. H. (2002b). Systemic tumor-targeted gene delivery by anti-transferrin receptor scFv-immunoliposomes. *Mol. Cancer Ther.* **1**, 337–346.

Xu, L., Tang, W. H., Huang, C. C., Alexander, W., Xiang, L. M., Pirollo, K. F., Rait, A., and Chang, E. H. (2001). Systemic p53 gene therapy of cancer with immunolipoplexes targeted by anti-transferrin receptor scFv. *Mol. Med.* **7**, 723–734.

Xu, Y., Hui, S. W., Frederik, P., and Szoka, F. C., Jr. (1999). Physicochemical characterization and purification of cationic lipoplexes. *Biophys. J.* **77**, 341–353.

Yanagihara, K., Cheng, H., and Cheng, P. W. (2000). Effects of epidermal growth factor, transferrin, and insulin on lipofection efficiency in human lung carcinoma cells. *Cancer Gene Ther.* **7**, 59–65.

Yanagihara, K., and Cheng, P. W. (1999). Lectin enhancement of the lipofection efficiency in human lung carcinoma cells. *Biochim. Biophys. Acta* **1472**, 25–33.

Yonemitsu, Y., Alton, E. W., Komori, K., Yoshizumi, T., Sugimachi, K., and Kaneda, Y. (1998). HVJ (Sendai virus) liposome-mediated gene transfer: Current status and future perspectives (review). *Int. J. Oncol.* **12**, 1277–1285.

Yonemitsu, Y., Kaneda, Y., Muraishi, A., Yoshizumi, T., Sugimachi, K., and Sueishi, K. (1997). HVJ (Sendai virus)-cationic liposomes: A novel and potentially effective liposome-mediated technique for gene transfer to the airway epithelium. *Gene Ther.* **4**, 631–638.

Young, J. L., Benoit, J. N., and Dean, D. A. (2003). Effect of a DNA nuclear targeting sequence on gene transfer and expression of plasmids in the intact vasculature. *Gene Ther.* **10**, 1465–1470.

Yu, W., Pirollo, K. F., Rait, A., Yu, B., Xiang, L. M., Huang, W. Q., Zhou, Q., Ertem, G., and Chang, E. H. (2004a). A sterically stabilized immunolipoplex for systemic administration of a therapeutic gene. *Gene Ther.* **11**, 1434–1440.

Yu, W., Pirollo, K. F., Yu, B., Rait, A., Xiang, L., Huang, W., Zhou, Q., Ertem, G., and Chang, E. H. (2004b). Enhanced transfection efficiency of a systemically delivered tumor-targeting immunolipoplex by inclusion of a pH-sensitive histidylated oligolysine peptide. *Nucleic Acids Res.* **32**, e48.

Zabner, J., Fasbender, A. J., Moninger, T., Poellinger, K. A., and Welsh, M. J. (1995). Cellular and molecular barriers to gene transfer by a cationic lipid. *J. Biol. Chem* **270**, 18997–19007.

Zanta, M. A., Belguise-Valladier, P., and Behr, J. P. (1999). Gene delivery: A single nuclear localization signal peptide is sufficient to carry DNA to the cell nucleus. *Proc. Natl. Acad. Sci. USA* **96**, 91–96.

Zhang, Y., Boado, R. J., and Pardridge, W. M. (2003a). *In vivo* knockdown of gene expression in brain cancer with intravenous RNAi in adult rats. *J. Gene Med.* **5,** 1039–1045.

Zhang, Y., Jeong Lee, H., Boado, R. J., and Pardridge, W. M. (2002a). Receptor-mediated delivery of an antisense gene to human brain cancer cells. *J. Gene Med.* **4,** 183–194.

Zhang, Y., Zhu, C., and Pardridge, W. M. (2002b). Antisense gene therapy of brain cancer with an artificial virus gene delivery system. *Mol. Ther.* **6,** 67–72.

Zhang, Y. F., Boado, R. J., and Pardridge, W. M. (2003b). Absence of toxicity of chronic weekly intravenous gene therapy with pegylated immunoliposomes. *Pharm. Res.* **20,** 1779–1785.

Zhang, Y. P., Sekirov, L., Saravolac, E. G., Wheeler, J. J., Tardi, P., Clow, K., Leng, E., Sun, R., Cullis, P. R., and Scherrer, P. (1999). Stabilized plasmid-lipid particles for regional gene therapy: Formulation and transfection properties. *Gene Ther.* **6,** 1438–1447.

Zhou, X., and Huang, L. (1994). DNA transfection mediated by cationic liposomes containing lipopolylysine: Characterization and mechanism of action. *Biochim. Biophys. Acta* **1189,** 195–203.

5

Lipoplex Structures and Their Distinct Cellular Pathways

Kai Ewert, Heather M. Evans, Ayesha Ahmad, Nelle L. Slack, Alison J. Lin, Ana Martin-Herranz, and Cyrus R. Safinya
Materials Department, Physics Department, and Molecular, Cellular and Developmental Biology Department, University of California, Santa Barbara Santa Barbara, California 93106

0065-2660/05 $35.00
DOI: 10.1016/S0065-2660(05)53005-0

ABSTRACT

Cationic liposomes (CLs) are used as non-viral vectors in worldwide clinical trials of gene therapy. Among other advantages, CL-DNA complexes have the ability to transfer very large genes into cells. However, since the understanding of their mechanisms of action is still incomplete, their transfection efficiencies remain low compared to those of viruses. We describe recent studies which have started to unravel the relationship between the distinct structures and physicochemical properties of CL-DNA complexes and their transfection efficiency by combining several techniques: synchrotron X-ray diffraction for structure determination, laser-scanning confocal microscopy to probe the interactions of CL-DNA particles with cells, and luciferase reporter-gene expression assays to measure transfection efficiencies in mammalian cells. Most CL-DNA complexes form a multilayered structure with DNA sandwiched between the cationic lipids (lamellar complexes, L_α^C). Much more rarely, an inverted hexagonal structure (H_{II}^C) with single DNA strands encapsulated in lipid tubules is observed. An important recent insight is that the membrane charge density σ_M of the CL-vector, rather than, for example, the charge of the cationic lipid, is a universal parameter governing the transfection efficiency of L_α^C complexes. This has led to a new model of the intracellular release of L_α^C complexes, through activated fusion with endosomal membranes. In contrast to L_α^C complexes, H_{II}^C complexes exhibit no dependence on σ_M, since their structure leads to a distinctly different mechanism of cell entry. Surface-functionalized complexes with poly(ethyleneglycol)-lipids (PEG-lipids), potentially suitable for transfection *in vivo*, have also been investigated, and the novel aspects of these complexes are discussed. © 2005, Elsevier Inc.

I. INTRODUCTION

Gene delivery by synthetic (non-viral) vectors continues to attract the interest of a large number of research groups, motivated mainly by the promises of gene therapy (Chesnoy and Huang, 2000; Clark and Hersh, 1999; De Smedt *et al.*, 2000; Ewert *et al.*, 2004; Felgner and Rhodes, 1991; Ferber, 2001; Friedmann, 1997; Henry, 2001; Huang *et al.*, 1999; Kumar *et al.*, 2003; Mahato and Kim, 2002; Miller, 1998; Niidome and Huang, 2002; Wagner, 2004). Following initial landmark studies (Felgner *et al.*, 1987; Nabel *et al.*, 1993; Wolff, 1994), cationic liposomes (CLs, closed bilayer membrane shells of lipid molecules) have been established as one of the most prevalent synthetic vectors. They are already used widely for *in vitro* transfection of mammalian cells in research applications, with numerous formulations commercially available (Miller, 2003). To enable the use of CL-DNA complexes in gene therapy, their mechanism of action is extensively investigated in many laboratories, concurrently with ongoing, mostly empirical, clinical trials (Edelstein *et al.*, 2004), for example, to develop cancer vaccines. Extensive and current information on clinical trials in the field of gene therapy is collected and made available at http://www.wiley.co.uk/genetherapy/clinical/ (The Journal of Gene Medicine Clinical Trial Web site, 2004).

A key advantage of CLs over viral methods is the lack of a specific immune response, due to the absence of viral proteins. This was tragically emphasized by a recent, much publicized complication in clinical trials of viral vectors: a patient died due to an unanticipated, severe inflammatory response to an adenoviral vector (Raper *et al.*, 2003). Further emphasizing safety concerns, a successful trial that used retroviral vectors to treat children with severe combined immunodeficiency (SCID) (Cavazzana-Calvo *et al.*, 2000) faced a major setback when two patients (out of ten) developed a leukemia-like disease (Hacein-Bey-Abina *et al.*, 2003a), later confirmed to have resulted from insertional mutagenesis (Hacein-Bey-Abina *et al.*, 2003b). In addition, while viral capsids have a maximum DNA-carrying capacity of about 40 kbp (Friedmann, 1997), CL-DNA complexes place no limit on the size of the DNA, since the vector is formed by self-assembly (Koltover *et al.*, 1998; Lasic *et al.*, 1997; Rädler *et al.*, 1997). Thus, optimally designed CL-vectors offer the potential of delivering multiple human genes and regulatory sequences, extending over hundreds of thousands of DNA base-pairs. In fact, fractions of an artificial human chromosome, about 1 million base-pairs in size, have been transferred into cells using cationic lipids as a vector, albeit extremely inefficiently (Harrington *et al.*, 1997; Willard, 2000).

The main disadvantage of CL-based gene vectors is that their transfection efficiency (TE), which is a measure of the amount of successfully transferred and expressed DNA, remains much lower than that of viral vectors, particularly for *in vivo* applications (Miller, 2003). This has spurred intense research activity aimed at enhancing TE. For further improvement of the transfection efficiency of non-viral methods, it will be crucial to gain deep insight into their mechanism of transfection on the molecular and self-assembled level. This requires a complete understanding of the supramolecular structures of CL-DNA complexes, their interactions with cell membranes, the intracellular events leading to release of DNA for possible expression as well as the physical and chemical basis of these.

Our work investigates the mechanisms of transfection by CL-DNA complexes at the molecular to cellular level, focusing on the influence of the nanostructure of the complex and, within a given structure (L_α^C or H_{II}^C), the physical (e.g., mechanical properties) and physico-chemical (e.g., charge density) parameters of the membranes forming the complex. To this end, a variety of techniques are employed: synchrotron X-ray diffraction (XRD) for structure determination (Koltover *et al.*, 1998, 1999, 2000; Lasic *et al.*, 1997; Lin *et al.*, 2000, 2003; Safinya, 2001; Salditt *et al.*, 1997; Rädler *et al.*, 1997), three-dimensional laser scanning confocal microscopy (LSCM) for imaging pathways and interactions of complexes within cells (Lin *et al.*, 2000, 2003) and reporter gene assays in animal cell culture (Lin *et al.*, 2000, 2003; Ewert *et al.*, 2002) for quantifying TE. By combining these characterization methods, the distinct interactions between cells and high and low transfection efficiency complexes, as observed in confocal microscopy, can be rationalized on the basis of the structure and properties of the complexes (Safinya and Koltover, 1999; Safinya *et al.*, 2002; Tranchant *et al.*, 2004). Unless otherwise stated, experiments reviewed in this chapter used mouse fibroblast L-cells, a luciferase reporter gene assay and complexes at a lipid-to-DNA charge ratio (ρ_{chg}) of 2.8, which corresponds to the middle of a plateau region observed when plotting TE as a function of increasing ρ_{chg} above the isoelectric point (Lin *et al.*, 2003).

The structures of most lipids mentioned in this chapter are shown in Fig. 5.1. As neutral lipids, 1,2-dioleoyl-*sn*-glycerophosphatidylethanolamine (DOPE) and 1,2-dioleoyl-*sn*-glycerophosphatidylcholine (DOPC) were employed. DOPE is one of the main neutral lipids currently in use in gene therapy applications of CLs. The cationic lipids were either commercially available (DOTAP), gifts from other investigators (DMRIE, DOSPA) or synthesized in our group (PEG-lipids, MVLs (Table 5.1)).

The *in vitro* studies described here apply mainly to TE optimization in *ex vivo* cell transfection, where cells are removed and returned to patients after transfection. In particular, an improved understanding of the key mechanisms of transfection in continuous cell lines should aid clinical efforts to develop efficient CL-vector cancer vaccines in *ex vivo* applications, which induce transient expression of genes encoding immuno-stimulatory proteins (Clark and Hersh, 1999; Ferber, 2001; Nabel *et al.*, 1993; Rinehart *et al.*, 1997; Stopeck *et al.*, 1998).

II. FORMATION, STRUCTURES, AND STABILITY OF CL-DNA COMPLEXES

CL-DNA complexes readily form for a large variety of lipids. This ease of preparation and the variability of the lipid composition constitute two of their main advantages. In this section, we discuss findings relevant to understanding the formation, nanostructure and thermodynamic stability of CL-DNA complexes.

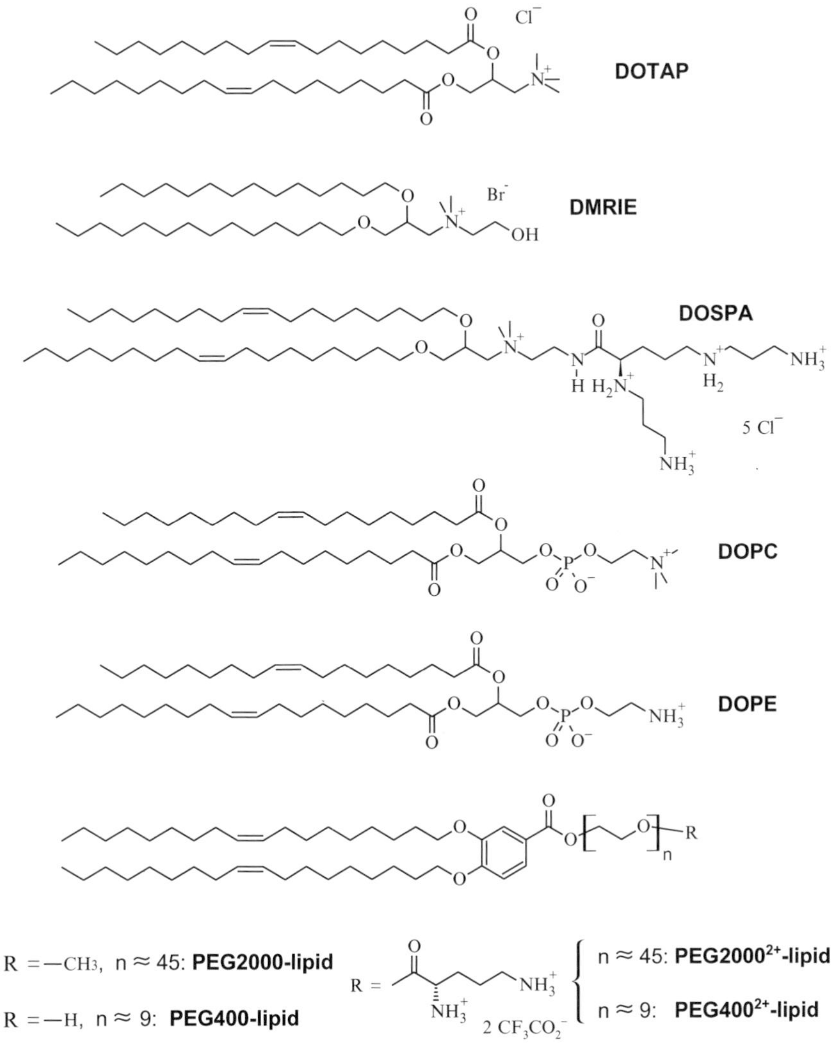

Figure 5.1. Chemical structures and abbreviated names of cationic and neutral lipids mentioned in this chapter.

A. Complex formation and structures

Combining solutions of cationic liposomes and DNA results in their spontaneous self-assembly into small (0.2 μm diameter) globular particles of CL-DNA complexes. This is schematically illustrated in Fig. 5.2. Depending on the conditions of formation, the initially formed particles may aggregate to form much larger assemblies. The driving force for complex formation is a large

Table 5.1. Newly Synthesized Multivalent Cationic Lipids. Their General Structure is Shown in Fig. 5.10

Headgroup	Spacer (EO)n	Maximum charge	Lipid name
	0	+1	MVL1
	0	+2	MVL2
	0	+3	MVL3
	0	+4	MVL4
	0, 2	+5	MVL5 / TMVL5 (n=2)

increase in entropy resulting from the release of tightly bound counter-ions (Le Bret and Zimm, 1984; Manning, 1978) from the cationic lipid membranes and the anionic DNA rods (Harries *et al.*, 1998; Koltover *et al.*, 1999; May and Ben-Shaul, 2004). The liposomes typically contain at least two components: a cationic lipid and a neutral lipid, which is sometimes called helper lipid. Depending on the ratio of charges on the cationic lipid and the DNA (ρ_{chg}), anionic, neutral or cationic complexes are obtained. We refer to neutral complexes, where the charges on the DNA exactly match those on the cationic lipids ($\rho_{chg} = 1$), as isoelectric. For transfection, positively charged ($\rho_{chg} > 1$) complexes are used, since the initial attachment of CL-DNA complexes to mammalian cells is mediated by electrostatic attractions between the complexes and negatively charged cell surface proteoglycans (Kopatz *et al.*, 2004; Mislick and Baldeschwieler, 1996; Mounkes *et al.*, 1998).

Synchrotron X-ray diffraction experiments have solved the two types of structures observed in CL-DNA complexes. Shown schematically on the right in

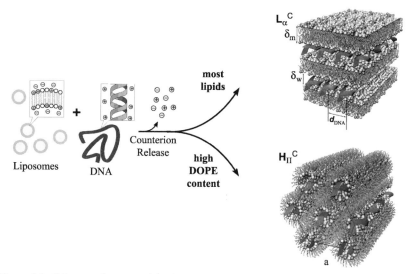

Figure 5.2. Schematic depiction of the formation and the nanostructures of CL–DNA complexes. Counterions are tightly bound to the charged surfaces of DNA and CLs (Manning condensation: Manning, 1978). The entropy gained from their release provides the main driving force for the self-assembly into distinctly structured complexes. The schematics on the right show the local interior structure of CL–DNA complexes for the two structures observed to date, as derived from synchrotron X-ray diffraction data. Neutral and cationic lipids are depicted as having white and gray headgroups, respectively. Top: The most commonly observed lamellar phase (denoted L_α^C), consisting of alternating lipid bilayers and DNA monolayers. The interlayer spacing is $d = \delta_w + \delta_m$. Bottom: Few lipid mixtures produce the inverted hexagonal phase of CL–DNA complexes (denoted H_{II}^C), which is comprised of lipid-coated DNA strands arranged on a hexagonal lattice. Schematics reprinted with permission from Koltover et al. (1998).

Fig. 5.2, these are a multilamellar structure with DNA monolayers sandwiched between cationic membranes (L_α^C) (Lasic et al., 1997; Rädler et al., 1997) and an inverted hexagonal structure with DNA encapsulated within cationic lipid monolayer tubes (H_{II}^C) (Koltover et al., 1998). Earlier work succeeded in elucidating the parameters which control the structure of the complex: the inverted hexagonal phase is observed for higher contents of DOPE, which has a negative spontaneous curvature (Israelachvili, 1992), as well as for complexes whose membranes have been softened by the addition of cosurfactant (Koltover et al., 1998).

In lamellar CL-DNA complexes, the DNA monolayers, sandwiched between lipid bilayers, are ordered with a well-defined separation distance d_{DNA} between adjacent DNA rods (Koltover et al., 2000; Rädler et al., 1997). By tuning the charge density of the membranes or introducing PEG-lipids, this DNA distance can be adjusted between 25 Å and 60 Å (see below).

B. Overcharging and stability of lamellar CL-DNA complexes

Koltover *et al.* (1999) carried out a comprehensive study of 2,3-dioleyloxy-propyltrimethylammonium chloride (DOTAP)/DOPC CL-DNA complexes as a function of lipid composition and lipid/DNA ratio (ρ:cationic lipid/DNA weight ratio), aimed at elucidating the interactions determining their structure, charge, and thermodynamic stability (Koltover *et al.*, 1999). These lamellar complexes incorporate excess DNA and lipid for $\rho_{chg} < 1$ and $\rho_{chg} > 1$, respectively, but only to a saturation level (Bruinsma, 1998). This "overcharging" and its limits are clearly evident from analysis of their structures and measurements of their zeta potential, shown in Fig. 5.3. The upper and lower saturation levels are strongly dependent on the membrane charge density (σ_M, the average charge per area of the membrane) and the salt concentration. Beyond them, complexes coexist with excess free DNA or liposomes. The addition of salt, which is invariably present in cell culture media, destabilizes complexes and reduces the achievable degree of overcharging, particularly for complexes with a larger content of neutral lipid (Koltover *et al.*, 1999).

The stability of CL-DNA complexes is an important parameter, since DNA eventually needs to be released in order to become transcriptionally active. The biogenic polycations spermidine[3+] and spermine[4+], which are known to be present at mM conditions in the cytosol during the cell cycle, are able to remove DNA from the CL-DNA complexes to form hexagonally packed DNA-polyamine particles in bulk solution (Koltover *et al.*, 2000). This suggests one possible mechanism for intracellular release of DNA from CL-DNA complexes during transfection. Once removed from the CL-DNA complex, the DNA can easily dissociate from the short polyamine molecules, since the probability of DNA dissociating from an oppositely charged oligo-electrolyte increases with decreasing oligo-electrolyte length. Not surprisingly, *in-vitro* X-ray studies also show that histones are able to remove DNA from CL-DNA complexes (Evans H., and Safinya C. R., 2001, unpublished results).

Negatively charged proteins are another cellular component that is likely involved in the dissociation of CL-DNA complexes in the cytoplasm. Synchrotron X-ray diffraction has solved the structure of CL-complexes of poly (L-glutamic acid) (PGA) (Subramanian *et al.*, 2000), a negatively charged model polypeptide, as well as F-actin, a highly abundant anionic cytoskeletal protein (Wong *et al.*, 2000). The data for the lipid-PGA complexes are consistent with a "pinched lamellar" phase, shown schematically in Fig. 5.4 (left). In this structure, PGA and the cationic lipid associate to form localized "pinched" regions. Between "pinches", pockets of water are stabilized with a large equilibrium spacing of 60 Å due to hydration repulsion of neutral bilayers.

CL-F-actin complexes consist of a network of flattened multilamellar tubules with an average width of $\approx 0.25\ \mu$m. The tubules are formed from stacks of triple-layers consisting of a lipid bilayer sandwiched between two monolayers

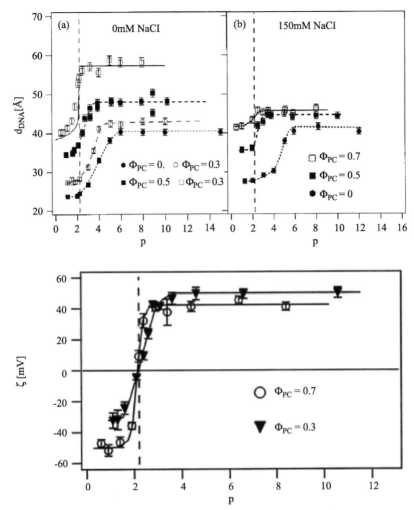

Figure 5.3. Top: (a) Variation of the DNA interaxial distance d_{DNA} with lipid/DNA weight ratio (ρ) in DOPC/DOTAP/DNA-complexes at a constant weight fraction of DOPC (Φ_{PC}), prepared in deionized water. The vertical dashed line indicates the isoelectric point ($\rho = 2.2$). The solid line through the data at $\Phi_{DOPC} = 0.7$ is the result of nonlinear Poisson–Boltzman theory for complexes with low membrane charge density (Bruinsma, 1998). The dashed lines are guides to the eye. The complexes are single-phase in the region of increasing d_{DNA}, coexisting with DNA at lower ρ and with lipid at higher ρ. (b) Same as (a) at 150 mM NaCl. All lines are guides to the eye. Bottom: Variation of the complexes' ζ-potential with changing ρ. The vertical line marks the isoelectric point. Lines through the data are a guide to the eye. Reprinted with permission from Koltover *et al.* (1999). © 1999 Biophysical Society.

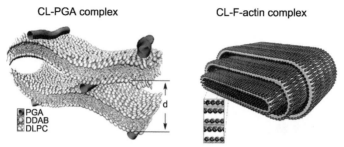

Figure 5.4. Left: Schematic of the local CL–PGA complex structure, showing the "pinching" mechanism. At larger length scales, the X-ray data is consistent with a model where the PGA macromolecules are positionally and orientationally disordered. DLPC: 1,2-Dilauroyl-*sn*-glycerophosphatidylcholine; DDAB: Didodecyl dimethyl ammonium bromide. Reprinted with permission from Subramanian *et al.* (2000). © 2000 American Chemical Society. Right: A schematic model of CL–F-actin complexes showing tubules formed by 3-layer composite membranes (inset) made up of a lipid bilayer sandwiched between two layers of F-actin. Reprinted with permission from Wong *et al.* (2000).

of F-actin, as shown in the schematic in Fig. 5.4 (right). The F-actin rods on the membrane are positionally locked into a 2-dimensional lattice (Wong *et al.*, 2000).

III. STRUCTURE-TRANSFECTION EFFICIENCY RELATIONSHIPS OF CL-DNA COMPLEXES

In contrast to typical molecular therapeutics, the efficiency of cationic lipids does not depend directly on their chemical structure. Rather, the mechanism and efficiency of transfection of CL-DNA complexes depend on properties of the whole self-assembled system. Examples are the complex structure (L_α^C or H_{II}^C) and, within a distinct structure, physical and chemical parameters of the complexes, such as the lipid/DNA charge ratio or the lipid composition. A lipid-independent measure of the ratio of neutral to cationic lipid is the membrane charge density (σ_M), the average charge per unit area of membrane. For isoelectric CL-DNA complexes in the L_α^C phase and d_{DNA} larger than the diameter of DNA with one hydration layer, d_{DNA} can directly be calculated from the average distance per anionic charge along the DNA backbone (l_0), and the membrane charge density σ_M: $d_{DNA} = e/(l_0\sigma_M)$ (Koltover *et al.*, 1999; Rädler *et al.*, 1997). Thus, the DNA interaxial spacing, which can be measured by X-ray diffraction, is a measure of σ_M, decreasing as the membrane charge density increases. For positively charged complexes ($\rho_{chg} > 1$), σ_M can no longer be directly calculated from d_{DNA}, but the same qualitative relationship still holds.

Early experiments in our group varied σ_M systematically for lamellar and inverted hexagonal phases of CL-DNA complexes to investigate the effect on TE and, via confocal microscopy, the pathways of gene delivery. Prior to those studies, TE measurements by other groups had shown that in mixtures of the monovalent cationic lipid DOTAP and neutral lipids, typically at a weight ratio between 1:1 and 1:3, DOPE aided transfection while DOPC severely suppressed it (Farhood et al., 1995; Hui et al., 1996). Therefore, it seemed that DOPE-based H_{II}^C complexes always transfect more efficiently than L_α^C complexes. The findings described below, from investigations on DNA complexes covering the whole range of compositions of DOTAP with neutral DOPC and DOPE, show that the difference in performance between the two types of complexes is more subtle than previously believed and that there are regimes of composition where the TE of complexes in the L_α^C phase rivals that of the best H_{II}^C complexes.

The striking difference in TE between L_α^C and H_{II}^C complexes at a cationic to neutral lipid ratio of 1:2 is illustrated in Fig. 5.5. Lamellar and inverted hexagonal complexes were prepared from DOTAP/DOPC and DOTAP/DOPE mixtures, respectively. Synchrotron XRD scans of complexes with plasmid DNA, measured in Dulbecco's Modified Eagle Medium (DMEM), a common cell culture medium, are shown in Fig. 5.5. Schematic views of the structures are also pictured (Lin et al., 2000, 2003). The DOTAP/DOPC complexes at a molar fraction $M_{DOPC} = 0.67$ (Fig. 5.5, left)

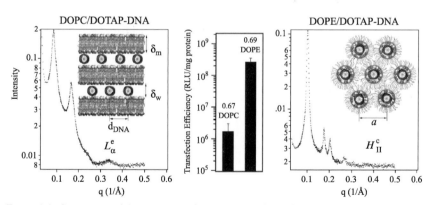

Figure 5.5. Comparison of CL–DNA complex structure and transfection efficiency for a cationic to neutral lipid ratio of $\approx$ 1:2. Complexes were prepared using pGL3 plasmid DNA and DOTAP as the cationic lipid. Left: Small angle X-ray scattering pattern and schematic view (inset) of lamellar L_α^C complexes (molar fraction $M_{DOPC} = 0.67$). Right: Small angle X-ray scattering pattern and schematic view (inset) of inverted hexagonal H_{II}^C complexes ($M_{DOPE} = 0.69$). Middle: TE of the complexes. Note the logarithmic scale. Reprinted with permission from Lin et al. (2003). © 2003 Biophysical Society.

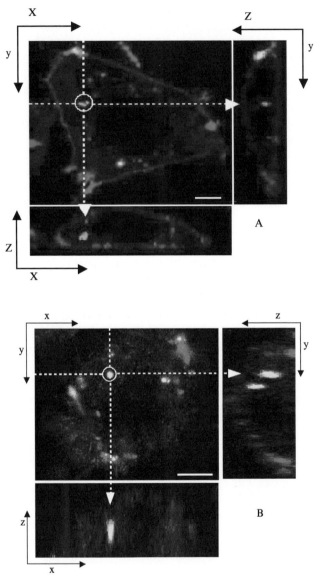

Figure 5.6. Laser scanning confocal microscopy images of transfected mouse L cells, fixed six hours after incubation with complexes. For each set of images, the center image is the x–y (top) view at a given z; the right shows a y–z side along the vertical dotted line; and the bottom a x–z side view along the horizontal dotted line. Red and green fluorescence corresponds to lipid and DNA labels, respectively; yellow, the overlap of the two, denotes CL–DNA complexes. Scale bars are 5 μm. (A): Cells transfected with H_{II}^{C} complexes ($M_{DOPE} = 0.69$) show transfer of fluorescent lipid to the cell plasma

showed sharp peaks at $q_{001} = 0.083$ Å^{-1}, $q_{002} = 0.166$ Å^{-1}, with a shoulder peak at $q_{003} = 0.243$ Å^{-1} and $q_{004} = 0.335$ Å^{-1}, resulting from the layered structure of the L_α^C phase. The interlayer spacing is $d = \delta_m + \delta_w = 2\pi/q_{001} = 75.70$ Å (cf. Fig. 5.2). For DOTAP/DOPE complexes at $M_{DOPE} = 0.69$, small angle X-ray scattering (Fig. 5.5, right) revealed four orders of Bragg peaks at $q_{10} = 0.103$ Å^{-1}, $q_{11} = 0.178$ Å^{-1}, $q_{20} = 0.205$ Å^{-1} and $q_{21} = 0.270$ Å^{-1}, denoting the H_{II}^C phase with a unit cell spacing $a = 4\pi/[(3)^{1/2}q_{10}] = 70.44$ Å (cf. Fig. 5.2). The results of TE experiments, shown in Fig. 5.5 (middle), demonstrate that the TE attainable with complexes in the H_{II}^C phase at $M_{DOPE} = 0.69$ is by more than two decades higher than that of L_α^C complexes at $M_{DOPC} = 0.67$.

A. Three-dimensional confocal imaging of CL-DNA complexes interacting with cells

To explore the structure–function correlation in more detail, we examined the transfer of CL-DNA complexes into cells and the subsequent DNA release by LSCM, which allows determining the position of objects relative to cells with certainty. The complexes were doubly tagged with fluorescent labels, Texas Red-DHPE for lipid and a covalently attached green for DNA (Mirus Label IT) (Lin *et al.*, 2003). Fig. 5.6 shows LSCM micrographs of mouse L cells, fixed six hours after the addition of complexes. In Fig. 5.6A, a typical image of a cell transfected with H_{II}^C complexes at $M_{DOPE} = 0.69$ is displayed. The lipid fluorescence clearly outlines the plasma membrane. Thus, either spontaneous transfer of labeled lipid or fusion of lipid with the plasma membrane has taken place. This may have occurred before or after entry through the endocytic pathway (Wrobel and Collins, 1995; Xu and Szoka 1996; Zabner *et al.* 1995, 1996). Inside the cell, both an aggregate of complexes (yellow) as well as lipid-free DNA (green) are visible. Thus, the interaction between H_{II}^C complexes and cells leads to the release of DNA from the CL-vector, consistent with the measured high TE.

Fig. 5.6B shows a corresponding confocal image for L_α^C complexes at $M_{DOPC} = 0.67$. In contrast to H_{II}^C complexes, no free DNA is visible. Instead, several intact CL-DNA complexes are found inside of the cell. One of these is highlighted in Fig. 5.6B. The complexes likely entered the cells through

membrane and the release of DNA (green; in the circle) within the cell. (B): LSCM image of cells transfected with L_α^C complexes at $M_{DOPC} = 0.67$, where TE is low, as shown in Figure 5.5. The cell outline was observed in reflection mode, appearing in blue. No evidence for fusion is visible and intact CL–DNA complexes such as the one marked by a circle are observed inside the cells. This observation implies that DNA remains trapped within the complexes, consistent with the observed low transfection efficiency. Reprinted with permission from Lin *et al.* (2003). © 2003 Biophysical Society. (See Color Insert.)

endocytosis: there is no fluorescent lipid observed in the cell membrane, ruling out cell entry via fusion. Cell entry by endocytosis has also been observed by others (Wrobel and Collins, 1995; Xu and Szoka 1996; Zabner *et al.* 1995). This was corroborated by LSCM images of cells prepared at $4°C$, where endocytosis is inhibited. In this case, complexes were found attached to the outside cell surface, but not within the cell body (Lin *et al.*, 2000). Thus, LSCM shows that at $M_{DOPC} = 0.67$, most of the DNA remains trapped in the endosome with the CL-vector, consistent with the observed low TE.

B. The relationship between the membrane charge density of lamellar CL-DNA complexes and transfection efficiency

In further transfection experiments, an unexpected enhancement of TE by two decades took place as the concentration of DOPC in lamellar CL-DNA complexes was decreased. The dependence of TE on M_{DOPC} for the whole range of compositions resulting in single-phase DOPC/DOTAP-DNA complexes is shown in Fig. 5.7A (diamonds). It is important to note that all TE measurements were done with 2 μg of plasmid DNA at constant $\rho_{chg} = 2.8$. Thus, every data point used an identical amount of charged species (DNA and cationic lipid) and M_{DOPC} was varied solely by adjusting the amount of neutral lipid. The efficiency starts low for $0.5 < M_{DOPC} < 0.7$, increasing dramatically to a value, at $M_{DOPC} = 0.2$, which rivals that achieved by DOPE/DOTAP-DNA H_{II}^C complexes. Similar results were obtained for another univalent cationic lipid, 2,3-di(myristyloxy)propyl(2-hydroxyethyl)dimethylammonium bromide (DMRIE), as also shown in Fig. 5.7A (triangles). However, a striking difference, which holds the key to a deeper understanding of the observed TE trends, was seen for the multivalent cationic lipid 2,3-Dioleyloxy-N-[2-(sperminecarboxamido)ethyl]-N,N-dimethyl-1-propylammonium chloride (DOSPA) as the cationic lipid (Fig. 5.7A, squares). A qualitatively similar curve is obtained, with TE decreasing rapidly above a critical M_{DOPC}^*. However, this M_{DOPC}^* is shifted from ≈ 0.2, as observed for DOTAP and DMRIE complexes, to 0.7 ± 0.1 for DOSPA.

The main difference between DOSPA and the other lipids are their headgroups (Fig. 5.7, inset): DOSPA has a charge of up to $+5$ at a headgroup size that is only somewhat larger than that of DOTAP. Thus, the charge density (charge per area) of the headgroup is much higher for DOSPA. Consequently, the membrane charge density σ_M at a given M_{DOPC} is also significantly larger in complexes containing DOSPA than in complexes prepared from DOTAP or DMRIE.

Utilizing this insight, Fig. 5.7B shows the same TE data as Fig. 5.7A, now plotted against σ_M. Given the complexity of the CL-DNA-cell system and the fact that all other parameters are fixed by the experiment or the chemical structure, it is remarkable that the data for monovalent and multivalent lipids,

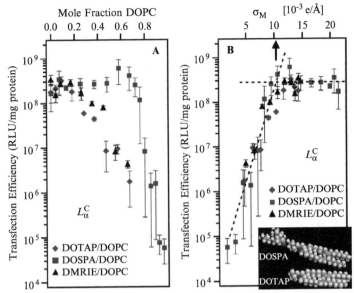

Figure 5.7. (A) Transfection efficiency, plotted as a function of varying molar fraction DOPC for complexes of the cationic lipids DOSPA, DOTAP, and DMRIE. (B) TE plotted versus the membrane charge density σ_M, demonstrating universal behavior of complexes containing cationic lipids with different charge and head group area (inset). For all three lipid systems, TE increases with σ_M up to an optimal σ_M^*, where it plateaus. An arrow marks $\sigma_M^* \approx 0.0104$ e/Å^2, determined by the intersection of two straight lines that were fit to the data (dashed lines). Note that the amount of transferred DNA was constant for all data points. Due to the fixed ρ_{chg} of 2.8, the same is true for the number of cationic charges on the lipids. The charge density was varied solely by varying the amount of neutral lipid. Reprinted with permission from Lin *et al.* (2003). © 2003 Biophysical Society.

spread out when plotted as a function of molar fraction of DOPC, coalesce into a single, universal curve as a function of σ_M. TE varies exponentially over nearly four decades as σ_M increases by a factor of ≈ 8 (Fig. 5.7B, σ_M between 0.0015 e/Å^2 and 0.012 e/Å^2) and the universal TE curve saturates beyond a common optimal $\sigma_M^* \approx 0.0104 \pm 0.0017$ e/$\text{Å}^2 \approx$ e/(100 Å^2), indicated by an arrow in Fig. 5.7B. This clearly implies that σ_M is an important and universal parameter for transfection by lamellar CL-DNA complexes. The membrane charge density was calculated as

$$\sigma_M = \text{total charge/total area} = eZN_{cl}/(N_{nl}A_{nl} + N_{cl}A_{cl})$$
$$= [1 - M_{nl}/(M_{nl} + rM_{cl})]\sigma_{cl} \qquad (5.1)$$

Here, $r = A_{cl}/A_{nl}$ is the ratio of the headgroup areas of cationic and neutral lipid; $\sigma_{cl} = eZ/A_{cl}$ is the charge density of the cationic lipid with valence Z; $M_{nl} = N_{nl}/$

$(N_{nl} + N_{cl})$ and $M_{cl} = N_{cl}/(N_{nl} + N_{cl})$ are the molar fractions of the neutral and cationic lipids, respectively. In Fig. 5.7B, we used $A_{nl} = 72$ Å^2 (Gruner et al., 1988; Tristram-Nagle et al., 1998), $r_{DOTAP} = r_{DMRIE} = 1$, $r_{DOSPA} = 2$, $Z_{DOTAP} = Z_{DMRIE} = 1$, $Z_{DOSPA} = 4$ (expected for pH ≈ 7 (Remy et al., 1994)). Since standard methods failed to yield well-defined values for the headgroup area of cationic lipids, r is an adjustable parameter. However, the values used in the plots agree well with chemical intuition (cf. also the inset in Fig. 5.7). Interestingly, the membrane charge density σ_M also controls the average DNA interaxial spacing d_{DNA} (see above), which decreases as σ_M increases and levels off for $\sigma_M > \sigma_M^*$ (Lin et al., 2003).

The markedly different transfection behavior of L_α^C CL-DNA complexes at low and high σ_M prompted further investigations with confocal microscopy (Lin et al., 2003). LSCM images of cells transfected with L_α^C complexes at high σ_M show a path of complex uptake and DNA release distinct from both L_α^C complexes at low σ_M (Fig. 5.6B) and H_{II}^C complexes (Fig. 5.6A). A typical confocal image of a cell incubated for 6 h with L_α^C complexes at $M_{DOPC} = 0.18$ ($\sigma_M \approx 0.012$ e/Å^2) is shown in Fig. 5.8. Since there is no indication of transfer of fluorescent lipid to the plasma membrane, the lamellar complexes with high σ_M must have entered the cells through endocytosis, similar to their low-σ_M counterparts. A few intact complexes are visible inside the cell: Fig. 5.8 (label 2; box 2) shows the equal green (DNA) and red (lipid) fluorescence intensity along the dotted line in the inset. In addition, a mass of exogenous DNA successfully transferred into the cytoplasm is evident (Fig. 5.8, label 1: box 1 shows the much larger green (DNA) fluorescence intensity along the x–y diagonal). The integrated fluorescence intensity of the observed DNA (Fig. 5.8, box 1) is comparable to that of DNA complexed with lipids (Fig. 5.8, box 2). This indicates that the released DNA is in the form of aggregates. Since there are no DNA-condensing molecules (Bloomfield, 1991) in the endosome, these aggregates must reside in the cytoplasm. The presence of lipid-free DNA in the cytoplasm after endocytosis of complexes is in agreement with the measured high TE and, moreover, implies fusion between CL-DNA lipids and endosomal membranes, enabling escape from the endosome. The confocal image also shows a large aggregate of complexes in one part of the cell (Fig. 5.8, label 3). Comparing the changes in fluorescence intensity along the x–y diagonal (Fig. 5.8, box 3) and z-axis (Fig. 5.8, box 4), from the outside toward the inside of the cell, we see an aggregate of complexes caught in the process of dissociation after endocytosis, with released DNA toward the inside of the cell.

Since LSCM suggested endocytosis as the mechanism of cell entry for L_α^C CL-DNA complexes and DNA appeared trapped within complexes at low σ_M, transfection experiments in the presence of chloroquine were performed. This is a well-established bioassay known to enhance the release of material trapped within endosomes (Felgner, 1990) by osmotically bursting late-stage

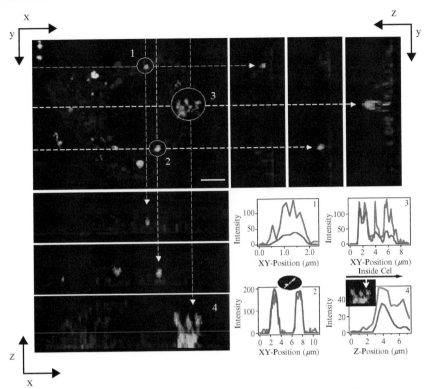

Figure 5.8. A typical LSCM image of a mouse L cell transfected with L_α^C complexes at $M_{DOPC} = 0.18$, corresponding to cationic membranes with a high charge density $\sigma_M \approx 0.012 \, e/\mathring{A}^2$ and high TE (cf. Figure 5.7). Red and green fluorescence corresponds to lipid and DNA labels, respectively; yellow, the overlap of the two, denotes CL–DNA complexes. The cell outline was observed in reflection mode, appearing in blue. Scale bars are 5 μm. The center image is the x–y (top) view at a given z; on the right are y–z side views along the vertical dotted lines; at the bottom are x–z side views along the horizontal dotted lines. In the boxes in the lower right corner, plots of lipid and DNA fluorescence intensity along the x–y diagonal or z-axis are shown for objects labeled with numbers. Although the lamellar complexes used here show high TE, no lipid transfer to the cell plasma membrane is seen in contrast to high-transfecting H_{II}^C complexes (Figure 5.6A). Both released DNA (1) and intact complexes (2) are observed inside the cell. Labels (3) and (4): A complex in the process of releasing its DNA into the cytoplasm. Reprinted with permission from Lin *et al.* (2003). © 2003 Biophysical Society. (See Color Insert.)

endosomes. The fractional increase (TE$_{chloroquine}$/TE; note the logarithmic scale) for the DOSPA/DOPC and DOTAP/DOPC lipid systems with added chloroquine, plotted as a function of σ_M (Fig. 5.9), shows a large increase by as much as a factor of 60 as σ_M decreases. This indicates that lamellar L_α^C complexes are trapped within endosomes at low σ_M, consistent both with the

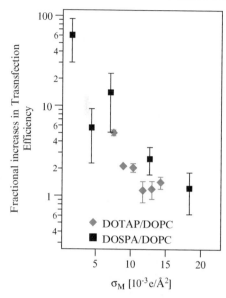

Figure 5.9. The fractional increase of TE upon addition of chloroquine, $TE_{chloroquine}/TE$, for DOSPA/DOPC and DOTAP/DOPC mixtures, plotted as a function of σ_M. Note the logarithmic scale. The fractional increase is substantial for low membrane charge densities, indicating that endosomal escape is limiting in this regime. Reprinted with permission from Lin *et al.* (2003). © 2003 Biophysical Society.

confocal images (Fig. 5.6B) and the measured low TE without chloroquine. At high σ_M, chloroquine has a much smaller effect on TE, with the fractional increase of order unity. Therefore, endosomal entrapment is not a significant limiting factor in this regime.

C. Transfection properties of new multivalent cationic lipids

To further investigate the finding of a well-defined relationship between transfection efficiency and membrane charge density in lamellar L_α^C complexes, and to more broadly explore the relevance of σ_M as a key chemical parameter, new lipids with multivalent cationic headgroups have been prepared (Ewert *et al.*, 2002; Schulze *et al.*, 1999). Their general structure is shown in Fig. 5.10, and their headgroups are compiled in Table 5.1. The headgroups are derived from the amino acid ornithine. By the addition of 0 to 3 propyl-amine groups, we have generated headgroup charges from +2 (MVL2) to +5 (MVL5). For comparison, a glycine-based monovalent lipid (MVL1) was prepared as well.

Earlier results for one of the lipids (MVL5) showed structurally stable MVL5/DOPC/DNA complexes, forming the L_α^C phase throughout the lipid composition range (Ewert *et al.*, 2002). As depicted in Fig. 5.11, X-ray

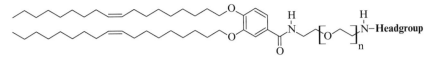

Figure 5.10. General structure of the new, ornithine-based multivalent lipids.

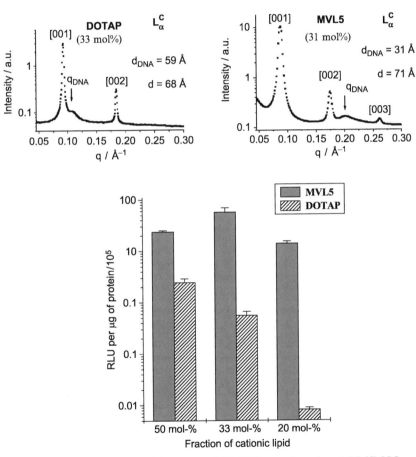

Figure 5.11. Top: Small angle XRD patterns of CL–DNA complexes from MVL5/DOPC and DOTAP/DOPC lipid mixtures at equivalent molar fractions of cationic lipid (31% MVL5; 33% DOTAP; complexes prepared in DMEM). The complexes are in the lamellar phase. The DNA interaxial spacing is reduced from 59 Å in the complexes with DOTAP to 31 Å in the complexes with MVL5, reflecting their higher membrane charge density. Bottom: Transfection efficiencies for cationic lipids DOTAP and MVL5 in mixtures with DOPC. Reprinted with permission from Ewert *et al.* (2002). © 2002 American Chemical Society.

diffraction shows the drastic effect of the increased head group charge density on σ_M, evident in the DNA spacing. The scans were taken for complexes at equivalent lipid molar fractions of 31% (MVL5) and 33% (DOTAP). The spacings d_{DNA}, extracted from the broad DNA correlation peaks (Rädler *et al.*, 1997), are $d_{DNA} = 2\pi/q_{DNA} = 31.2$ Å and $d_{DNA} = 59.5$ Å for the complexes prepared with pentavalent MVL5 and univalent DOTAP, respectively. At this composition, the transfection efficiency of the MVL5 complexes is approximately 100-fold higher than that of the DOTAP complexes. This is also shown in Fig. 5.11, demonstrating again the importance of high membrane charge density for optimal transfection efficiency. The data shows that MVL5 consistently gives higher transfection efficiencies than DOTAP at several molar ratios of neutral to cationic lipid. However, the difference in TE increases dramatically from one to three orders of magnitude as the amount of cationic lipid is reduced from 50 to 20 mol-%. While the TE for MVL5 remains high, that for DOTAP drops rapidly. As already seen in Fig. 5.7 for DOSPA, multivalent lipids maintain a $\sigma_M > \sigma_M^*$, and thus high TE, over a large range of cationic to neutral lipid molar ratios. This provides a rationale for the frequently reported superior TE of multivalent lipids, particularly in cases where only few neutral/cationic lipid ratios were tested.

The results of more recent investigations on the series of MVLs show that at very high membrane charge densities, which these lipids allowed us to access for the first time, the transfection efficiency drops again, rather than staying at a saturated level (Ahmad *et al.*, 2005). A universal maximum in TE at $\sigma_M \approx 17.10^{-3}$ e/Å^2 is observed for all lipids, demonstrating again the universality of σ_M as a parameter that predicts the transfection efficiency of lamellar complexes.

D. A model of cell entry by L_α^C CL carriers: activated fusion, dependent on σ_M and elastic moduli of complex membranes

The combined X-ray diffraction, LSCM, and TE data have led to a model of cellular entry via L_α^C CL carriers (Lin *et al.*, 2003). The key features of this model are shown in the schematic in Fig. 5.12. Previous work by others has established that the initial electrostatic attraction and attachment of cationic CL-DNA complexes to mammalian cells is mediated by negatively charged cell surface proteoglycans (Fig. 5.12a) (Kopatz *et al.*, 2004; Mislick and Baldeschwieler, 1996; Mounkes *et al.*, 1998). LSCM at room temperature and 4 °C, transfection in the presence of chloroquine as well as the results from other groups indicate that L_α^C complexes enter via the endocytic pathway (Fig. 5.12b,c) (Lin *et al.*, 2003; Wrobel and Collins, 1995; Xu and Szoka 1996; Zabner *et al.* 1995). As seen in LSCM, the further intracellular fate of the complexes strongly depends on their membrane charge density: at low σ_M, complexes remain intact and trapped in the endosomes (Fig. 5.12c), whereas they successfully escape at high

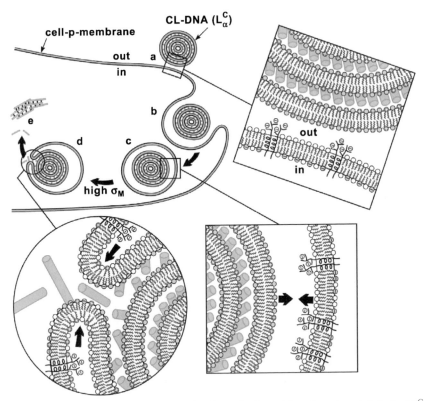

Figure 5.12. Model of cellular uptake and endosomal release (through activated fusion) of L_α^C complexes. (a) Cationic complexes adhere to cells due to electrostatic attraction between positively charged CL–DNA complexes and negatively charged cell-surface proteoglycans (shown in expanded views). (b and c) After attachment, complexes enter through endocytosis. (d) Only complexes with sufficiently high membrane charge density escape from the endosome through activated fusion with the endosomal membrane. (e) Confocal microscopy shows that lipid-free DNA inside the cell exists primarily in the form of aggregates. These DNA aggregates must reside in the cytoplasm because oppositely charged cellular biomolecules able to condense DNA are not present in the endosome. Arrows in the expanded view of (c) indicate the electrostatic attraction between the oppositely charged membranes of the complex and endosome, which enhances adhesion and fusion. Arrows in the expanded view of (d) indicate the bending of the membranes required for fusion, which constitutes the main barrier for the process. Reprinted with permission from Lin *et al.* (2003). © 2003 Biophysical Society. (See Color Insert.)

σ_M. Therefore, Lin *et al.* (2003) proposed that escape from the endosome is the process limiting TE for $\sigma_M < \sigma_M^*$. The most likely mechanism for this is through fusion with the endosomal membrane (Fig. 5.12d), with a kinetic barrier that has to be overcome (activated fusion). An Arrhenius-type equation then yields the exponential increase of the universal TE curve (Fig. 5.7B) with σ_M, where TE is proportional to the rate of fusion:

$$\text{Transfection Efficiency} \propto \text{rate of fusion} = \tau^{-1}[\exp(-\delta E/k_B T)], \qquad (5.2)$$

with δE = kinetic energy barrier height = $a \cdot \kappa - b \cdot \sigma_M$. Here, a,b > 0 are constants, and τ^{-1} is the collision rate between the trapped CL-DNA particle and the endosomal wall (Fig. 5.12c); κ is the membrane-bending modulus, which is a measure of the flexibility of the membrane (Helfrich, 1973; Israelachvili, 1992; Janiak *et al.*, 1979; Seddon, 1990). Bending of a membrane requires energy proportional to κ and provides the main barrier to fusion, because bending is inevitable during the fusion of membranes. This is shown schematically by arrows in the expanded view of Fig. 5.12d. On the other hand, electrostatic attraction favors membrane adhesion because the fusing membranes of complex and endosome are oppositely charged (Fig. 5.12c, arrows in expanded view). This leads to the term "$- b \cdot \sigma_M$", which describes the fact that σ_M lowers the barrier height. The fact that higher σ_M lowers the barrier height and thus appears in the exponent in Eq. (5.2) is responsible for the observed exponential increase of TE (Fig. 5.7b). In agreement with this model, a recent theoretical study, which considers fusion between two neutral lipid bilayers (i.e., pores forming between neighboring membranes), has found that the main energy barrier against fusion is proportional to κ (Gompper and Goos, 1995).

    It is important to note that if cellular attachment and uptake were limiting TE via a σ_M-dependent mechanism, a linear increase of TE with σ_M would be predicted. Thus, the TE data, which shows an exponential increase, excludes this possibility. In addition, the observed effect of chloroquine on TE can not be explained by assuming uptake-limited transfection.

E. Transfection properties of inverted hexagonal H_{II}^C complexes

A comparison of TE as a function of σ_M for DOTAP/DOPC-DNA and DOTAP/DOPE-DNA complexes is shown in Fig. 5.13 (left). The DOTAP/DOPE system goes through 2 phase transitions: From L_α^C (filled squares) at high amounts of DOTAP to coexisting $L_\alpha^C + H_{II}^C$ (squares with cross) to H_{II}^C (open squares). In contrast to DOPC containing complexes, for which TE increases exponentially over nearly four decades with increasing $\sigma_M < \sigma_M^*$, the TE of DOPE-containing H_{II}^C complexes is independent of σ_M: at high $M_{DOPE} > 0.56$, DOTAP/DOPE-DNA complexes are in the H_{II}^C phase and continue to exhibit high TE. We can thus conclude that σ_M is an essential parameter for transfection with L_α^C complexes but not H_{II}^C complexes. The mechanism of transfection by DOPE containing H_{II}^C complexes is dominated by other effects as we describe below.

F. A model of cell entry by CL-vectors with the H_{II}^C structure: relevance of the outermost lipid layer

In LSCM, mixing of the lipids of H_{II}^C complexes with the plasma membrane is observed as lipid fluorescence outlining the plasma membrane (Lin et al., 2003). Confocal imaging further shows that the interaction between H_{II}^C complexes and cells leads to release of DNA from the lipid vectors, consistent with their high TE. Therefore, H_{II}^C complexes must undergo rapid fusion with cellular membranes. Fig. 5.13 (right) schematically shows a mechanism that may be responsible for this. The top part (a) pictures a H_{II}^C complex, approaching either the plasma or the endosomal membrane. The cell-surface

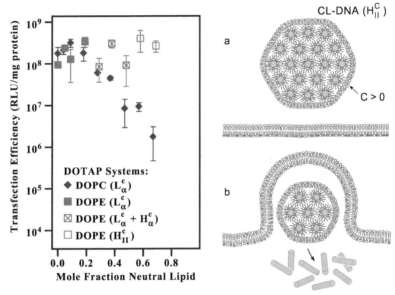

Figure 5.13. Left: TE and complex structure as a function of molar fraction of neutral lipid for complexes prepared from DOTAP/DOPC and DOTAP/DOPE mixtures. Right: Schematic of the proposed mechanism of cell entry/endosomal fusion of inverted hexagonal complexes. An H_{II}^C CL–DNA complex is shown interacting with the plasma membrane or the endosomal membrane. The cell-surface proteoglycans of the cellular membrane have been omitted for clarity. The outer lipid monolayer covering the H_{II}^C CL–DNA complex has a positive curvature. However, the preferred curvature of the lipids forming the complex membrane is negative, as realized in the tubules coating DNA within the complex. Thus, the outer layer is energetically costly. This results in a driving force, independent of the cationic membrane charge density, for rapid fusion of the H_{II}^C complex with the bilayer of the cell plasma membrane or the endosomal membrane. Reprinted with permission from Lin et al. (2003). © 2003 Biophysical Society. (See Color Insert.)

proteoglycans, which again mediate the attraction between the complex and the membrane (cf. Fig. 5.12), have been omitted for clarity. The preferred (natural) membrane curvature for lipids forming H_{II}^C complexes is negative, as opposed to ≈ 0 for lipids that form the L_{α}^C structure. Such curvature is realized for the lipids coating DNA inside the H_{II}^C complex, but the curvature of the outermost lipid monolayer, which must cover the H_{II}^C complex to provide a hydrophilic surface, is positive. This elastically frustrated state of the outer monolayer, which is independent of σ_M, drives the rapid fusion with the plasma or endosomal membrane. The fusion releases a layer of DNA and a smaller H_{II}^C complex, as shown in Fig. 5.13 (right, bottom). Speaking in terms of the model for cell entry of L_{α}^C complexes, the activation energy for fusion of H_{II}^C complexes with the endosomal membrane is negligible and this step no longer limits TE. The outer layer of the released, smaller H_{II}^C complex is again elastically frustrated and will drive quick intracellular release of the remaining DNA through interactions with negatively charged membranes or proteins. It is interesting to speculate that a similar mechanism might also be responsible for the observed poor serum stability of DOPE-containing complexes. By comparison, the onion-like lamellar complexes are inherently more stable. They are expected to dissociate much more slowly, layer-by-layer, through interactions with anionic components of the cell or DNA-condensing biomolecules (see above).

IV. PEGYLATED CL-DNA COMPLEXES: SURFACE FUNCTIONALIZATION AND DISTINCT DNA–DNA INTERACTION REGIMES

CL-DNA complexes of the type currently used in *ex vivo* and *in vivo* clinical trials (the latter involving intratumoral injection methods (Mahato and Kim, 2002; Ferber, 2001)) are not suitable for systemic applications. Cationic lipids and their complexes with DNA activate the complement system (Plank *et al.*, 1996), which results in their rapid removal from circulation through opsonization. Poly(ethyleneglycol) (PEG) conjugation to non-viral vectors can reduce the activation of the complement system, as is well known for liposomes (Bradley *et al.*, 1998; Lasic and Martin, 1995; Lasic and Papahadjopoulos, 1995; Woodle, 1995). The presence of a hydrophilic polymeric shell on liposomes provides a repulsive barrier and results in vastly increased circulation lifetimes, a phenomenon referred to as steric stabilization. Thus, incorporating PEG-lipids is an essential step in making CL-DNA complexes a viable option for systemic gene delivery. For future applications, it is also important to have the ability to use polymer chains of variable length as tethers for target-specific ligands (e.g., peptides), which will add functionality and specificity to the complexes. Therefore, it is crucial to gain an understanding of the effects of incorporating PEG-lipids into CL-DNA complexes.

A. DNA–DNA interaction regimes in PEGylated CL-DNA complexes

Recent experiments in our group have probed the structure, morphology and function of CL-DNA complexes consisting of a ternary mixture of DOTAP, DOPC, and PEG-lipids (Martin-Herranz et al., 2004; Schulze et al., 1999). This work has shown that a critical value of PEG chain length exists, above which steric stabilization and other polymer-specific effects become evident. The structures of the investigated lipids are displayed in Fig. 5.1. X-ray diffraction of isoelectric ($\rho_{chg} = 1$) complexes revealed one phase of stable complexes with a well-ordered lamellar structure for the PEG400-lipids. The lamellar structure was also observed for the PEG2000-lipids, but phase separation occurs at higher molar ratios of PEG-lipid (>7 mol-% for PEG2000^{2+}-lipid; >10 mol-% for PEG2000-lipid). Complete incorporation of PEG-lipid into the complexes is not possible beyond these limits.

The effect of incorporating PEG400^{2+}-lipid into isoelectric CL-DNA complexes was studied by monitoring the distance between DNA chains (d_{DNA}; see Fig. 5.2) while increasing the molar fraction of PEG400^{2+}-lipid at various constant molar fractions of DOTAP (M_{DOTAP}). The results reveal that PEG400^{2+}-lipid simply acts as a cationic co-lipid, leading to condensation of DNA through an increase in the membrane charge density. As with DOTAP (Salditt et al., 1998), two distinct DNA interaction regimes are observed: in the electrostatic regime ($4 < \sigma_M/10^{-3}$ e/Å^2 < 8.5), d_{DNA} depends purely on the membrane charge density. In the regime of $\sigma_M > 8.5 \cdot 10^{-3}$ e/Å^2, a strong repulsive hydration barrier between DNA rods dominates (Salditt et al., 1998), preventing further condensation of DNA by added PEG400^{2+}-lipid. Neutral PEG400-lipid also shows no polymer-specific behavior, its exchange with DOPC resulting in no effect on the DNA spacing.

A distinctly different picture arises for PEG-lipids with chains of molecular weight 2000 g/mol. Fig. 5.14A shows XRD data from single-phase PEG-CL-DNA complexes. Increasing amounts of neutral PEG2000-lipid were added to the membranes of DOTAP/DOPC-DNA complexes at constant $M_{DOTAP} = 30\%$. The DNA interaxial spacing d_{DNA}($d_{DNA} = 2\pi/q_{DNA}$) decreases from 53.7 Å (0% PEG-lipid) to 49.1 Å (1.6 mol-% PEG-lipid) to 41.6 Å (6.7 mol-% PEG-lipid) with increasing molar fraction of PEG2000-lipid, indicating the existence of an additional attractive force. This attraction is due to the presence of a polymer chain in the confined space between the lipid bilayers where the DNA chains reside (depletion attraction force; see below). The fact that addition of neutral as well as cationic (data not shown) PEG2000-lipid decreases the DNA spacing confirms that the polymer

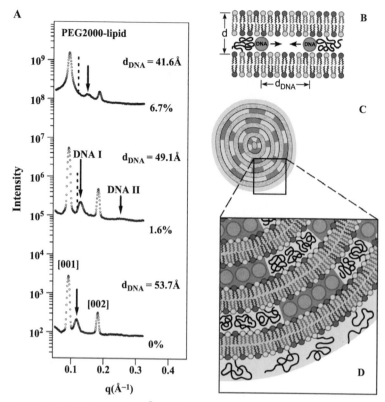

Figure 5.14. (A): XRD scans of L_α^C CL–DNA complexes with PEG$_{2000}$-lipid incorporated within their membranes. Arrows point to the DNA interaxial peak which moves to larger q_{DNA} (corresponding to a decrease of d_{DNA}) as PEG$_{2000}$-lipid is added to the membranes of the complex. The dashed line marks the position of q_{DNA} for 0% PEG-lipid. (B–D): Schematics of a lamellar CL–DNA complex containing long-chain PEG-lipids. (B) The presence of the polymer is forcing the DNA rods closer together than predicted by electrostatics (depletion attraction force). (C) Cross-section of a PEG–CL–DNA complex with DNA rich domains (dark gray) and polymer rich domains (light gray) in-between lipid bilayers (gray) (D) Enlarged view showing the internal phase separation as well as the outer shell of polymer chains. Reprinted in part with permission from Martin-Herranz *et al.* (2004). © 2004 Biophysical Society.

chain-DNA interaction is the dominating effect of adding PEG2000-lipids, as opposed to electrostatics in the case of PEG400-lipids.

Figure 5.14B schematically shows the origin of the polymer-induced depletion attraction force between DNA strands, which is well known in bulk solution (where much larger PEG molecular weights are required) (Vasilevskaya *et al.*, 1995). Since PEG2000 has a radius of gyration of $\approx35\,\text{Å}$

(Devanand and Selser, 1991; Warriner *et al.*, 1996), the PEG2000 part of the PEG2000-lipid will be excluded from regions between DNA rods (Israelachvili, 1992). This causes a phase separation between the polymer and the DNA within the layers of the complex, as shown schematically in Fig. 5.14C. The decreased DNA spacing then is a result of osmotic stress, exerted on the DNA domains by the PEG2000 chains confined to the outside of these domains, which increases with the concentration of polymer. As depicted in Fig. 5.14D, the resulting complex has DNA-rich domains, shown in dark gray, and polymer-rich domains, shown in light gray.

B. Surface functionalization of CL-DNA complexes with PEG-lipids

Optical microscopy of PEG-CL-DNA complexes at $\rho_{chg} = 2.8$ and $M_{DOTAP} = 0.82$ in DMEM was performed to demonstrate coverage of their surface with the PEG-lipids and further pinpoint their distinct, chain length-dependent behavior. Figure 5.15A shows complexes without PEG-lipid. Some degree of aggregation of the complexes is observed, due to the presence of salts in DMEM. Figure 5.15B shows complexes at $M_{PEG400\text{-lipid}} = 10\%$. Again, the aggregation of the complexes is clearly evident. However, complexes prepared using the long-chain PEG-lipid at $M_{PEG2000\text{-lipid}} = 10\%$ exhibit a strong shielding effect of the polymer (Fig. 5.15C). No aggregation of complex particles occurs, due to steric repulsion conferred by the shell of PEG2000-lipid polymer chains of thickness ≈ 35 Å (Devanand and Selser, 1991; Warriner *et al.*, 1996). As mentioned earlier, this stabilization is important for developing a viable *in vivo* gene delivery system (Harvie *et al.*, 2000; Pitard *et al.*, 2001; Wheeler *et al.*, 1999).

Figure 5.15D shows transfection results for positively charged PEG-CL-DNA complexes ($\rho_{chg} = 2.8$). At $M_{DOTAP} = 0.80$, transfection efficiency is high without PEG-lipid, but the addition of 6 mol-% PEG2000-lipid or PEG2000^{2+}-lipid strongly suppresses TE, by about 2 orders of magnitude. This indicates that the electrostatic binding of the cationic CL-DNA complexes to cells is efficiently reduced due to shielding by the PEG2000 polymer coat. In contrast, the addition of cationic or neutral PEG400-lipid only negligibly affects TE, even at 20 mol-% PEG400-lipid. Thus, no shielding occurs with these shorter amphiphiles and the further addition of cationic co-lipid (PEG400^{2+}) does not aid transfection in this regime of high σ_M, where TE is at a maximum (Lin *et al.*, 2003). Note that 6 and 20 mol-% of PEG2000-lipid and PEG400-lipid, respectively, correspond to an approximately equal total weight of PEG.

In summary, microscopy and transfection experiments show that the PEG-lipid is coating the surfaces of the complexes, whereas X-ray diffraction results clearly demonstrate that the PEG-lipid is also located internally.

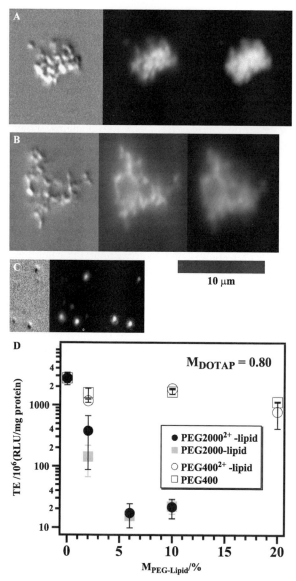

Figure 5.15. (A–C): Microscopy images of DOTAP/DOPC CL–DNA complexes at $M_{DOTAP} =$ 0.33 with plasmid DNA in the presence of DMEM, taken in DIC (left), lipid fluorescence (center) and DNA fluorescence modes. The images show complexes prepared (A) without PEG-lipid, (B) with 10 mol-% PEG400-lipid and (C) with 10 mol-% PEG2000-lipid. The complex particles aggregate when no PEG-lipid or PEG400-lipid are added but are sterically stabilized by 10 mol-% PEG2000-lipid. (D): Transfection efficiency of PEG-lipid/DOTAP/DOPC–DNA complexes as a function of increasing molar fraction of PEG-lipid ($M_{PEG\text{-lipid}}$) added to a DOTAP/

V. CONCLUSION

The structure–function data obtained from the reported work has propelled us further toward the long-term goal of our research: unraveling the mechanisms of transfection of CL-DNA complexes at the molecular to cellular level to enable the rational design of optimal non-viral transfection reagents for gene therapy.

CL-DNA complexes of distinct structure (i.e. L_α^C versus H_{II}^C) differ widely in their interactions with cells and their ability to deliver exogenous DNA to the cytoplasm. We have demonstrated that the membrane charge density of the CL-vector, σ_M, is a key universal parameter for transfection with L_α^C complexes. The universal TE curve for DOPC-containing L_α^C complexes increases exponentially with $\sigma_M < \sigma_M^*$ (the optimal membrane charge density). The TE level of L_α^C complexes at σ_M^* is comparable to the high TE of DOPE-containing H_{II}^C complexes, which exhibit no dependence on σ_M. This is an important finding: by not having to rely on DOPE as the neutral lipid in order to induce the H_{II}^C phase, a compositional degree of freedom is gained, and neutral lipids with various functionalities may be introduced while maintaining high TE.

PEG-lipids of various chain lengths are readily incorporated into CL-DNA complexes. The PEG-chains are present on the inside as well as the outside of the complexes, giving rise to polymer-specific effects such as depletion attraction forces between DNA rods and steric stabilization of complex particles above a critical PEG chain length of molecular weight >400 g/mol and ≤2000 g/mol. Sterical stabilization achieved by PEG2000-lipids reduces TE significantly but is a necessary first step on the way towards CL-vectors viable for *in vivo* applications.

VI. FUTURE DIRECTIONS

The capabilities of CL-DNA complexes as non-viral vectors and the understanding of their structure and mechanism of action have improved vastly over the last decade. Out of a large variety of lipids and formulations, vectors suitable

---

DOPC liposome mixture ($M_{DOTAP} = 0.80$ for all complexes). Adding PEG2000^{2+}-lipid or PEG2000-lipid leads to a decrease in TE by nearly two orders of magnitude with only 6 mol-% added PEG-lipid. By contrast, addition of PEG400^{2+}-lipid or PEG400-lipid does not modify TE significantly, even at 20 mol-% PEG-lipid. As a reference, typical TE for naked DNA is on the order of 300,000 RLU/mg protein. Note that 20 mol-% PEG400-lipid and 6 mol-% PEG2000-lipid correspond to approximately equal total weight of PEG. Reprinted with permission from Martin-Herranz et al. (2004). © 2004 Biophysical Society.

for *in vitro* transfection in research settings, some general mechanisms, and a few unifying themes such as the relevance of the membrane charge density have emerged. More such unifying themes are likely to surface, and will help to rationally design more efficient lipids. In standard CL-DNA complexes, the cationic lipids perform multiple functions along the transfection pathway. They condense DNA and provide for electrostatic attachment to the cell surface; they help with escape of the complex from the endosome via activated fusion into the cytoplasm, where they need to dissociate from the DNA. In the future, dividing these responsibilities between several components and enabling specific interactions between complexes and cells, for example, by the incorporation of peptide-ligands, will be an important and exciting avenue of research.

The universality of σ_M as a parameter for transfection by lamellar complexes is best explored with a series of lipids bearing headgroups of systematically varied size and charge (MVLs, Table 5.1). These lipids open access to a new regime of very high membrane charge density. Endosomal escape is not limiting in this regime, which shows an unexpected decrease in TE with increasing σ_M. The full universal curve fits a simple Gaussian bell curve, with an optimal, intermediate membrane charge density σ_M^* resulting from the compromise between the opposing requirements for high TE: higher σ_M improves TE by enhancing the activated fusion with the endosomal membrane, but itself becomes an obstacle to transfection at very high σ_M (Ahmad et al., 2005).

Future studies will also move towards revealing structure–function correlations for transfection *in vivo*. Preparing PEG-coated CL-DNA complexes with strongly reduced electrostatic binding to cells has been a crucial initial step in the development of competent vectors for systemic gene delivery. The next step will be to re-equip these complexes with means of attaching to the targeted cells, preferably via specific, ligand-receptor-based binding. The feasibility of this concept has been demonstrated for liposomes with antibodies as targeting moieties (Allen, 1994), and some work along these lines has also been done with non-viral vectors. However, the need for a thorough and systematic mechanistic investigation has not been met.

Currently, a major continuing focus of work on non-viral gene delivery systems is on optimizing transfection at the cell level in continuous (dividing) cell lines with potential relevance for the clinical development of cancer vaccines (Ferber, 2001; Mahato and Kim, 2002; Nabel et al., 1993; Rinehart et al., 1997; Stopeck et al., 1998). Since the nuclear membrane dissolves during mitosis, it is not a major barrier for delivery and expression of DNA in these experiments (Mortimer et al., 1999). However, even cancer cells are dividing much more slowly than the cells used for *in vitro* experiments and some very interesting target cells for gene delivery, such as nerve cells, are non-dividing cells. Thus, optimizing TE for complexes that can cross the nuclear membrane is of great interest. This constitutes another application for functionalized CL-DNA complexes, since the nucleus may be targeted with a nuclear localization sequence (NLS) (Jans et al., 1998; Kalderon et al., 1984). Existing work employing NLS peptides to non-viral gene delivery is still ambiguous, and systematic mechanistic studies are lacking (Cartier and Reszka, 2002; Escriou et al., 2003; Hebert, 2003).

Another future avenue of research is exploiting the predictions of our current model of transfection by lamellar complexes. In this model, TE is limited by the rate of the activated fusion of the oppositely charged membranes of CL-DNA complexes and endosome (Ewert et al., 2002; Lin et al., 2003). Since the elastic cost of fusion provides the main barrier for this process, the model suggests that softening of the membranes of CL-DNA complexes should enhance transfection efficiency. The bending modulus of membranes κ scales with lipid chain length l and the area per lipid chain A_L as $\kappa \propto l^3/A_L^5$ (Safinya et al., 1989; Szleifer et al., 1990). Thus, the effect of κ on the TE of CL-DNA complexes can be probed by using lipids of varied chain length.

We can only briefly mention several other exciting new directions in the research on lipid-mediated delivery of nucleic acids here, such as imaging of the transfection process in live cells. Cholesterol has become a highly popular choice of neutral co-lipid and a thorough investigation of its effects on the properties of CL-DNA complexes is required. A rapidly growing and important field is that of delivering small interfering RNA (siRNA). Simultaneously, an ongoing long term objective is to improve efficiency for delivering large pieces of DNA containing human genes and their regulatory sequences (>100 kbp). With such a broad range of cargo, a diversification of vectors, tailored to their respective delivery tasks, seems a logical consequence. Lipid vectors, with their great compositional variability and fundamentally unlimited capacity, are ideally suited for these challenges.

VII. MAIN ABBREVIATIONS

CL   Cationic liposome
DDAB   Didodecyl dimethyl ammonium bromide
DLPC   1,2-Dilauroyl-sn-glycerophosphatidylcholine
DMEM   Dulbecco's modified Eagle's medium
DMRIE   2,3-Di(myristyloxy)propyl(2-hydroxyethyl)
          dimethylammonium bromide

DOPC   1,2-Dioleoyl-sn-glycerophosphatidylcholine
DOPE   1,2-Dioleoyl-sn-glycerophosphatidylethanolamine
DOSPA   2,3-Dioleyloxy-N-[2-(sperminecarboxamido)ethyl]-N,N-dimethyl-1-propylammonium chloride
DOTAP   2,3-Dioleyloxypropyltrimethylammonium chloride
H_{II}^C Inverted hexagonal phase of lipid-DNA complexes
LSCM   Laser scanning confocal microscopy
L_α^C Lamellar phase of lipid-DNA complexes
MVL   Multivalent lipid
MVL5   N1-[2-((1S)-1-[(3-Aminopropyl)amino]
          -4-[di(3-aminopropyl)-amino] butyl-carbox-amido)
          ethyl]-3,4-di[oleyloxy]-benzamide
NLS   Nuclear localization sequence
PEG   Poly(ethylene glycol)
PGA   Poly(L-glutamic acid)
RLU   Relative light units
TE   Transfection efficiency
XRD   X-ray diffraction
κ Membrane bending rigidity
σ_M Membrane charge density

Acknowledgments

We gratefully acknowledge useful discussions with Charles Samuel, Cyril George, Robijn Bruinsma and Avinoam Ben-Shaul. We have also benefited over the years through extensive discussions with Phillip Felgner and Leaf Huang. The reported work was supported by grant GM-59288 on DNA-lipid gene delivery studies from the National Institute of General Medical Sciences of the National Institute of Health. Support for the structure studies was also provided by the National Science Foundation DMR 0203755. DOSPA and DMRIE were gifts from Phillip Felgner for which we are grateful. The synchrotron X-ray diffraction experiments were carried out at the Stanford Synchrotron Radiation Laboratory, which is supported by the U.S. Department of Energy. This work made use of MRL Central Facilities supported by the MRSEC Program of the National Science Foundation under award No. DMR00-80034.

References

Ahmad, A., Evans, H. M., Ewert, K., George, C. X., Samuel, C. E., and Safinya, C. R. (2005). New multivalent lipids reveal bell curve for transfection versus membrane charge density: Non-viral lipid-DNA complexes for gene delivery. *J. Gene Med.* **7,** in press.
Allen, T. M. (1994). Long-circulating (sterically stabilized) liposomes for targeted drug-delivery. *Trends Pharmacol. Sci.* **15,** 215–220.
Bloomfield, V. A. (1991). Condensation of DNA by multivalent cations: Considerations on mechanism. *Biopolymers* **31,** 1471–1481.

Bradley, A. J., Devine, D. V., Ansell, S. M., Janzen, J., and Brooks, D. E. (1998). Inhibition of liposome-induced complement activation by incorporated poly(ethylene glycol) lipids. *Arch. Biochem. Biophys.* **357,** 185–194.

Bruinsma, R. (1998). Electrostatics of DNA cationic lipid complexes: Isoelectric instability. *Eur. Phys. J. B* **4,** 75–88.

Cartier, R., and Reszka, R. (2002). Utilization of synthetic peptides containing nuclear localization signals for non-viral gene transfer systems. *Gene Ther.* **9,** 157–167.

Cavazzana-Calvo, M., Hacein-Bey, S., de Saint Basile, G., Gross, F., Yvon, E., Nusbaum, P., Selz, F., Hue, C., Certain, S., Casanova, J. L., Bousso, P., Le Deist, F., and Fischer, A. (2000). Gene therapy of human severe combined immunodeficiency (SCID)-X1 disease. *Science* **288,** 669–672.

Chesnoy, S., and Huang, L. (2000). Structure and function of lipid-DNA complexes for gene delivery. *Annu. Rev. Biophys. Biomol. Struct.* **29,** 27–47.

Clark, P. R., and Hersh, E. M. (1999). Cationic lipid-mediated gene transfer: Current concepts. *Curr. Opin. Mol. Ther.* **1,** 158–176.

De Smedt, S. C., Demeester, J., and Hennink, W. E. (2000). Cationic polymer-based gene delivery systems. *Pharm. Res.* **17,** 113–126.

Devanand, K., and Selser, J. C. (1991). Asymptotic-behavior and long-range interactions in aqueous-solutions of poly(ethylene oxide). *Macromolecules* **24,** 5943–5947.

Edelstein, M. L., Abedi, M. R., Wixon, J., and Edelstein, R. M. (2004). Gene therapy clinical trials worldwide 1989–2004—an overview. *J. Gene Med.* **6,** 597–602.

Escriou, V., Carriere, M., Scherman, D., and Wils, P. (2003). NLS bioconjugates for targeting therapeutic genes to the nucleus. *Adv. Drug Deliv. Rev.* **55,** 295–306.

Ewert, K., Ahmad, A., Evans, H. M., Schmidt, H.-W., and Safinya, C. R. (2002). Efficient synthesis and cell-transfection properties of a new multivalent cationic lipid for non-viral gene delivery. *J. Med. Chem.* **45,** 5023–5029.

Ewert, K., Slack, N. L., Ahmad, A., Evans, H. M., Lin, A. J., Samuel, C. E., and Safinya, C. R. (2004). Cationic lipid-DNA complexes for gene therapy: Understanding the relationship between complex structure and gene delivery pathways at the molecular level. *Curr. Med. Chem.* **11,** 133–149 and other recent reviews in this issue of *Curr. Med. Chem.* 133–223.

Farhood, H., Serbina, N., and Huang, L. (1995). The role of dioleoyl phosphatidylethanolamine in cationic liposome mediated gene transfer. *Biochim. Biophys. Acta* **1235,** 289–295.

Felgner, P. L., Gadek, T. R., Holm, M., Roman, R., Chan, H. W., Wenz, W., Northrop, J. P., Ringold, G. M., and Danielsen, M. (1987). Lipofection: A highly efficient, lipid-mediated DNA-transfection procedure. *Proc. Natl. Acad. Sci. USA* **84,** 7413–7417.

Felgner, P. L. (1990). Particulate systems and polymers for *in vitro* and *in vivo* delivery of polynucleotides. *Adv. Drug Deliv. Rev.* **5,** 163–187.

Felgner, P. L., and Rhodes, G. (1991). Gene therapeutics. *Nature* **349,** 351–352.

Ferber, D. (2001). Gene therapy: Safer and virus-free? *Science* **294,** 1638–1642.

Friedmann, T. (1997). Overcoming the obstacles to gene therapy. *Sci. Am.* **276,** 96–101.

Gompper, G., and Goos, J. (1995). Fluctuations and phase behavior of passages in a stack of fluid membranes. *J. Phys. II (France)* **5,** 621–634.

Gruner, S. M., Tate, M. W., Kirk, G. L., So, P. T. C., Turner, D. C., Keane, D. T., Tilcock, C. P. S., and Cullis, P. R. (1988). X-ray-diffraction study of the polymorphic behavior of N-methylated dioleoylphosphatidylethanolamine. *Biochemistry* **27,** 2853–2866.

Hacein-Bey-Abina, S., von Kalle, C., Schmidt, M., Le Deist, F., Wulffraat, N., McIntyre, E., Radford, I., Villeval, J. L., Fraser, C. C., Cavazzana-Calvo, M., and Fischer, A. (2003a). A serious adverse event after successful gene therapy for X-linked severe combined immunodeficiency. *N. Engl. J. Med.* **348,** 255–256.

Hacein-Bey-Abina, S., von Kalle, C., Schmidt, M., McCormack, M. P., Wulffraat, N., Leboulch, P., et al. (2003b). LMO2-associated clonal T cell proliferation in two patients after gene therapy for SCID-X1. Science **302,** 415–419.

Harries, D., May, S., Gelbart, W. M., and Ben-Shaul, A. (1998). Structure, stability, and thermodynamics of lamellar DNA-lipid complexes. Biophys. J. **75,** 159–173.

Harrington, J. J., van Bokkelen, G., Mays, R. W., Gustashaw, K., and Willard, H. F. (1997). Formation of de novo centromeres and construction of first-generation human artificial microchromosomes. Nat. Genet. **15,** 345–355.

Harvie, P., Wong, F. M. P., and Bally, M. B. (2000). Use of poly(ethylene glycol)-lipid conjugates to regulate the surface attributes and transfection activity of lipid-DNA particles. J. Pharm. Sci. **89,** 652–663.

Hebert, E. (2003). Improvement of exogenous DNA nuclear importation by nuclear localization signal-bearing vectors: A promising way for non-viral gene therapy? Biol. Cell **95,** 59–68.

Helfrich, W. (1973). Elastic properties of lipid bilayers: Theory and possible experiments. Z. Naturforsch. **28c,** 693–703.

Henry, C. M. (2001). Gene delivery—without viruses. Chem. Eng. News **79,** 35–41.

Huang, L., Hung, M.-C., and Wagner, E. (eds.) (1999). "Non-viral Vectors for Gene Therapy." Academic Press, San Diego.

Hui, S., Langner, M., Zhao, Y., Ross, P., Hurley, E., and Chan, K. (1996). The role of helper lipids in cationic liposome-mediated gene transfer. Biophys. J. **71,** 590–599.

Israelachvili, J. N. (1992). "Intermolecular and Surface Forces." Academic Press, London.

Janiak, M. J., Small, D. M., and Shipley, G. G. (1979). Temperature and compositional dependence of the structure of hydrated dimyristoyl lecithin. J. Biol. Chem. **254,** 6068–6078.

Jans, D. A., Chan, C. K., and Hübner, S. (1998). Signals mediating nuclear targeting and their regulation: Application in drug delivery. Med. Res. Rev. **18,** 189–223.

Kalderon, D., Roberts, B. L., Richardson, W. D., and Smith, A. E. (1984). A short amino acid sequence able to specify nuclear localization. Cell **39,** 499–509.

Koltover, I., Salditt, T., Rädler, J. O., and Safinya, C. R. (1998). An inverted hexagonal phase of cationic liposome-DNA complexes related to DNA release and delivery. Science **281,** 78–81.

Koltover, I., Salditt, T., and Safinya, C. R. (1999). Phase diagram, stability and overcharging of lamellar cationic lipid—DNA self-assembled complexes. Biophys. J. **77,** 915–924.

Koltover, I., Wagner, K., and Safinya, C. R. (2000). DNA condensation in two dimensions. Proc. Natl. Acad. Sci. USA **97,** 14046–14052.

Kopatz, I., Remy, J.-S., and Behr, J.-P. (2004). A model for non-viral gene delivery: Through syndecan adhesion molecules and powered by actin. J. Gene Med. **6,** 769–776.

Kumar, V. V., Singh, R. S., and Chaudhuri, A. (2003). Cationic transfection lipids in gene therapy: Successes, setbacks, challenges and promises. Curr. Med. Chem. **10,** 1297–1306 and the other recent reviews in this issue of Curr. Med. Chem. 1185–1315.

Lasic, D. D., and Martin, F. J. (eds.) (1995). "Stealth Liposomes." CRC Press, Boca Raton.

Lasic, D. D., and Papahadjopoulos, D. (1995). Liposomes revisited. Science **267,** 1275–1276.

Lasic, D. D., Strey, H., Stuart, M. C. A., Podgornik, R., and Frederik, P. M. (1997). The structure of DNA-liposome complexes. J. Am. Chem. Soc. **119,** 832–833.

Le Bret, M., and Zimm, B. H. (1984). Distribution of counterions around a cylindrical polyelectrolyte and Manning's condensation theory. Biopolymers **23,** 287–312.

Lin, A. J., Slack, N. L., Ahmad, A., Koltover, I., George, C. X., Samuel, C. E., and Safinya, C. R. (2000). Structure–function studies of lipid-DNA non-viral gene delivery systems. J. Drug Targeting **8,** 13–27.

Lin, A. J., Slack, N. L., Ahmad, A., George, C. X., Samuel, C. E., and Safinya, C. R. (2003). Three-dimensional imaging of lipid gene-carriers: Membrane charge density controls universal transfection behavior in lamellar cationic liposome-DNA complexes. *Biophys. J.* **84,** 3307–3316.

Mahato, R. I., and Kim, S. W. (eds.) (2002). "Pharmaceutical Perspectives of Nucleic Acid-Based Therapeutics." Taylor & Francis, London.

Manning, G. S. (1978). Limiting laws and counterion condensation in polyelectrolyte solutions. I. Colligative properties. *J. Chem. Phys.* **51,** 924–933.

Martin-Herranz, A., Ahmad, A., Evans, H. M., Ewert, K., Schulze, U., and Safinya, C. R. (2004). Surface functionalized cationic lipid-DNA complexes for gene delivery: PEG-ylated lamellar complexes exhibit distinct DNA–DNA interaction regimes. *Biophys. J.* **86,** 1160.

May, S., and Ben-Shaul, A. (2004). Modeling of cationic lipid-DNA complexes. *Curr. Med. Chem.* **11,** 151–167.

Miller, A. D. (1998). Cationic liposomes for gene therapy. *Angew. Chem., Int. Ed.* **37,** 1768–1785.

Miller, A. D. (2003). The Problem with cationic liposome/micelle-based non-viral vector systems for gene therapy. *Curr. Med. Chem.* **10,** 1195–1211.

Mislick, K. A., and Baldeschwieler, J. D. (1996). Evidence for the role of proteoglycans in cation-mediated gene transfer. *Proc. Natl. Acad. Sci. USA* **93,** 12349–12354.

Mortimer, I., Tam, P., Maclachlan, I., Graham, R. W., Saravolac, E. G., and Joshi, P. B. (1999). Cationic lipid-mediated transfection of cells in culture requires mitotic activity. *Gene Ther.* **6,** 403–411.

Mounkes, L. C., Zhong, W., Ciprespalacin, G., Heath, T. D., and Debs, R. J. (1998). Proteoglycans mediate cationic liposome-DNA complex-based gene delivery *in vitro* and *in vivo. J. Biol. Chem.* **273,** 26164–26170.

Nabel, G., Nabel, E., Yang, Z.-Y., Fox, B. A., Plautz, G. E., Gao, X., Huang, L., Shu, S., Gordon, D., and Chang, A. E. (1993). Direct gene transfer with DNA-liposome complexes in melanoma: Expression, biologic activity, and lack of toxicity in humans. *Proc. Natl. Acad. Sci. USA* **90,** 11307–11311.

Niidome, T., and Huang, L. (2002). Gene therapy progress and prospects: Non-viral vectors. *Gene Ther.* **9,** 1647–1652.

Pitard, B., Oudrhiri, N., Lambert, O., Vivien, E., Masson, C., Wetzer, B., Hauchecorne, M., Scherman, D., Rigaud, J.-L., Vigneron, J.-P., Lehn, J.-M., and Lehn, P. (2001). Sterically stabilized BGTC-based lipoplexes: Structural features and gene transfection into the mouse airways *in vivo. J. Gene Med.* **3,** 478–487.

Plank, C., Mechtler, K., Szoka, F. C., and Wagner, E. (1996). Activation of the complement system by synthetic DNA complexes, a potential barrier for intravenous gene delivery. *Human Gene Ther.* **7,** 1437–1446.

Rädler, J. O., Koltover, I., Salditt, T., and Safinya, C. R. (1997). Structure of DNA-cationic liposome complexes: DNA intercalation in multilamellar membranes in distinct interhelical packing regimes. *Science* **275,** 810–814.

Raper, S. E., Chirmule, N., Lee, F. S., Wivel, N. A., Bagg, A., Gao, G. P., Wilson, J. M., and Batshaw, M. L. (2003). Fatal systemic inflammatory response syndrome in an ornithine trans-carbamylase-deficient patient following adenoviral gene transfer. *Mol. Genet. Metab.* **80,** 148–158.

Remy, J.-S., Sirlin, C., Vierling, P., and Behr, J.-P. (1994). Gene transfer with a series of lipophilic DNA-binding molecules. *Bioconjugate Chem.* **5,** 647–654.

Rinehart, J., Hersh, E., Issell, B., Triozzi, P., Buhles, W., and Neidhart, J. (1997). Phase 1 trial of recombinant human interleukin-1-beta (rhIL-1-beta), carboplatin, and etoposide in patients with solid cancers: Southwest Oncology Group Study 8940. *Cancer Invest.* **15,** 403–410.

Safinya, C. R. (2001). Structures of lipid-DNA complexes: Supramolecular assembly and gene delivery. *Curr. Opin. Struct. Biol.* **11,** 440–448.

Safinya, C. R., and Koltover, I. (1999). Self assembled structures of lipid-DNA non-viral gene delivery systems from synchrotron X-ray diffraction. In "Non-Viral Vectors for Gene Therapy" (L. Huang, M.-C. Hung, and E. Wagner, eds.), pp. 92–117. Academic Press, San Diego.

Safinya, C. R., Sirota, E. B., Roux, D., and Smith, G. S. (1989). Universality in interacting membranes: The effect of cosurfactants on the interfacial rigidity. *Phys. Rev. Lett.* **62**, 1134–1137.

Safinya, C. R., Slack, N. L., Lin, A. J., and Koltover, I. (2002). Cationic lipid-DNA complexes for gene delivery: Structure-function correlations. In "Pharmaceutical Perspectives of Nucleic Acid-Based Therapeutics" (R. I. Mahato and S. W. Kim, eds.), pp. 190–209. Taylor & Francis, London.

Salditt, T., Koltover, I., Rädler, J. O., and Safinya, C. R. (1997). Two-dimensional smectic ordering of linear DNA chains in self-assembled DNA-cationic liposome mixtures. *Phys. Rev. Lett.* **79**, 2582–2585.

Salditt, T., Koltover, I., Rädler, J. O., and Safinya, C. R. (1998). Self-assembled DNA-cationic-lipid complexes: Two-dimensional smectic ordering, correlations, and interactions. *Phys. Rev. E* **58**, 889–904.

Schulze, U., Schmidt, H.-W., and Safinya, C. R. (1999). Synthesis of novel cationic poly(ethylene glycol) containing lipids. *Bioconjugate Chem.* **10**, 548–552.

Seddon, J. M. (1990). Structure of the inverted hexagonal (HII) phase, and non-lamellar phase transitions of lipids. *Biochim. Biophys. Acta* **1031**, 1–69.

Stopeck, A. T., Hersh, E. M., Brailey, J. L., Clark, P. R., Norman, J., and Parker, S. E. (1998). Transfection of primary tumor cells and tumor cell lines with plasmid DNA/lipid complexes. *Cancer Gene Ther.* **5**, 119–126.

Subramanian, G., Hjelm, R. P., Deming, T. J., Smith, G. S., Li, Y., and Safinya, C. R. (2000). Structure of complexes of cationic lipids and poly(glutamic acid) polypeptides: A pinched lamellar phase. *J. Am. Chem. Soc.* **122**, 26–34.

Szleifer, I., Ben-Shaul, A., and Gelbart, W. M. (1990). Chain packing statistics and thermodynamics of amphiphile monolayers. *J. Phys. Chem.* **94**, 5081–5089.

Tranchant, I., Thompson, B., Nicolazzi, C., Mignet, N., and Scherman, D. (2004). Physicochemical optimization of plasmid delivery by cationic lipids. *J. Gene Med.* **6**, S24–S35.

Tristram-Nagle, S., Petrache, H. I., and Nagle, J. F. (1998). Structure and interactions of fully hydrated dioleoylphosphatidylcholine bilayers. *Biophys. J.* **75**, 917–925.

Vasilevskaya, V. V., Khokhlov, A. R., Matsuzawa, Y., and Yoshikawa, K. (1995). Collapse of single DNA molecule in poly(ethylene glycol) solutions. *J. Chem. Phys.* **102**, 6595–6602.

Wagner, E. (2004). Strategies to improve DNA polyplexes for *in vivo* gene transfer: Will "Artificial Viruses" be the answer? *Pharm. Res.* **21**, 8–14.

Warriner, H. E., Idziak, S. H. J., Slack, N. L., Davidson, P., and Safinya, C. R. (1996). Lamellar biogels: Fluid-membrane-based hydrogels containing polymer lipids. *Science* **271**, 969–973.

Wheeler, J. J., Palmer, L., Ossanlou, M., MacLachlan, I., Graham, R. W., Zhang, Y. P., Hope, M. J., Scherrer, P., and Cullis, P. R. (1999). Stabilized plasmid-lipid particles: Construction and characterization. *Gene Ther.* **6**, 271–281.

Willard, H. F. (2000). Genomics and gene therapy: Artificial chromosomes coming to life. *Science* **290**, 1308–1309.

Wolff, J. A. (ed.) (1994). "Gene Therapeutics: Methods and Applications of Direct Gene Transfer." Birkhäuser, Boston.

Wong, G. C. L., Tang, J. X., Lin, A. J., Li, Y., Janmey, P. A., and Safinya, C. R. (2000). Hierarchical self-assembly of F-Actin and cationic lipid complexes: Stacked three-layer tubule networks. *Science* **288**, 2035–2039.

Woodle, M. C. (1995). Sterically stabilized liposome therapeutics. *Adv. Drug Deliv. Rev.* **16,** 249–265.

Wrobel, I., and Collins, D. (1995). Fusion of cationic liposomes with mammalian cells occurs after endocytosis. *Biochim. Biophys. Acta* **1235,** 296–304.

Xu, Y. H., and Szoka, F. C. (1996). Mechanism of DNA release from cationic liposome/DNA complexes used in cell transfection. *Biochemistry* **35,** 5616–5623.

Zabner, J., Fasbender, A. J., Moninger, T., Poellinger, K. A., and Welsh, M. J. (1995). Cellular and molecular barriers to gene transfer by a cationic lipid. *J. Biol. Chem.* **270,** 18997–19007.

6

"Diffusible-PEG-Lipid Stabilized Plasmid Lipid Particles"

Ian MacLachlan* and Pieter Cullis*,[†]

*Protiva Biotherapeutics Incorporated, Burnaby, BC, Canada V5G 4Y1
[†]Inex Pharmaceuticals Inc., Burnaby, BC, Canada V5J 5J8

ABSTRACT

Many viral and non-viral gene transfer systems suffer from common pharmacological issues that limit their utility in a systemic context. By application of the liposomal drug delivery paradigm, many of the limitations of the first generation non-viral delivery systems can be overcome. Encapsulation in small, long-circulating particles called stabilized plasmid lipid particles (SPLP) results in enhanced accumulation at disease sites and selective protein expression. This work compares the detergent dialysis method of SPLP manufacture with an alternative method, spontaneous vesicle formation by ethanol dilution. The

Advances in Genetics, Vol. 53
Copyright 2005, Elsevier Inc. All rights reserved.

0065-2660/05 $35.00
DOI: 10.1016/S0065-2660(05)53006-2

pharmacology of SPLP, as determined by monitoring lipid label and quantitative real time PCR, is also presented.   © 2005, Elsevier Inc.

I. INTRODUCTION

Current efforts in gene transfer research focus on the development of genetic drugs capable of treating acquired diseases such as cancer, inflammation, viral infection or cardiovascular disease. The disseminated nature of these diseases requires the development of vector systems capable of accessing distal sites following systemic or intravenous administration. Unfortunately, most vectors have limited utility for systemic applications. Viral vectors, for example, are rapidly cleared from the circulation, limiting transfection to "first-pass" organs such as the lungs, liver and spleen. In addition, many viruses induce immune responses that compromise potency upon subsequent administration. In the case of most non-viral vectors such as plasmid DNA-cationic lipid complexes (lipoplexes), the large size and positively charged nature of these systems also results in rapid clearance upon systemic administration with the highest expression levels observed in first-pass organs, particularly the lungs (Huang and Li, 1997; Hofland et al., 1997; Templeton et al., 1997; Thierry et al., 1995). In addition, lipoplexes often give rise to significant toxicities both in vitro and in vivo (Harrison et al., 1995; Li and Huang, 1997; Tousignant et al., 2000, 2003). In spite of these limitations, non-viral gene transfer systems offer specific clinical and commercial advantages as therapeutics. Because non-viral systems use synthetic or highly purified components, they are chemically defined and free of adventitious agents. Non-viral systems can be manufactured under controlled conditions, relatively unconstrained by the biological considerations that define the scale-up of viral production in mammalian cell culture. These advantages have encouraged a number of investigators to focus on the development of non-viral gene transfer systems that have utility in a systemic context (Dzau et al., 1996; Li and Huang, 1997; Templeton et al., 1997; Wheeler et al., 1999; Zhu et al., 1993). Here we will describe one system that specifically attempts to address the inability of current vector systems to overcome the first barrier to systemic gene delivery, delivery to the disease site and the target cell.

II. PROPERTIES OF A PLASMID DELIVERY SYSTEM FOR THE TREATMENT OF SYSTEMIC DISEASE

A. Definition of an appropriate vector

We propose the following definition of an ideal carrier for systemic gene transfer: The ideal vector will (i) be safe and well tolerated upon systemic administration;

(ii) have the appropriate pharmacokinetic attributes to ensure delivery to disseminated disease sites; (iii) deliver intact DNA to target tissue and mediate transfection of that tissue; (iv) be non-immunogenic; and (v) be stable upon manufacture to facilitate production at commercial scale with uniform, reproducible performance specifications.

Gene-based drugs must maximize the benefit to patient health while minimizing the risks associated with treatment. Accordingly, gene transfer systems must be safe and well tolerated. Attempts to bypass the inherent pharmacology of a given vector by invoking elaborate or invasive treatment methodologies are likely to result in an increased, potentially unacceptable, risk to the patient. Methods such as 'hydrodynamic injection' or direct portal vein infusion may continue to generate exciting preclinical results, but translation of these methods to a clinical setting will be limited. Gene-based drugs will be adopted more readily if they can be delivered in a manner analogous to conventional medicines, for example by intravenous injection or in oral form.

The toxicity associated with systemic administration of poorly tolerated compounds is exacerbated by accumulation in non-target tissue and can be reduced by optimizing delivery to the target site. In the case of gene-based drugs, 'delivery' is determined by physical and biochemical properties including stability, size, charge, hydrophobicity, interaction with serum proteins and non-target cell surfaces, as well as the mechanism of action of the nucleic acid payload. In the context of a disease site, effective delivery requires that a vector overcome obstacles associated with heterogeneous cell populations that are often proliferating rapidly, at different stages of the cell cycle and not conforming to the patterns of organization established during the development of normal tissue. As demonstrated in this work, these challenges, and other potential barriers to transfection, can represent opportunities for conferring a degree of selectivity greater than that associated with the use of conventional therapeutics.

B. Overcoming the barriers to transfection

The barriers to transfection include the pharmacological barriers inhibiting delivery to the target cell, and the intracellular barriers that inhibit nuclear delivery and expression of the plasmid DNA construct. An effective delivery system must be able to confer stability to the nucleic acid payload in the blood despite the presence of serum nucleases and membrane lipases. Systemic delivery requires the use of a 'stealthy' delivery system, since indiscriminate interaction with blood components, lipoproteins or serum opsonins, can cause aggregation before the carrier reaches the disease site. This is especially important in the case of systemic delivery systems containing large polyanionic molecules such as plasmid DNA, which have a greater potential for inducing toxicity through interaction with complement and coagulation pathways (Chonn et al., 1991).

Other barriers to gene delivery may include the microcapillary beds of the "first pass" organs, the lung and the liver, and the phagocytic cells of the reticuloendothelial system. Accessing target cell populations requires extravasation from the blood compartment to the disease site. Carriers of appropriate size can pass through the fenestrated epithelium of tumor neovasculature and accumulate at the tumor site via the "enhanced permeation and retention" (EPR) effect (Mayer et al., 1990), also referred to as "passive" targeting or "disease site" targeting. In order to take advantage of the EPR effect, which can result in accumulation of up to 10% of the injected dose per gram of tumor tissue, the gene carriers must be small (diameter on the order of 100 nm) and long-circulating (circulation lifetimes of 5 h or more following intravenous injection in mice). Clearly, nucleic acids require pharmaceutical enablement in the form of appropriate carriers that confer: protection from degradation, an extended circulation lifetime, appropriate biodistribution and delivery facilitation of the nucleic acid payload to the disease site.

While delivery of intact plasmid DNA to a target cell is a *prerequisite*, it in no way *guarantees* transfection. Once at the cell surface, vectors are confronted with a number of physical and biochemical barriers, each of which must be overcome in order to effect transfection and transgene expression. The first physical barrier to transfection is the plasma membrane, protected by the carbohydrate coating, or glycocalyx, formed by the post-translational glycosylation of transmembrane proteins. Although early models of lipid-mediated transfection invoked a putative fusion event between the plasma membrane and the membrane of the lipid vesicle, it is now generally agreed that the majority of intracellular delivery occurs through endocytosis.

Endocytosis is a complex process by which cells take up extracellular material. This occurs through a number of discrete pathways, reviewed elsewhere in this volume. While there is some evidence that non-viral vectors may be taken up by caveolae, syndecan-mediated endocytosis or other clathrin-independent pathways, the classical endocytic pathway involves the activity of cell surface clathrin-coated pits, invaginations in the plasma membrane that are subsequently pinched off into the cytoplasm (Goldstein et al., 1985). When this occurs, internalized material remains trapped on the exoplasmic side of the internalized vesicle, without direct access to the cytoplasm or the nucleus. Endocytic vesicles undergo a series of biochemical changes that represent escape opportunities for a non-viral vector. The first such change occurs within 5 min of uptake as internalized vesicles form the early endosome containing the "Compartment of Uncoupling of Receptor and Ligand" (CURL) (Geuze et al., 1983). Early endosomes are transiently fusogenic (Dunn and Maxfield, 1992) with a pH close to that of the exoplasm, while late endosomes have a significantly lower luminal pH (Murphy et al., 1993). As endosomes mature to form lysosomes they experience a further decrease in internal pH and an increase in fusogenicity.

Although the process of clatharin-dependent endocytosis has been well characterized, the processing and release of internalized non-viral vectors or their DNA payload is not well understood. Even less clear is the relative import of clathrin-independent uptake through mechanisms that share some, but not all of the features of the classical pathway. Improvements in our understanding of these alternative pathways, and their role in non-viral gene transfer, will be important for the rational design of more effective intracellular delivery strategies for non-viral vectors.

Following uptake, plasmid DNA spends some indeterminate residency time in the cytoplasm prior to gaining entry to the nucleus. Unlike viral systems that have evolved specific mechanisms to traverse this barrier, untargeted non-viral vectors rely on diffusion to facilitate interaction with the nuclear envelope (Kopatz *et al.*, 2004). However the cytoplasm, rather than an empty space, is a highly organized compartment containing networks of cytoskeletal elements and membrane-bound organelles that have the potential to interact with and accumulate vector systems that arrive at the cytosol intact. When plasmid DNA is delivered by direct microinjection into the cytosol of mammalian cells it is rapidly degraded by divalent-cation-dependent cytosolic nucleases (Howell *et al.*, 2003; Lechardeur *et al.*, 1999). This has implications for vector design. Vector systems that either protect the DNA payload from degradation following endosome release or effectively minimize the cytoplasmic residency time are to be expected to yield improved transfection efficiencies.

The final physical barrier to transfection is delivery to the nucleus. The nucleus has evolved as a means of organizing, isolating and protecting the genome of eukaryotic cells from adventitious agents such as viruses or transposons. The nuclear uptake of DNA is limited by the presence of an intact nuclear envelope and as such non-viral transfection is considerably more efficient in highly mitotic cells (Mortimer *et al.*, 1999; Wilke *et al.*, 1996). Strategies to overcome this barrier to transfection take one of two forms: either targeting transfection reagents to cell populations with a high degree of mitotic activity, such as tumor tissue; or enhancing the low level of transfection that occurs in quiescent cells by using either nuclear targeting technologies or condensing agents that compact plasmid DNA to a size more amenable to uptake through the nucleopore complex (Blessing *et al.*, 1998; Sebestyen *et al.*, 1998).

C. Proposed mechanism of stabilized plasmid lipid particle mediated transfection

1. Delivery to the target cell

The demands imposed upon vectors used for systemic applications are conflicting. First, the carrier must be stable and long-circulating, circulating long enough to facilitate accumulation at disease sites via the EPR effect.

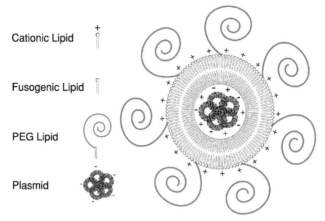

Figure 6.1. Structure of stabilized plasmid lipid particles.

Second, the carrier must interact with—and be taken up by—target cells following arrival at the target site in order to facilitate gene expression. The "stabilized plasmid-lipid particle" (SPLP) attempts to satisfy both of these requirements.

SPLPs consist of a single plasmid encapsulated in a lipid bilayer containing a *diffusible* polyethylene glycol (PEG)-lipid conjugate (Fig. 6.1). The PEG-lipid conjugates in the SPLP play an essential role during the formulation process, stabilizing the nascent particle and preventing aggregation of the particles in the vial. In the blood, the PEG-lipid shields the positive surface charge, preventing rapid clearance following intravenous injection. Following administration, at $37°C$ and in the presence of sufficient lipid sink, the PEG conjugate dissociates from the SPLP, revealing the positive charge and an increasingly fusogenic lipid bilayer, transforming the particle into a transfection-competent entity. The residency of the PEG conjugate in the SPLP bilayer is determined by the length of the lipid anchor. PEGs with shorter lipid anchors, such as ceramide-C_8 or dimyristoyl-glycerol, dissociate more quickly from the bilayer, quickly 'activating' the SPLP into which they are incorporated. As a result, particles incorporating PEG-lipids with shorter lipid anchors show higher transfection potency *in vitro* than those containing longer lipid anchors (e.g., ceramide-C_{20} or distearoyl-glycerol) (Mok *et al.*, 1999). When injected systemically, PEG conjugates with a larger, more securely fastened anchor and will confer greater stability and extended circulation lifetimes, leading to greater levels of accumulation at disseminated disease sites (Monck *et al.*, 2000; Tam *et al.*, 2000).

2. The role of cationic lipids in promoting intracellular delivery

While the factors facilitating intracellular delivery of non-viral vectors are poorly understood, it is believed that both polycation and cationic lipid-containing systems function, at least in part, by coating plasmid DNA with a positive charge that enables binding of the DNA complex to anionic cell surface molecules, such as cell surface proteoglycans that appear to facilitate transfection both *in vitro* (Mislick and Baldeschwieler, 1996) and *in vivo* (Mounkes et al., 1998). Inhibition of the interaction between the positively charged lipoplex and negatively charged cell surface molecules by pretreatment with polyanionic compounds greatly inhibits lipoplex-mediated transfection while having no effect on electroporation or adenoviral transfection. Intravenous administration of heparinase I, an enzyme specific for the cleavage of heparan sulfate proteoglycans, also inhibits cationic lipoplex-mediated transfection. Given that the basis of the interaction between proteoglycans and cationic vectors appears to be electrostatic, differences in charge and charge density between vector systems could yield differences in transfection efficiency. This has certain implications for the design of vector systems for systemic gene therapy. SPLP are transiently charge shielded due to the incorporation of diffusible PEG-lipids. As the PEG-lipid leaves the particle in the blood compartment, the positive charge conferred by the cationic lipid component is revealed. Although systems with a lower surface charge might be expected to benefit from increased circulation time, incorporation of additional cationic lipid in the SPLP lipid bilayer yields an appreciable gain in potency *in vitro* (Zhang et al., 1999).

It is also believed that cationic lipids play a direct role to facilitate intracellular delivery following internalization in the endosome. The proposed mechanism involves the ability of cationic lipids to promote formation of the H_{II} phase in combination with anionic lipids (Hafez et al., 2001), thereby destabilizing the bilayer structure of the endosomal membrane, encouraging fusion with the SPLP bilayer and facilitating the cytoplasmic translocation of the associated plasmid DNA. Clearly, the cationic lipid content of systemic carrier systems must be optimized with a view towards achieving both an extended circulation lifetime and effective intracellular delivery. In particular, a compromise must be made between incorporating high amounts of cationic lipid, which facilitates transfection and the fact that high amounts of cationic lipid result in shorter circulation lifetimes, reducing the amount of the material that arrives at the disease site.

3. The role of helper lipids in promoting intracellular delivery

The majority of cationic lipids require the addition of a fusogenic 'helper' lipid for efficient *in vitro* gene transfer (Farhood et al., 1995; Felgner et al., 1994; Gao and Huang, 1995; Hui et al., 1996). Inclusion of lipids, such as unsaturated phosphatidylethanolamines like dioleoylphosphatidylethanolamine (DOPE), promote destabilization of the lipid bilayer and fusion (Farhood et al., 1995; Hui et al., 1981; Litzinger and Huang, 1992). The fusogenicity of DOPE-containing bilayers is thought to be due to their polymorphic nature. Upon formulation, most lipids adopt the bilayer-forming Lamellar Phase (L_α), while DOPE has a tendency to form the inverse hexagonal (H_{II}) phase (Cullis PR and B., 1978; Koltover et al., 1998). Several researchers have noted that increasing the degree of unsaturation of the lipid hydrophobic domain increases the affinity for the H_{II} phase (Cullis PR and B., 1979; Dekker et al., 1983; Epand et al., 1991; Sankaram et al., 1989; Szule et al., 2002). As a result, the fusogenicity of an SPLP bilayer can be increased by increasing the degree of unsaturation in the hydrophobic domain of either the helper lipid or cationic lipid components (Heyes et al., 2004). Furthermore, certain cationic lipids can function in the absence of fusogenic helper lipids, either alone (Felgner et al., 1994; Gao and Huang, 1995) or in the presence of the non-fusogenic lipid cholesterol (Liu et al., 1995).

The specific role of fusogenic helper lipids in the transfection process, and whether this role is conserved between lipoplex and systems such as SPLP which fully encapsulate plasmid DNA, is not clear. Membrane fusion events could theoretically occur at a number of different stages in the gene delivery process, either at the plasma membrane, endosome or nuclear envelope. In order for fusion with the plasma membrane to occur, positively charged lipid particles must first bypass the negatively charged glycocalyx. Fusion of lipoplex systems with the plasma membrane would be expected to be a particularly inefficient method of introducing DNA into the cytosol since lipoplex fusion events may resolve with plasmid DNA, formerly attached to the cationic liposome surface, deposited on the outside surface of the plasma membrane. Encapsulated systems differ from lipoplex in this respect. Fusion with the plasma membrane could result in an encapsulated carrier delivering its contents into the cytosol. However, the bulk of both lipoplex- and SPLP-mediated transfection is thought to be by fusion with the endosomal membrane of particles that are taken up intact by endocytosis (Wrobel and Collins, 1995). There is considerable biochemical evidence to support an endosomal route for internalized plasmid DNA. One example is the transient inhibition of endocytosis and concomitant transfection upon treatment of cells with cytochalasin-B, an inhibitor of actin polymerization required in the endocytic process (Hui et al., 1996). Another example utilizes fluorescently labeled lipids to track the fate of lipoplex or SPLP upon delivery to the cell. Fusion of labeled liposomes with the plasma membrane would result in the transfer of lipid label to the membrane. Cells exposed to lipoplex or SPLP containing rhodamine-phosphatidyl-ethanolamine accumulate fluorescent label in endocytic granules, well before plasma membranes become fluorescent (Hui et al., 1996; Palmer et al., 2003). In the absence of a fusion-induced translocation event, fusion of lipoplex systems with endosomal compartments results in a gradual destabilization and disruption of the endosomal membrane. Encapsulated systems have an advantage over lipoplex in that a single fusion event within an endosomal compartment would be expected to result in efficient delivery of the DNA payload to the cytosol.

The role of fusogenic lipids *in vivo* remains unclear. A number of investigators have reported that replacement of fusogenic DOPE with the less fusogenic lipid cholesterol yields higher levels of gene expression upon systemic administration of either lipoplex or encapsulated systems (Sakurai *et al.*, 2001; Templeton *et al.*, 1997). However, it is important to distinguish the effect of helper lipids on biodistribution from the effect on intracellular delivery. The enhanced gene expression observed upon incorporation of cholesterol in lipoplex or SPLP formulations may be a result of either an increase in transfection efficiency or improved pharmacokinetics and delivery to the target cell. Fusogenic formulations are more likely to interact with the vascular endothelium, blood cells, lipoproteins and the fixed and free macrophages of the mononuclear phagocyte system while in the blood compartment, leading to rapid clearance and decreasing the proportion of carriers that reach target tissue. Incorporation of cholesterol may simply render vectors less promiscuous and thereby improve delivery to the target cell. The implication is that there is further rationale for transiently shielding the fusogenic potential of systemic carriers through the use of diffusible PEG-lipids.

A variety of approaches can be considered for enhancing the endosomal release of internalized liposomes. In addition to the use of fusogenic lipids that are thought to facilitate endosome release, another strategy involves the incorporation of specific lipids that render the liposome pH-sensitive such that it becomes more fusogenic in low pH compartments such as the late endosome and lysosome (Wang and Huang, 1987, 1989; Lee and Huang, 1996). One example of this approach utilizes titratable cationic lipids that become positively charged at the reduced pH values that may be encountered in endosomes. Cationic lipids such as 1,2-dioleoyl-3-(N,N-dimethylamino)propane (AL1) that exhibit pK of approximately 6.6 (Bailey and Cullis, 1994) confer no significant positive charge to carriers at neutral pH yet are fully positively charged at the pH values commonly encountered in endosomes.

4. Nuclear delivery

Attempts to improve the nuclear uptake of plasmid DNA must take into consideration the physical constraints of the nucleopore complex that mediates the uptake of plasmid DNA into the intact nucleus. When fully condensed by monovalent detergent counterions, a 5.5 kb supercoiled plasmid DNA molecule becomes a sphere of about 25 nm in diameter (Blessing *et al.*, 1998) while the passive diffusion channel of the nuclear pore complex has an internal diameter of 9 nm (Ohno *et al.*, 1998). The diameter of the activated nuclear pore complex through which active transport occurs, and therefore the size limit for signal mediated nuclear import, is 25 nm. Although there does appear to be considerable potential for improving the nuclear uptake of supercoiled plasmids through attachment of nuclear localization peptides or other nuclear import signals, it remains to be seen if this can be accomplished in a manner that is compatible with large-scale formulation and systemic gene delivery (Sebestyen *et al.*, 1998).

III. METHODS OF ENCAPSULATING PLASMID DNA

In order to capitalize on the pharmacology and disease site targeting demonstrated by liposomal drug carriers it is necessary to completely entrap plasmid DNA within the contents of a liposome. Unlike small molecule drugs, plasmid DNA cannot easily be "loaded" into preformed liposomes using pH gradients or other similar strategies. Lipid encapsulation of high molecular weight DNA was first demonstrated in the late 1970s, prior to the development of cationic lipid-containing lipoplex (Hoffman *et al.*, 1978; Mannino *et al.*, 1979; Mukherjee *et al.*, 1978). Plasmid DNA has subsequently been encapsulated by reverse-phase evaporation (Cudd and Nicolau, 1985; Fraley *et al.*, 1980; Nakanishi *et al.*, 1985; Soriano *et al.*, 1983), ether injection (Fraley *et al.*, 1979; Nicolau and Rottem, 1982), lipid hydration-dehydration techniques (Alino *et al.*, 1993; Baru *et al.*, 1995; Lurquin, 1979) sonication (Jay and Gilbert, 1987; Ibanez *et al.*, 1997; Puyal *et al.*, 1995), spontaneous internalization into pre-formed liposomes (Templeton *et al.*, 1997) and others methods (Monnard *et al.*, 1997; Szelei and Duda, 1989) (summarized in Table 6.1). Early attempts to encapsulate plasmid DNA yielded mostly large multilamellar vesicles with poor transfection efficiency (Baru *et al.*, 1995; Nicolau *et al.*, 1983; Scaefer-Ridder *et al.*, 1982), while more recently, improvements in formulation technology have resulted in the production of cationic lipid-containing particles with a much greater transfection potential. SPLP initially utilized detergent dialysis, a process in which unilamellar vesicles are formed upon removal of detergent from a DNA:lipid solution. While early efforts to encapsulate plasmid DNA using detergent dialysis yielded low encapsulation efficiencies (Fraley *et al.*, 1979; Nakanishi *et al.*, 1985), these results were significantly improved upon through the use of PEG lipids to stabilize the vesicles during the formulation process (Wheeler *et al.*, 1999). In this way plasmid-containing cationic liposomes are stabilized in a manner analogous to PEGylated liposomal drug formulations that exhibit extended circulation lifetimes (Allen and Chong, 1987; Klibanov *et al.*, 1990; Needham *et al.*, 1992; Papahadjopoulos *et al.*, 1991; Wu *et al.*, 1993). PEG conjugates sterically stabilize liposomes by forming a protective hydrophilic layer that shields the hydrophobic lipid layer, preventing the association of serum proteins and resulting uptake by the reticuloendothelial system (Gabizon and Papahadjopoulos, 1988; Senior et al., 1991). Although this approach has been investigated with a view towards improving the stability and pharmacokinetics of lipoplex (Hong et al., 1997), lipoplex incorporating PEG-lipids systems suffer from the heterogeneity common to most complexes of plasmid DNA and cationic lipid.

The detergent dialysis method of plasmid encapsulation involves the simultaneous solubilization of hydrophobic (cationic and helper lipid) and hydrophilic (PEG lipid and plasmid DNA) components in a single detergent-containing phase (Fenske et al., 2002;Wheeler et al., 1999). Particle formation occurs spontaneously upon removal of the detergent by dialysis. This technique can result in the formation of small (approximately 100 nm diameter) 'stabilized plasmid lipid particles' (SPLP) containing one plasmid per vesicle in combination with optimized plasmid trapping efficiencies approaching 70%. The SPLP protocol results in stable particles with low levels of cationic lipids, high levels of fusogenic lipids and high DNA-to-lipid ratios. SPLP can be concentrated to achieve plasmid DNA concentrations of >5 mg/ml. These attributes compare favorably with the previously reported plasmid encapsulation processes (Table 6.1). The SPLP method yields the highest plasmid DNA-to-lipid ratio of any method and SPLP are remarkably stable when compared to other encapsulated systems. Although the detergent dialysis process results in 30–50% unencapsulated DNA, free plasmid DNA can be removed by simple ion-exchange chromatography.

Although SPLP shows considerable potential as systemic gene transfer agents, the detergent dialysis method suffers from a number of limitations. Detergent dialysis is exquisitely sensitive to minor changes in the ionic strength of the formulation buffer. Changes as small as 10 mM result in dramatic decrease in encapsulation efficiency (Fenske et al., 2002). Even when SPLP are formed under ideal conditions, the detergent dialysis method results in the formation of large numbers of empty vesicles that are usually separated from SPLP by gradient ultracentrifugation (Fenske et al., 2002). The detergent dialysis method is also difficult to scale to the size required to support preclinical and clinical development of the technology. For these reasons, alternative methods of preparing stable plasmid lipid particles have been explored.

One such method uses ethanol-destabilized cationic liposomes (Maurer et al., 2001). Though this method does not require gradual detergent removal or ultracentrifugation steps, it does require the formation of cationic vesicles prior to the encapsulation of pDNA. Once cationic liposomes of the desired size have been prepared, they are destabilized by ethanol addition to 40% v/v. Destabilization of vesicles with ethanol requires very slow addition of ethanol to a rapidly

Table 6.1. Procedures for Encapsulating Plasmid in Lipid-Based Systems

Procedure	Lipid composition	Length of DNA	Trapping efficiency[a]	DNA-to-lipid ratio[a]	Diameter
Reverse-phase evaporation (Fraley et al., 1980)	PS or PS:Chol (50:50)	SV40 DNA	30 to 50%	<4.2 μg/μmol	400 nm
Reverse-phase evaporation (Soriano et al., 1983)	PC:PS:Chol (40:10:50)	11.9 kb plasmid	13 to 16%	0.23 μg/μmol	100 nm to 1 μm
Reverse-phase evaporation (Nakanishi et al., 1985)	PC:PS:Chol (50:10:40)	8.3 kb, 14.2 kbp plasmid	10%	0.97 μg/μmol	ND
Reverse-phase evaporation (Cudd and Nicolau, 1985)	EPC:PS:Chol (40:10:50)	3.9 kb plasmid	12%	0.38 μg/μmol	400 nm
Ether injection (Fraley et al., 1979)	EPC:EPG (91:9)	3.9 kb plasmid	2 to 6%	<1 μg/μmol	0.1 to 1.5 μm; Aug = 230 nm
Ether injection (Nicolau and Rottem, 1982)	PC:PS:Chol (40:10:50) PC:PG:Chol (40:10:50)	3.9 kb plasmid	15%	15 μg/μmol	ND
Detergent dialysis (Stavridis et al., 1986)	EPC:Chol:stearylamine (43.5:43.5:13)	sonicated genomic DNA (approximately 250,000 MW)	11%	0.26 μg/μmol	50 nm
Detergent dialysis, extrusion (Wang and Huang, 1987)	DOPC:Chol:oleic acid or DOPE:Chol:oleic acid (40:40:20)	4.6 kb plasmid	14 to 17%	2.25 μg/μmol	180 nm (DOPC) 290 nm (DOPE)
Lipid hydration (Lurquin, 1979)	EPC:Chol (65:35) or EPC	3.9 kb, 13 kb plasmid	ND	ND	0.5 to 7.5 μm
Dehydration-rehydration, extrusion (400 or 200 nm filters) (Alino et al., 1993)	Chol:EPC:PS (50:40:10)	ND	ND	0.83 μg/μmol (200 nm) 1.97 μg/μmol (400 nm)	142.5 nm (200 nm filter) 54.6 nm (400 nm filter, ultracentrifugation)

Dehydration-rehydration (Baru et al., 1995)	EPC	2.96 kb, 7.25 kb plasmid	35 to 40%	2.65 to 3.0 μg/μmol	1 to 2 μm
Sonication (in the presence of lysozyme) (Jay and Gilbert, 1987)	Asolectin (soybean phospholipids)	1.0 kb linear DNA	50%	0.08 μg/μmol	100 to 200 nm
Sonication (Puyal et al., 1995)	EPC:Chol:lysine-DPPE (55:30:15)	6.3 kb ssDNA 1.0 kb dsRNA	60 to 95% ssDNA 80 to 90% dsRNA	13 μg/μmol ssDNA; 14 μg/μmol dsRNA	100 to 150 nm
Spermidine-condensed DNA, sonication, extrusion (Ibanez et al., 1997)	EPC:Chol:PS (40:50:10) EPC:Chol:EPA (40:50:10) or EPC: Chol:CL (50:40:10)	4.4 kb, 7.2 kb plasmid	46 to 52%	2.53 to 2.87 μg/μmol	400 to 500 nm
Ca^{2+}-EDTA entrapment of DNA-protein complexes (Szelei and Duda, 1989)	PS:Chol (50:50)	42.1 kbp bacteriophage	52 to 59%	22 μg/μmol	ND
Freeze-thaw, extrusion (Monnard et al., 1997)	POPC:DDAB (99:1)	3.4 kb linear plasmid	17 to 50%	ND	80 to 120 nm
SPLP – Detergent Dialysis (Wheeler et al., 1999)	Various	4.4 to 15 kb plasmid	60 to 70%	62.5 μg/μmol	75 nm (QELS); 65 nm (freeze-fracture)
SPLP – Ethanol Dilution (Jeffs et al., 2005)	Various	4.4 to 15 kb plasmid	80 to 95%	70 μg/μmol	100–150 nm (QELS)

[a]Some values calculated based on presented data.
ND = Not determined.

mixing aqueous suspension of vesicles, to avoid localized areas of high ethanol concentration (>50% v/v) that promote fusion and conversion of liposomes into large lipid structures (Maurer et al., 2001). The addition of plasmid must also be accomplished slowly in a drop-wise manner to the destabilized vesicle. The uncontrolled nature of both the vesicle destabilization and nucleic acid addition steps poses challenges for reproducibly preparing SPLP at a scale suitable for clinical evaluation.

A more simple, robust and fully scalable method for the encapsulation of plasmid DNA in stable plasmid lipid particles has been developed. This method, termed 'stepwise ethanol dilution,' produces SPLP with the same desirable properties as those prepared by detergent dialysis (Jeffs et al., 2005). Lipid vesicles encapsulating plasmid DNA are formed instantaneously by mixing lipids dissolved in ethanol with an aqueous solution of DNA in a controlled, stepwise manner (Fig. 6.2). Combining DNA and lipid flow streams results in rapid dilution of ethanol below the concentration required to support lipid solubility. Using this method, vesicles are prepared with particle sizes less than 150 nm and DNA encapsulation efficiencies as high as 95%. Although analysis by transmission electron microscopy shows that SPLPs prepared by stepwise ethanol dilution are a more heterogenous population of unilamellar, bilamellar and oligolamellar vesicles than those prepared by detergent dialysis, extensive analysis reveals that this more diverse morphology has little effect on the stability or activity of SPLPs *in vitro* or *in vivo*.

The ethanol dilution method represents an effective solution to the issues confounding SPLP preparation by detergent dialysis (Wheeler et al., 1999). The efficiency of the ethanol dilution method obviates the requirement for an ultracentrifugation purification step (Fig. 6.3). Another benefit of the method is that it facilitates the rapid preparation of SPLP samples, allowing formulation development to proceed at a much faster pace. An SPLP formulation can be prepared in a few hours with ethanol dilution, whereas it takes days to prepare SPLP by detergent dialysis (Fenske et al., 2002). Furthermore, the improved method enables the accelerated optimization of formulation composition (e.g., lipid molar ratios) and process parameters (e.g., buffer concentration or pH). The ethanol dilution method has been used to formulate a 4.5 g batch of plasmid DNA under current Good Manufacturing Practices (cGMP) (Fig. 6.4). The properties of SPLP from this batch were identical to batches prepared at 1/10 and 1/100 of this scale using smaller process steps. Until now, an issue plaguing the development of non-viral vectors has been their less than predictable formulation and performance characteristics, particularly when manufactured at large scale. Preparation of SPLP batches at a scale suitable for clinical evaluation or commercialization is possible using this novel approach.

The ability of the ethanol dilution method to rapidly prepare liposomes of desirable size and encapsulate plasmid DNA with high efficiency is thought to

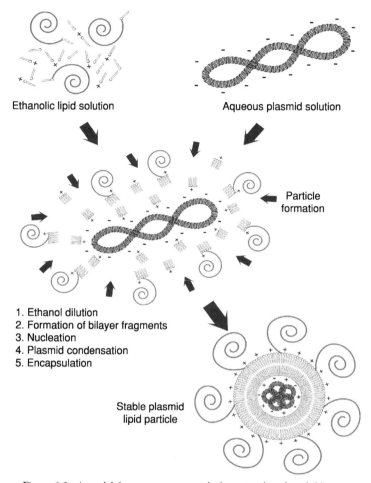

Ethanolic lipid solution

Aqueous plasmid solution

Particle
formation

1. Ethanol dilution
2. Formation of bilayer fragments
3. Nucleation
4. Plasmid condensation
5. Encapsulation

Stable plasmid
lipid particle

Figure 6.2. A model for spontaneous vesicle formation by ethanol dilution.

result from the precise control of the conditions under which the lipids enter the aqueous environment, self arrange into lipid bilayer fragments and then form liposomes. Several parameters have been shown to be critical for SPLP formation and plasmid encapsulation when using detergent dialysis (Wheeler *et al.*, 1999; Zhang *et al.*, 1999). Ionic strength, cationic lipid and PEG lipid content must be optimized to maximize plasmid entrapment and minimize aggregation or the formation of empty vesicles (Wheeler *et al.*, 1999). The first stage of dialysis is proposed to result in the formation of macromolecular intermediates, possibly lamellar lipid sheets or micelles. Plasmid DNA is recruited to these bilayer

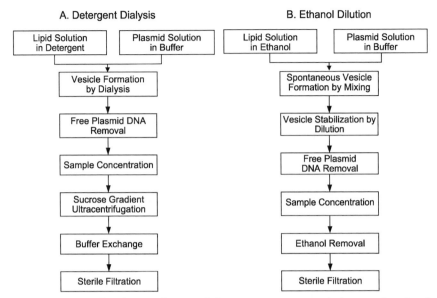

A. Detergent Dialysis   B. Ethanol Dilution

Figure 6.3. Process flow diagram: detergent dialysis vs. spontaneous vesicle formation by ethanol dilution.

fragments by electrostatic attraction. If the cationic lipid content is too low, the plasmid fails to associate with these intermediates, favoring the formation of empty vesicles. If the cationic lipid concentration is too high, the surface charge on the lipid intermediate attracts excess plasmid DNA leading to the formation of polydisperse aggregates. At optimal cationic lipid concentrations, plasmid DNA is proposed to associate with the lipid intermediates in such a way as to reduce the net positive charge on the lipid surface. Association of additional lipid leads to the formation of vesicles containing encapsulated plasmid (Wheeler *et al.*, 1999). Similar to detergent dialysis, SPLP formation by ethanol dilution is optimized by balancing ionic strength, cationic lipid and PEG lipid content. However the ethanol dilution method appears much more robust than detergent dialysis, with optimal results achieved through a wide range of formulation conditions.

In summary, while a variety of techniques are available for encapsulating plasmid DNA into lipid-based systems, only the SPLP approach—employing PEG-lipid conjugates during formulation—satisfies the demands of generating small (diameter ~100 nm), well-defined (one plasmid per particle) stable systems with high encapsulation efficiencies (>50%) and high plasmid-to-lipid ratios (>0.1 mg plasmid DNA/mg lipid) that exhibit the extended circulation lifetimes required to achieve preferential accumulation at disease sites such as

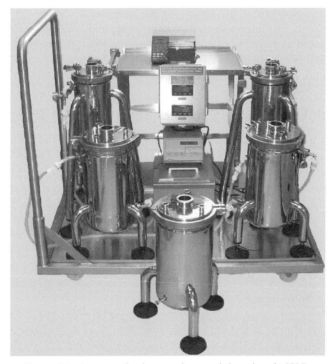

Figure 6.4. Apparatus for the manufacture of clinical grade SPLP.

solid tumors. Among the methods for producing SPLP, the stepwise ethanol dilution approach most adequately satisfies demands related to scalability and reproducibility.

IV. PHARMACOLOGY OF ENCAPSULATED PLASMID DNA

As described above, SPLP were designed to take advantage of the EPR effect whereby small, long-circulating particulate carriers preferentially accumulate at disease sites such as tumors. The importance of such an approach is profound. For example, liposomal vincristine formulations that have these characteristics facilitate the accumulation of 50- to 100-fold higher amounts of drug at a tumor site compared to injection of the same amount of free drug (Boman et al., 1994; Mayer et al., 1993) resulting in significantly improved efficacy (Webb et al., 1995). In the case of SPLP, accumulation of the plasmid carrier at a tumor site contributes to a remarkable and unexpected benefit, in that preferential gene expression is observed at the tumor site as compared to expression in normal tissues resulting in tumor-selective protein expression. The pharmacokinetics, tumor accumulation and transfection properties of SPLP have been extensively characterized in several murine models and will be described here in more detail (Fenske *et al.*, 2001, 2002; Monck *et al.*, 2000; Tam *et al.*, 2000).

A. Biodistribution following systemic administration of SPLP

Following intravenous injection into mice, the clearance of SPLP can be assessed by lipid and/or DNA markers (Monck *et al.*, 2000; Tam *et al.*, 2000). Previous experience shows that, due to the stability of the SPLP, the lipid and DNA components are cleared from the blood compartment at the same rate and the plasmid DNA remains intact while encapsulated within the SPLP lipid bilayer (Tam *et al.*, 2000). Because the SPLP remains intact in the blood compartment, the biodistribution of a non-exchangeable lipid marker (Stein *et al.*, 1980) incorporated into an SPLP is representative of the biodistribution of the entire particle, including the plasmid DNA component, at early time points. This finding is applied to analysis of SPLP clearance and biodistribution up to 24 h after administration.

An example of the clearance of an SPLP from the circulation of tumor-bearing mice is shown in Fig. 6.5. SPLP was formulated containing DSPC, cholesterol, DODMA and PEG-disterylglycerol (20:55:15:10 mol%, respectively), encapsulating plasmid DNA containing the luciferase reporter

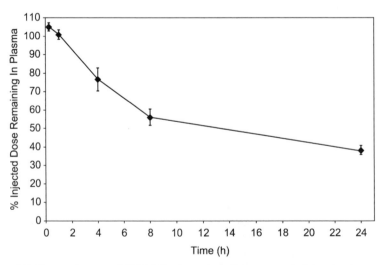

Figure 6.5. Plasma clearance of SPLP following a single intravenous administration in neuro-2a tumor-bearing male A/J mice.

gene. Trace amounts of radiolabelled lipid marker ([³H]-cholesteryl hexadecyl ether (CHE)) were included in the SPLP to quantify SPLP in the plasma and tissue samples.

SPLP was administered intravenously and blood was subjected to analysis for ³H-CHE lipid. Twenty-four hours after intravenous administration of SPLP, forty percent of the injected dose remains in the plasma with a half-life of 13 h. This result is typical of numerous such experiments in normal and tumor-bearing mice (Monck et al., 2000). SPLP, regardless of the DNA encapsulated, remain in the circulation for many hours with 15–40% of the injected dose in circulation at 24 h following injection. The serum half-life of unprotected plasmid DNA is known to be <5 min (Monck et al., 2000; Thierry et al., 1997).

The extended circulation of SPLP results in the accumulation of particles in tumors following intravenous administration. The accumulation of SPLP at a distal Neuro-2a tumor site is shown in Fig. 6.6. The amount of SPLP delivered to the tumor is substantial, in this experiment corresponding to >8% of the total injected dose per gram of tumor at 24 h. This result is also typical of numerous experiments in tumor-bearing mice. SPLP, regardless of the DNA encapsulated, achieve significant levels of accumulation, with 8–15% of the injected dose per gram of tumor at 24 h following injection (Fenske et al., 2001, 2002; Monck et al., 2000; Tam et al., 2000).

In addition to tumor accumulation, the biodistribution of SPLP in various other tissues has been studied. The accumulation of SPLP in the tumor, liver, spleen, lymph nodes and small intestine increases in the first 24 h after

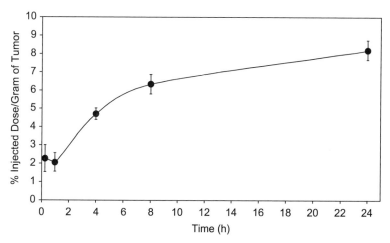

Figure 6.6. Tumor accumulation of SPLP following a single intravenous administration in neuro-2a tumor-bearing male A/J mice.

Table 6.2. Biodistribution of [³H]-CHE Labeled SPLP Following A Single Intravenous Administration in Neuro-2a Tumor-Bearing Male A/J Mice. Data is expressed as percent injected dose per gram of tissue. SEM is the standard error of the mean

Tissue	0.25 h	1 h	4 h	8 h	24 h
Spleen	7.94 ± 2.70	10.6 ± 1.86	10.1 ± 0.95	0.97	44.1 ± 1.28
Liver	4.16 ± 0.30	6.18 ± 0.90	10.7 ± 1.01	19.0 ± 1.90	25.8 ± 2.67
Adrenal Glands	7.92 ± 1.99	11.5 ± 1.92	20.8 ± 2.25	19.9 ± 3.88	14.1 ± 0.99
Small Intestine	1.92 ± 0.42	2.65 ± 0.40	5.35 ± 1.58	8.60 ± 1.20	12.4 ± 1.61
Tumor	2.25 ± 0.74	2.05 ± 0.51	4.70 ± 0.31	6.32 ± 0.52	8.19 ± 0.54
Mesenteric L.N.	1.56 ± 0.24	2.19 ± 0.27	5.03 ± 0.55	6.85 ± 1.65	7.23 ± 0.80
Thymus	9.31 ± 1.38	4.87 ± 0.54	4.28 ± 0.87	4.87 ± 0.15	5.72 ± 1.90
Large Intestine	1.81 ± 0.27	1.46 ± 0.14	1.86 ± 0.46	2.31 ± 0.56	4.47 ± 0.75
Kidneys	9.41 ± 1.84	9.81 ± 0.37	7.17 ± 0.48	7.36 ± 1.09	4.27 ± 0.13
Lungs	3.76 ± 1.27	5.10 ± 1.91	4.21 ± 1.96	4.62 ± 1.46	4.14 ± 0.39
Heart	5.02 ± 1.03	3.88 ± 0.43	4.23 ± 1.03	3.14 ± 0.45	3.54 ± 0.35
Bone Marrow	1.77 ± 0.21	2.10 ± 0.22	2.54 ± 0.20	3.23 ± 0.24	2.44 ± 0.41
Testes	0.87 ± 0.11	0.95 ± 0.21	0.82 ± 0.26	1.02 ± 0.23	1.65 ± 0.20
Brain	1.49 ± 0.20	0.89 ± 0.06	0.97 ± 0.23	0.69 ± 0.08	0.60 ± 0.08

administration. The amount of SPLP in other tissues, including the bone marrow, testes, and the brain, decreases or is maintained at low levels during this period (Table 6.2). The spleen and liver consistently demonstrate the highest levels of SPLP accumulation (44% and 26% of the injected dose per gram, respectively); whereas, the testes and brain accumulate the least amount of SPLP (Fig. 6.7).

To determine the pattern of biodistribution and rate of clearance of the plasmid DNA component of SPLP, pharmacokinetics have been evaluated based upon the detection of the plasmid DNA by quantitative real-time polymerase chain reaction (QPCR). This approach allows for analysis at later timepoints than would be appropriate using a lipid label. It also allows for the determination of intact plasmid payload, rather than a lipid marker for the SPLP.

Peak levels of plasmid accumulation occur on Day 1, 24 h after intravenous administration of SPLP. The greatest concentration of plasmid at this time is found in the blood (Fig. 6.8). At 6.5×10^{10} copies/μg, plasmid DNA accounts for greater than one-third of the total DNA extracted from whole blood (corresponding to a ratio of 300,000 plasmid copies per diploid cell). Plasmid, in both supercoiled and open circular conformations, was observed upon conventional electrophoretic analysis of DNA extracted from blood

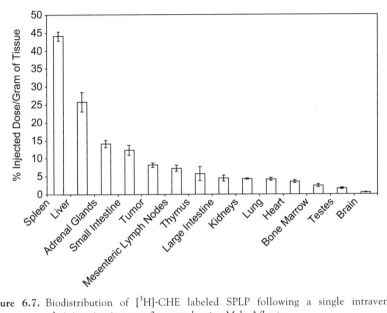

Figure 6.7. Biodistribution of [³H]-CHE labeled SPLP following a single intravenous administration in neuro-2a tumor-bearing Male A/J mice.

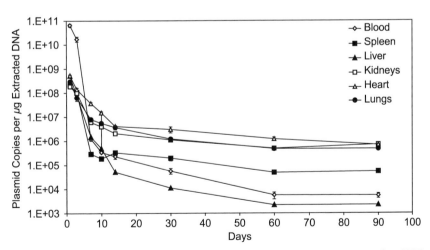

Figure 6.8. Clearance of plasmid DNA from the blood and first pass organs after SPLP administration in A/J mice.

samples. Plasmid concentrations in tumor and non-target tissues are generally two to three orders of magnitude lower than in blood, with the brain accumulating the least plasmid DNA at four orders of magnitude lower than in blood. Differences in plasmid concentration between male and female tissues are found only in the brain, femur (bone marrow) and reproductive organs. The target tissue in this test system is the subcutaneous Neuro-2a tumor. Tumor tissue is found to accumulate roughly the same relative plasmid concentration as the liver on Day 1, however, plasmid clearance from the tumor is much slower than from non-target tissues and blood (Fig. 6.9). By Day 7, plasmid concentration in tumor tissue has decreased, on average, 5.5-fold; whereas it has declined at least 21-fold and, on average, 196-fold in the non-target tissues. At Day 10, the most extended time-point for tumor-bearing mice, the tumor plasmid concentration has decreased, on average, 60-fold, compared to an average decline of 320-fold in the non-target tissues. Comparison of plasmid distribution in tumor-bearing and non-tumored animals shows that the pattern of tissue distribution is qualitatively similar yet plasmid concentrations in blood and non-target tissues of non-tumored animals are generally found to be higher than in the tumor-bearing animals (Lee and MacLachlan, 2004).

Ninety days after administration, plasmid concentrations in blood are 5.4×10^3 copies/μg, corresponding to 1–2 copies per 100 diploid cells (Fig. 6.8). The liver, followed by blood and brain tissue, contains the least plasmid DNA. The tissue containing the highest concentration of plasmid DNA after ninety days is the heart at 6.7×10^5 copies/μg, corresponding to 2 copies per diploid cell.

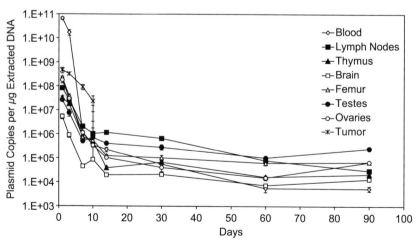

Figure 6.9. Clearance of plasmid DNA from the tumor and other tissues of interest after SPLP administration in A/J mice

This represents a decrease of less than three orders of magnitude from peak levels observed on Day 1, the slowest rate of clearance of all the tissues assayed.

The greatest change in plasmid concentration occurred in the blood compartment (decreasing seven orders of magnitude, Fig. 6.8) followed by liver tissue (decreasing five orders of magnitude), whereas plasmid levels decreased less than 1,000-fold between Day 1 and Day 90 in the kidneys, heart, lungs, brain and testes. The remaining analyzed tissues (lymph nodes, thymus, spleen, femur and ovaries) displayed moderate decreases in plasmid concentration of four orders of magnitude or less over the time period examined.

The rate of SPLP plasmid DNA clearance, from all tissues, is initially rapid between Day 1 and Day 7, and then slows considerably beyond Day 14. This pattern is reminiscent of that expected in a two-compartment model of pharmacokinetics. Two-compartment models assume that a drug, in this case plasmid DNA, distributes between two compartments, moves between the two compartments in proportion to its concentration, and is eliminated from the first compartment in proportion to its concentration (Fig. 6.10). Two-compartment models are often used to describe the behavior of a drug as it is distributed from a more accessible compartment, such as the systemic circulation or the more perfused tissues, to less perfused tissues such as the adipose tissue, skin or brain. In the case of SPLP, each individual tissue exhibits two-compartment behavior. One may speculate that each tissue contains a less accessible compartment, perhaps intracellular, that acts to sequester plasmid DNA delivered by SPLP and that this remarkably stable compartment is responsible for the more gradual clearance observed at later time-points. However, this is purely speculation. In compartmental analysis, the term 'compartment' refers to a mathematically distinct kinetic pool that does not necessarily correspond to any physical location or process. It remains to be determined experimentally if the second compartment responsible for the long-term sequestration of SPLP plasmid DNA following the initial distribution phase is a real physical compartment or merely a mathematical construct.

The extent of plasmid DNA distribution in tissues following SPLP administration is markedly greater than has been observed in other non-viral gene delivery systems (Lew *et al.*, 1995). This can be attributed to the extended blood circulation lifetimes of SPLP formulations and their ability to protect encapsulated plasmid DNA from degradation, greatly extending the available timeframe for plasmid delivery to—and accumulation within—tissues. While SPLP formulations provide plasma half-lives for intact plasmid DNA of 7 to 13 h (Tam *et al.*, 2000), 'naked' DNA and cationic lipoplex have half-lives of 30 min or less (Lew *et al.*, 1995; Ogris *et al.*, 1999; Thierry *et al.*, 1997). However the persistence of distributed plasmid DNA following SPLP administration is not entirely dissimilar to that observed in other non-viral systems. Plasmid has been detected by conventional PCR in mice up to six months after intravenous

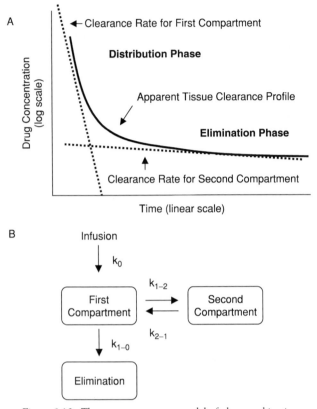

Figure 6.10. The two-compartment model of pharmacokinetics.

injection of lipoplex DNA (Lew *et al.*, 1995). Plasmid has also been demonstrated to persist in mice at least 19 months after intramuscular injection of naked DNA (Wolff *et al.*, 1992), and 1.5 years after three intraperitoneal doses of lipoplex (Xing *et al.*, 1998). With SPLP, residual plasmid DNA has been detected as long as 360 days after a single intravenous injection of SPLP at a dose level equivalent to 100 μg for a mouse weighing 20 grams. The implications of these findings should be assessed in conjunction with other relevant information. In particular, it is important to note that the initial plasmid distribution pattern and persistence of a gene delivery system is distinct from its gene expression pattern, as has been demonstrated for SPLP, DNA/lipid complexes (Lew *et al.*, 1995; Osaka *et al.*, 1996; Xing *et al.*, 1998) and naked DNA (Wolff *et al.*, 1992).

B. Biodistribution of protein expression following systemic administration of SPLP

Encapsulation of plasmid DNA coding for luciferase facilitates analysis of the biodistribution of protein expression resulting from intravenous administration of SPLP. Fig. 6.11 shows that the accumulation of SPLP following intravenous administration leads to significant levels of expression in Neuro-2a tumors, with other tissues yielding much lower levels of luciferase. With the exception of the adrenal glands, with one order of magnitude less than the tumor, all other tissues express 2 or 3 orders of magnitude less luciferase than the tumor. The liver and the brain demonstrate the least amount of expression. These results confirm that SPLP are capable of preferential disease-site targeting and expression, a conclusion supported by a number of preclinical studies in which expression has been observed in tumors following intravenous administration of various other SPLP formulations. In each of these studies, intravenous administration of SPLP leads to protein expression in the tumor on the hind flank of the animal.

Tumor-selective protein expression is achieved with SPLP in the absence of tumor-specific promoters, targeting ligands or any other so called 'targeting' technology. In fact, the results compare favorably with the degree of selectivity conferred to protein therapeutics, small molecule drugs or viral

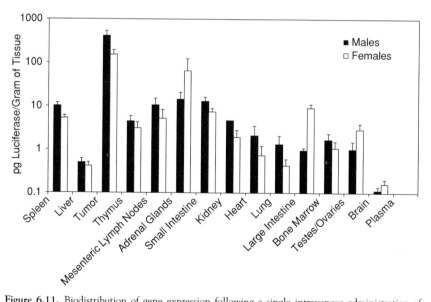

Figure 6.11. Biodistribution of gene expression following a single intravenous administration of SPLP in neuro-2a tumor bearing A/J mice.

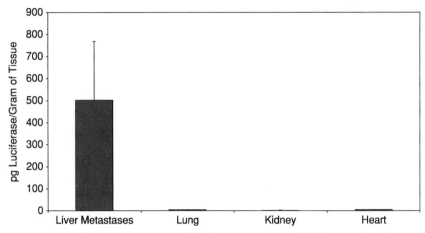

Figure 6.12. Luciferase gene expression 48 h after administration of SPLP in female balb/c mice bearing CT26 liver metastases.

vectors conjugated to monoclonal antibodies or peptides directed toward tumor cells in an attempt to convey tumor selectivity.

In an effort to further explore the potential of disease site targeting and tumor selective protein expression, the SPLP gene expression pattern has been evaluated in murine models of metastatic disease. One such model is the CT26 liver metastases model. Murine CT26 colon carcinoma cells are injected into the mouse spleen 10 min prior to splenectomy. Twenty-four days after tumor inoculation, mice are treated with SPLP by injection in the lateral tail vein. Forty-eight hours later, tissues are collected for luciferase analysis. In this experiment, and other models of liver metastasis, gene expression is greatest in the liver metastases (Fig. 6.12). Although previous studies have shown that significant amounts of SPLP are delivered to the liver in both normal and tumor-bearing mice, it is important to note that gene expression in this instance is increased by more than 100-fold in the presence of hepatic tumor nodules.

V. CONCLUSION

Many viral and non-viral gene transfer systems suffer from common pharmacological issues that limit their utility in a systemic context. Attempts to address these issues through manipulation of the virology or molecular biology of these systems have met with limited success. By application of the liposomal drug delivery paradigm, many of the limitations of the first generation non-viral delivery systems can be overcome. Encapsulation in small, long-circulating particles results in enhanced accumulation at disease sites. The selectivity of nucleic acid payloads is further enhanced by their mechanism of action. Plasmid DNA must overcome a number of intracellular barriers to reach the nucleus of the cell where gene expression occurs. Since these barriers are more readily overcome in transformed cells relative to their more quiescent, healthy counterparts the barriers to transfection actually represent opportunities to confer additional selectivity in the gene expression pattern.

Although lipid-mediated systemic gene delivery and expression was reported by Zhu et al in 1993 (Zhu et al., 1993), progress in developing lipoplex systems capable of delivering plasmid DNA to distal disease sites has been slow. The ability of SPLP to mediate tumor-selective protein expression in disseminated tumor sites represents a promising foundation upon which to build targeted molecular therapeutics for oncology, inflammation and infectious disease. The pharmacology of these systems is currently being evaluated in a Phase I clinical trial.

References

Alino, S. F., Bobadilla, M., Garcia-Sanz, M., Lejarreta, M., Unda, F., and Hilario, E. (1993). *In vivo* delivery of human alpha 1-antitrypsin gene to mouse hepatocytes by liposomes. *Biochem. Biophys. Res. Commun.* **192,** 174–181.

Allen, T. M., and Chong, A. (1987). Large unilamellar liposomes with low uptake into the reticuloendothelial system. *FEBS Lett.* **223,** 42–46.

Bailey, A., and Cullis, P. (1994). Modulation of membrane fusion by asymmetric transbilayer distributions of amino lipids. *Biochemistry* **33**(42), 12573–12580.

Baru, M., Axelrod, J. H., and Nur, I. (1995). Liposome-encapsulated DNA-mediated gene transfer and synthesis of human factor IX in mice. *Gene* **161,** 143–150.

Blessing, T., Remy, J. S., and Behr, J. P. (1998). Monomolecular collapse of plasmid DNA into stable virus-like particles. *Proc. Natl. Acad. Sci. USA* **95**(4), 1427–1431.

Boman, N. L., Masin, D., Mayer, L. D., Cullis, P. R., and Bally, M. B. (1994). Liposomal vincristine which exhibits increased drug retention and increased circulation longevity cures mice bearing P388 tumors. *Cancer Research* **54,** 2830–2833.

Chonn, A., Cullis, P. R., and Devine, D. V. (1991). The role of surface charge in the activation of the classical and alternative pathways of complement by liposomes. *J. Immunol.* **146**(12), 4234–4241.

Cudd, A., and Nicolau, C. (1985). Intracellular fate of liposome encapsulated DNA in mouse liver: Analysis using electron microscope autoradiography and subcellular fractionation. *Biochim. Biophys. Acta* **845,** 477–491.

Cullis, P. R., and de Kruiji, B. (1978). The polymorphic phase behaviour of phosphatidylethanolamines of natural and synthetic origin. A 31P NMR study. *Biochim. Biophys. Acta* **513**(1), 31–42.

Cullis, P. R., and de Kruiji, B. (1979). Lipid polymorphism and the functional roles of lipids in biological membranes. *Biochim. Biophys. Acta* **559**(4), 399–420.

Dekker, C. J., Vankessel, W., Klomp, J. P. G., Pieters, J., and de Kruiji, B. (1983). Synthesis and polymorphic phase-behavior of poly-unsaturated phosphatidylcholines and phosphatidylethanolamines. *Chem. Phys. Lipids* **33**(1), 93–106.

Dunn, K. W., and Maxfield, F. R. (1992). Delivery of ligands from sorting endosomes to late endosomes occurs by maturation of sorting endosomes. *J. Cell Biol.* **117**, 301–310.

Dzau, V. J., Mann, M. J., Morishita, R., and Kaneda, Y. (1996). Fusigenic viral liposome for gene therapy in cardiovascular diseases. *Proc. Natl. Acad. Sci. USA* **93**(21), 11421–11425.

Epand, R. M., Epand, R. F., Ahmed, N., and Chen, R. (1991). Promotion of hexagonal phase formation and lipid mixing by fatty-acids with varying degrees of unsaturation. *Chem. Phys. Lipids* **57**(1), 75–80.

Farhood, H., Serbina, N., and Huang, L. (1995). The role of dioleoylphosphatidylethanolamine in cationic liposome mediated gene transfer. *Biochim. Biophys. Acta* **1235**, 289–295.

Felgner, J. H., Kumar, R., Sridhar, C. N., Wheeler, C. J., Tsai, Y. J., Border, R., Ramsey, P., Martin, M., and Felgner, P. L. (1994). Enhanced gene delivery and mechanism studies with a novel series of cationic lipid formulations. *J. Biol. Chem.* **269**, 2550–2561.

Fenske, D. B., MacLachlan, I., and Cullis, P. (2001). Long circulating vectors for the systemic delivery of genes. *Curr. Opin. Mol. Thera.* **3**(2), 153–158.

Fenske, D. B., MacLachlan, I., and Cullis, P. R. (2002). Stabilized plasmid-lipid particles: A systemic gene therapy vector. *Met. Enzymol.* **346**, 36–71.

Fraley, R., Subramani, S., Berg, P., and Papahadjopoulos, D. (1980). Introduction of liposome-encapsulated SV-40 DNA into cells. *J. Biol. Chem.* **255**, 10431–10435.

Fraley, R. T., Fornari, C. S., and Kaplan, S. (1979). Entrapment of a bacterial plasmid in phospholipid vesicles: Potential for gene therapy. *Proc. Natl. Acad. Sci. USA* **76**, 3348–3352.

Gabizon, A., and Papahadjopoulos, D. (1988). Liposome formulations with prolonged circulation time in blood and enhanced uptake by tumors. *Proc. Natl. Acad. Sci. USA* **85**, 6949–6953.

Gao, X., and Huang, L. (1995). Cationic liposome-mediated gene transfer. *Gene Ther.* **2**, 710–722.

Geuze, H. J., Slot, J. W., Strous, G. J., Lodish, H. F., and Schwartz, A. L. (1983). Intracellular site of asialoglycoprotein receptor-ligand uncoupling: Double-label immunoelectron microscopy during receptor-mediated endocytosis. *Cell* **32**, 277–287.

Goldstein, J. L., Brown, M. S., Anderson, R. G. W., Russell, D., and Schneider, W. (1985). Receptor mediated endocytosis: Concepts emerging from the LDL receptor system. *Annu. Rev. Cell Biol.* **1**, 1–39.

Hafez, I. M., Maurer, N., and Cullis, P. (2001). On the mechanism whereby cationic lipids promote intracellular delivery of polynucleic acids. *Gene Thera.* **8**, 1188–1196.

Harrison, G., Wang, Y., Tomczak, J., Hogan, C., Shpall, E., Curiel, T., and Felgner, P. L. (1995). Optimization of gene transfer using cationic lipids in cell lines and primary human CD4+ and CD34+ hematopoietic cells. *Biotechniques* **19**, 816–823.

Heyes, J., Palmer, L., and MacLachlan, I. (2004). Degree of cationic lipid saturation influences fusogenicity and subsequent potency of stable plasmid lipid particles. In Press.

Hoffman, R. M., Margolis, L. B., and Bergelson, L. D. (1978). Binding and entrapment of high molecular weight DNA by lecithin liposomes. *FEBS Lett.* **93**, 365–368.

Hofland, H. E. J., Nagy, D., Liu, J. J., Spratt, K., Lee, Y. L., Danos, O., and Sullivan, S. M. (1997). *In vivo* gene transfer by intravenous administration of stable cationic lipid DNA complex. *Pharma. Res.* **14**(6), 742–749.

Hong, K., Zheng, W., Baker, A., and Papahadjopoulos, D. (1997). Stabilization of cationic liposome-plasmid DNA complexes by polyamines and poly(ethylene glycol)-phospholipid conjugates for efficient *in vivo* gene delivery. *FEBS Lett.* **400**, 233–237.

Howell, D., Krieser, R., Eastman, A., and MA, B. (2003). Deoxyribonuclease II is a lysosomal barrier to transfection. *Mol Ther.* **8**(6), 957–963.

Huang, L., and Li, S. (1997). Liposomal gene delivery: A complex package. *Nat. Biotech.* **15**, 620–621.

Hui, S. W., Langner, M., Zhao, Y. L., Ross, P., Hurley, E., and Chan, K. (1996). The role of helper lipids in cationic liposome-mediated gene transfer. *Biophys. J.* **71**, 590–599.

Hui, S. W., Stewart, T. P., Boni, L. T., and Yeagle, P. L. (1981). Membrane fusion through point defects in bilayers. *Science* **212,** 921–923.

Ibanez, M., Gariglio, P., Chavez, P., Santiago, R., Wong, C., and Baeza, I. (1997). Spermidine-condensed DNA and cone-shaped lipids improve delivery and expression of exogenous DNA transfer by liposomes. *Biochem. Cell Biol.* **74,** 633–643.

Jay, D. G., and Gilbert, W. (1987). Basic protein enhances the incorporation of DNA into lipid vesicles: Model for the formation of primordial cells. *Proc. Natl. Acad. Sci. USA* **84,** 1978–1980.

Jeffs, L., Palmer, L., Ambegia, E., Giesbrecht, C., Ewanick, S., and MacLachlan, I. (2005). A scalable, extrusion-free method for efficient liposomal encapsulation of plasmid DNA. *Pharm. Res.* **22**(3), 362–372.

Klibanov, A. L., Maruyama, K., Torchilin, V. P., and Huang, L. (1990). Amphipathic polyethyleneglycols effectively prolong the circulation time of liposomes. *FEBS Lett.* **268,** 235–237.

Koltover, I., Salditt, T., Radler, J. O., and Safinya, C. R. (1998). An inverted hexagonal phase of cationic liposome-DNA complexes related to DNA release and delivery. *Science* **281**(5373), 78–81.

Kopatz, I., Remy, J., and Behr, J. (2004). A model for non-viral gene delivery: Through syndecan adhesion molecules and powered by actin. *J. Gene Med.* **6**(7), 769–776.

Lechardeur, D., Sohn, K. J., Haardt, M., Joshi, P. B., Monck, M., Graham, R. W., Beatty, B., Squire, J., O'Brodovich, H., and Lukacs, G. L. (1999). Metabolic instability of plasmid DNA in the cytosol: A potential barrier to gene transfer. *Gene Thera.* **6**(4), 482–497.

Lee, R. J., and Huang, L. (1996). Folate-targeted, anionic liposome-entrapped polylysine-condensed DNA for tumor cell-specific gene transfer. *J. Biol. Chem.* **271,** 8481–8487.

Lew, D., Parker, S. E., Latimer, T., Abai, A. M., Kuwahara-Rundell, A., Doh, S. G., Yang, Z. Y., LaFace, D., Gromkowski, S. H., Nabel, G. J., Manthorpe, M., and Norman, J. (1995). Cancer gene therapy using plasmid DNA: Pharmacokinetic study of DNA following injection in mice. *Human Gene Thera.* **6,** 553–564.

Li, S., and Huang, L. (1997). *In vivo* gene transfer via intravenous administration of cationic lipid-protamine-DNA (LPD) complexes. *Gene Ther.* **4,** 891–900.

Litzinger, D. C., and Huang, L. (1992). Phosphatidylethanolamine Liposomes-Drug Delivery, Gene-Transfer and Immunodiagnostic Applications. *Biochim. Biophys. Acta* **1113**(2), 201–227.

Liu, Y., Liggitt, D., Zhong, W., Tu, G., Gaensler, K., and Debs, R. (1995). Cationic liposome mediated intravenous gene delivery. *J. Biol. Chem.* **270**(42), 24864–24870.

Lurquin, P. F. (1979). Entrapment of plasmid DNA by liposomes and their interactions with plant protoplasts. *Nucl. Acids Res.* **6,** 3773–3784.

Mannino, R. J., Allebach, E. S., and Strohl, W. A. (1979). Encapsulation of high molecular weight DNA in large unilamellar phospholipid vesicles. *FEBS Lett.* **101,** 229–232.

Maurer, N., Wong, K. F., Stark, H., Louie, L., McIntosh, D., Wong, T., Scherrer, P., Semple, S. C., and Cullis, P. R. (2001). Spontaneous entrapment of polynucleotides upon electrostatic interaction with ethanol-destabilized cationic liposomes. *Biophys. J.* **80,** 2310–2326.

Mayer, L. D., Bally, M. B., Loughrey, H., Masin, D., and Cullis, P. R. (1990). Liposomal vincristine preparations which exhibit decreased drug toxicity and increased activity against murine L1210 and P388 tumors. *Cancer Res.* **50,** 575–579.

Mayer, L. D., Nayar, R., Thies, R. L., Boman, N. L., Cullis, P. R., and Bally, M. B. (1993). Identification of vesicle properties that enhance the antitumour activity of liposomal vincristine against murine L1210 leukemia. *Cancer Chemother. Pharmacol.* **33,** 17–24.

Mislick, K. A., and Baldeschwieler, J. D. (1996). Evidence for the role of proteoglycans in cation mediated gene transfer. *Proc. Natl. Acad Sci. USA* **93,** 12349–12354.

Mok, K. W. C., Lam, A. M. I., and Cullis, P. R. (1999). Stabilized plasmid-lipid particles: Factors influencing plasmid entrapment and transfection properties. *Biochimi. Biophys. Acta-Biomembranes* **1419,** 137–150.

Monck, M. A., Mori, A., Lee, D., Tam, P., Wheeler, J. J., Cullis, P. R., and Scherrer, P. (2000). Stabilized plasmid-lipid particles: Pharmacokinetics and plasmid delivery to distal tumors following intravenous injection. *J. Drug Target.* **7**(6), 439–452.

Monnard, P. A., Oberholzer, T., and Luisi, P. L. (1997). Entrapment of nucleic acids in liposomes. *Biochim. Biophys. Acta-Biomembranes* **1329**(1), 39–50.

Mortimer, I., Tam, P., MacLachlan, I., Graham, R. W., Saravolac, E. G., and Joshi, P. B. (1999). Cationic lipid-mediated transfection of cells in culture requires mitotic activity. *Gene Thera.* **6**(3), 403–411.

Mounkes, L. C., Zhong, W., Cipres-Palacin, G., Heath, T. D., and Debs, R. J. (1998). Proteoglycans mediate cationic liposome-DNA complex-based gene delivery *in vitro* and *in vivo. J. Biol. Chem.* **273**(40), 26164–26170.

Mukherjee, A. B., Orloff, S., Butler, J. D., Triche, T., Lalley, P., and Schulman, J. D. (1978). Entrapment of metaphase chromosomes into phospholipid vesicles (lipochromosomes): Carrier potential in gene transfer. *Proc. Natl. Acad. Sci. USA* **75,** 1361–1365.

Murphy, R. F., Schmid, J., and Fuchs, R. (1993). Endosome maturation: Insights from somatic cell genetics and cell free analysis. *Biochem. Soc. Trans* **21,** 716–720.

Nakanishi, M., Uchida, T., Sugawa, H., Ishiura, M., and Okada, Y. (1985). Efficient introduction of contents of liposomes into cells using HVJ (sendai virus). *Exp. Cell Res.* **159,** 399–409.

Needham, D., McIntosh, T. J., and Lasic, D. D. (1992). Repulsive interactions and mechanical stability of polymer grafted lipid membranes. *Biochim. Biophys. Acta* **1108,** 40–48.

Nicolau, C., Le Pape, A., Soriano, P., Fargette, F., and Juhel, M. F. (1983). *In vivo* expression of rat insulin after intravenous administration of the liposome-entrapped gene for rat insulin. *Proc. Natl. Acad. Sci. USA* **80,** 1068–1072.

Nicolau, C., and Rottem, S. (1982). Expression of beta-lactamase activity in Mycoplasma carpicolum transfected with the liposome-encapsulated *E. coli* pBR32 plasmid. *Biochem. Biophys. Res. Commun.* **108,** 982–986.

Ogris, M., Brunner, S., Schüller, S., Kircheis, R., and Wagner, E. (1999). PEGylated DNA/transferrin-PEI complexes: Reduced interaction with blood components, extended circulation in blood and potential for systemic gene delivery. *Gene Thera.* **6,** 595–605.

Ohno, M., Fornerod, M., and Mattaj, I. W. (1998). Nucleocytoplasmic transport: The last 200 nanometers. *Cell* **92,** 327–336.

Osaka, G., Carey, K., Cuthbertson, A., Godowski, P., Patapoff, T., Ryan, A., Gadek, T., and Mordenti, J. (1996). Pharmacokinetics, tissue distribution, and expression efficiency of plasmid [33P]DNA following intravenous administration of DNA/cationic lipid complexes in mice: Use of a novel radionuclide approach. *J. Pharma. Sci.* **85**(6), 612–617.

Palmer, L., Chen, T., Lam, A., Fenske, D., Wong, K., MacLachlan, I., and PR, C. (2003). Transfection properties of stabilized plasmid-lipid particles containing cationic PEG lipids. *Biochim. Biophys. Acta* **1611**(1–2), 204–216.

Papahadjopoulos, D., Allen, T. M., Gabizon, A., Mayhew, E., Matthay, K, Huang, S. K., Lee, K. D., Woodle, M. C., Lasic, D. D., Redemann, C., and Martin, F. J. (1991). Sterically stabilized liposomes: improvements in pharmacokinetics and anti-tumor therapeutic efficacy. *Proc. Natl. Acad. Sci. USA* **88,** 11460–11464.

Puyal, C., Milhaud, P., Bienvenue, A., and Philippot, J. R. (1995). A new cationic liposome encapsulating genetic material. A potential delivery system for polynucleotides. *Eur. J. Biochem.* **228,** 697–703.

Sakurai, F., Nishioka, T., Saito, H., Baba, T., Okuda, A., Matsumoto, O., Taga, T., Yamashita, F., Takakura, Y., and M., H. (2001). Interaction between DNA-cationic liposome complexes and erythrocytes is an important factor in systemic gene transfer via the intravenous route in mice: The role of the neutral helper lipid. *Gene Ther.* **8**(9), 677–686.

Sankaram, M. B., Powell, G. L., and Marsh, D. (1989). Effect of acyl chain composition on salt-induced lamellar to inverted hexagonal phase-transitions in cardiolipin. *Biochim. Biophys. Acta* **980**(3), 389–392.

Scaefer-Ridder, M., Wang, Y., and Hofschneider, P. H. (1982). Liposomes as gene carriers: Efficient transformation of mouse L cells by thymidine kinase gene. *Science* **215**, 166–168.

Sebestyen, M. G., Ludtke, J. J., Bassik, M. C., Zhang, G., Budker, V., Lukhtanov, E. A., Hagstrom, J. E., and Wolff, J. A. (1998). DNA vector chemistry: The covalent attachment of signal peptides to plasmid DNA. *Nat. Biotech.* **16**(1), 80–85.

Senior, J., Delgado, C., Fisher, D., Tilcock, C., and Gregoriadis, G. (1991). Influence of surface hydrophilicity of liposomes on their interaction with plasma protein and their clearance from the circulation: Studies with poly(ethylene glycol)-coated vesicles. *Biochim. Biophys. Acta* **1062,** 77–82.

Soriano, P., Dijkstra, J., Legrand, A., Spanjer, H., Londos-Gagliardi, D., Roerdink, F., Scherphof, G., and Nicolau, C. (1983). Targeted and non-targeted liposomes for *in vivo* transfer to rat liver cells of plasmid containing the preproinsulin I gene. *Proc. Natl. Acad. Sci. USA* **80**, 7128–7131.

Stavridis, J. C., Deliconstantinos, G., Psallidopoulos, M. C., Armenakas, N. A., Hadjiminas, D. J., and Hadjiminas, J. (1986). Construction of transferrin-coated liposomes for *in vivo* transport of exogenous DNA to bone marrow erythroblasts in rabbits. *Exp. Cell Res.* **164**, 568–572.

Stein, Y., Halperin, G., and Stein, O. (1980). Biological stability of [3H]cholesteryl esther in cultured fibroblasts and intact rat. *FEBS Lett.* **111**(1), 104–106.

Szelei, J., and Duda, E. (1989). Entrapment of high molecular mass DNA molecules in liposomes for the genetic transformation of animal cells. *Biochem. J.* **259**, 549–553.

Szule, J. A., Fuller, N. L., and Rand, R. P. (2002). The effects of acyl chain length and saturation of diacylglycerols and phosphatidylcholines on membrane monolayer curvature. *Biophys. J.* **83**(2), 977–984.

Tam, P., Monck, M., Lee, D., Ludkovski, O., Leng, E., Clow, K., Stark, H., Scherrer, P., Graham, R. W., and Cullis, P. R. (2000). Stabilized plasmid lipid particles for systemic gene therapy. *Gene Thera.* **7**, 1867–1874.

Templeton, N. S., Lasic, D. D., Frederik, P. M., Strey, H. H., Roberts, D. D., and Pavlakis, G. N. (1997). Improved DNA: Liposome complexes for increased systemic delivery and gene expression. *Nat. Biotech.* **15**(July), 647–652.

Thierry, A. R., Lunardiiskandar, Y., Bryant, J. L., Rabinovich, P., Gallo, R. C., and Mahan, L. C. (1995). Systemic Gene-Therapy—Biodistribution and Long-Term Expression of a Transgene in Mice. *Proc. Nat. Acad. Sci. USA* **92**(21), 9742–9746.

Thierry, A. R., Rabinovich, P., Peng, B., Mahan, L. C., Bryant, J. L., and Gallo, R. C. (1997). Characterization of liposome-mediated gene delivery: Expression, stability and pharmacokinetics of plasmid DNA. *Gene Ther.* **4**, 226–237.

Tousignant, J. D., Gates, A. L., Ingram, L. A., Johnson, C. L., Nietupski, J. B., Cheng, S. H., Eastman, S. J., and Scheule, R. K. (2000). Comprehensive analysis of the acute toxicities induced by systemic administration of cationic lipid: Plasmid DNA complexes in mice. *Human Gene Thera.* **11**, 2493–2513.

Tousignant, J. D., Zhao, H., Yew, N. S., Cheng, S. H., Eastman, S. J., and Scheule, R. K. (2003). DNA sequences in cationic lipid: pDNA-mediated systemic toxicities. *Human Gene Thera.* **14,** 203–214.

Wang, C. Y., and Huang, L. (1987). pH-sensitive immunoliposomes mediate target cell-specific delivery and controlled expression of a foreign gene in mouse. *Proc. Natl. Acad. Sci. USA* **84**, 7851–7855.

Wang, C. Y., and Huang, L. (1989). Highly efficient DNA delivery mediated by pH sensitive immunoliposomes. *Biochemistry* **28**, 9508–9514.

Webb, M. S., Harasym, T. O., Masin, D., Bally, M. B., and Mayer, L. D. (1995). Sphingomyelin-cholesterol liposomes significantly enhance the pharmacokinetic and therapeutic properties of vincristine in murine and human tumour models. *B. J. Cancer* **72,** 896–904.

Wheeler, J. J., Palmer, L., Ossanlou, M., MacLachlan, I., Graham, R. W., Zhang, Y. P., Hope, M. J., Scherrer, P., and Cullis, P. R. (1999). Stabilized plasmid-lipid particles: Construction and characterization. *Gene Thera.* **6,** 271–281.

Wilke, M., Fortunati, E., vandenBroek, M., Hoogeveen, A. T., and Scholte, B. J. (1996). Efficacy of a peptide-based gene delivery system depends on mitotic activity. *Gene Thera.* 3(12), 1133–1142.

Wolff, J. A., Ludtke, J. J., Acsadi, G., Williams, P., and Jani, A. (1992). Long-term persistence of plasmid DNA and foreign gene expression in mouse muscle. *Hum. Mol. Genet.* 1(6), 363–369.

Wrobel, I., and Collins, D. (1995). Fusion of cationic liposomes with mammalian cells occurs after endocytosis. *Biochim. Biophys. Acta* **1235,** 296–304.

Wu, N. Z., Da, D., Rodolf, T. L., Needham, D., Whorton, A. R., and Dewhirst, M. W. (1993). Increased microvascular permeability contributes to preferential accumulation of stealth liposomes in tumor tissue. *Cancer Res.* **53,** 3765–3770.

Xing, X., Zhang, S., Chang, J. Y., Tucker, S. D., Chen, H., Huang, L., and Hung, M. C. (1998). Safety study and characterization of E1A-liposome complex gene-delivery protocol in an ovarian cancer model. *Gene Thera.* **5,** 1538–1544.

Zhang, Y. P., Sekirov, L., Saravolac, E. G., Wheeler, J. J., Tardi, P., Clow, K., Leng, E., Sun, R., Cullis, P. R., and Scherrer, P. (1999). Stabilized plasmid-lipid particles for regional gene therapy: Formulation and transfection properties. *Gene Thera.* **6,** 1438–1447.

Zhu, N., Liggitt, D., Liu, Y., and Debs, R. (1993). Systemic gene expression after intravenous DNA delivery into adult mice. *Science* **261,** 209–211.

7

Toxicity of Cationic Lipid-DNA Complexes

Nelson S. Yew and Ronald K. Scheule

Genzyme Corporation
Framingham, Massachusetts 01701

Advances in Genetics, Vol. 53
Copyright 2005, Elsevier Inc. All rights reserved.

0065-2660/05 $35.00
DOI: 10.1016/S0065-2660(05)53007-4

ABSTRACT

As with any conventional drug, the body's response to cationic lipid-DNA complexes is highly dependent on both the dose administered and the route of delivery. At relatively low doses there is little to no effect on organ function or tissue architecture, but at higher doses, acute inflammation and tissue damage can occur that is sometimes quite profound. Of the two most common routes of delivery, intravenous (IV) or intrapulmonary, IV administration tends to cause more severe adverse effects and can be lethal at higher doses of complex. Both routes activate an innate immune response that includes the induction of proinflammatory cytokines and immune cell activation, a major portion of which has been attributed to the presence of immunostimulatory CpG motifs within the plasmid DNA vector. Removing CpGs from the plasmid vector reduces several, but not all of the acute inflammatory responses to cationic lipid-DNA complexes. Therefore, other strategies are required to improve the therapeutic potential of these vectors, such as transient immune suppression, aerosolization of the complex, and novel formulations that have increased efficiency of transduction and decreased interaction with immune cells. © 2005, Elsevier Inc.

I. INTRODUCTION

The close relationship between safety and efficacy that is associated with any therapeutic agent is no less important for cationic lipid-mediated gene therapy. Non-viral vectors have long been considered to have a safety advantage over recombinant DNA viruses, since the absence of any viral antigens and the inevitable immune complications is a distinct plus, especially when re-administration of vector is required. However, the levels of expression that can be achieved with non-viral systems, with a few notable exceptions, are substantially lower than what is required for most disease applications. In the absence of therapeutic efficacy, the safety of current formulations is less meaningful. Reporter proteins such as luciferase are most commonly used when testing new cationic liposomes and formulations, and while useful and convenient for evaluating relative expression levels, reporter proteins are minimally informative when evaluating toxicity. Ideally, expressing a therapeutic protein in a relevant animal model of a given disease is required to understand fully the relationship between the efficacy, toxicity and dose of a given vector.

The physical properties of complexes contribute to many of their inflammatory effects. They are known to aggregate extensively *in vivo* and are avidly recognized by the reticuloendothelial system. Cells that endocytose complexes have been shown to contain large clumps of aggregated lipid-DNA structures in their cytoplasm (Zabner *et al.*, 1995). Cell function is likely compromised as a result, and the body's defense system responds accordingly to this unnatural deposition of lipid and DNA. Interestingly, cationic liposomes and DNA have little toxicity individually. Mice have tolerated IV injection of 2 mg of plasmid DNA (66 mg/kg for a 30 g mouse) and greater than 1 mg (33 mg/kg) of cationic liposomes (Liu *et al.*, 1995). In contrast, a maximum of about 3–4 mg/kg of cationic lipid-DNA complex can be administered intravenously, above which there is significant lethality. Liposomes are not totally benign, as lipofectamine and DOSPER have been shown to induce activation of interferon-stimulated genes in the absence of DNA (Li *et al.*, 1998). However, the overwhelming portion of the acute toxicity is due to the characteristics of the cationic lipid-DNA complex, rather than the toxicity of the individual components.

There are recent reviews that discuss the toxicity of cationic lipid-DNA and non-viral gene therapy (Audouy *et al.*, 2002; Niidome and Huang, 2002; Tan and Huang, 2002). This chapter will focus mainly on the acute inflammatory responses associated with intrapulmonary and IV delivery of cationic lipid-DNA complexes, with a bias toward the role that DNA plays in contributing to this toxicity, in particular the immunostimulatory CpG motifs within plasmid DNA. Strategies to reduce the effects of CpGs and other approaches to decrease dose-limiting toxicities will also be discussed.

II. ACUTE TOXICITY AFTER INTRAPULMONARY DELIVERY

A. Studies in mice and sheep

The conducting airways provide a convenient and noninvasive route for delivering genes into the body, and the lung is the primary target for treating pulmonary diseases such as α-1-antitrypsin deficiency and cystic fibrosis (Griesenbach *et al.*, 2004; Stecenko and Brigham, 2003). However, intratracheal

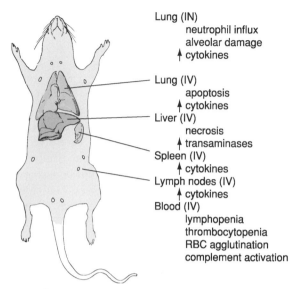

Lung (IN)
　　neutrophil influx
　　alveolar damage
　　↑ cytokines

Lung (IV)
　　apoptosis
　　↑ cytokines
Liver (IV)
　　necrosis
　　↑ transaminases
Spleen (IV)
　　↑ cytokines
Lymph nodes (IV)
　　↑ cytokines
Blood (IV)
　　lymphopenia
　　thrombocytopenia
　　RBC agglutination
　　complement activation

Figure 7.1. Spectrum of acute toxic responses in mice after intrapulmonary (IN) or intravenous (IV) delivery of cationic lipid DNA complex. The magnitude of these effects is dependent on the dose administered.

or intranasal instillation of lipid-DNA complexes in mice results in a dose-dependent, acute inflammatory response, characterized by an influx of neutrophils, induction of proinflammatory cytokines, and multifocal lesions within the alveolar region (Fig. 7.1; Scheule et al., 1997). The cytokines that are induced include TNF-α, IL-6, IL-12, IFN-γ and KC, their levels peaking around 1 day after instillation and returning to normal within a few days (Freimark et al., 1998; Scheule et al., 1997; Yew et al., 1999). Resolution of the histopathological damage takes longer, greater than 14 days in mice that received a high dose of complex (132 μg DNA). Instillation of cationic lipid GL-67 alone was found to induce almost as much alteration to the lung architecture as instillation of complex. It is thought that the pooling of complex in the lung and the detergent effect of the liposomes on surface of the epithelium may be responsible for this damage. Indeed, using fluorescently tagged complexes, Uyechi et al. (2001) observed that the distribution of the complex was nonuniform and concentrated around bronchioles and distal airways.

Cationic lipid-DNA complexes have also been instilled into sheep, a large animal model being used to develop non-viral vectors for cystic fibrosis and other lung diseases. Using cationic lipid GL-67, different doses of plasmid DNA (0.2, 1, and 5 mg) were complexed and instilled into spatially distinct lung segments (Emerson *et al.*, 2003). A dose-dependent inflammatory response was observed in the lung, characterized by an influx of neutrophils. Complexes have also been administered into fetal sheep in an effort to develop a method of *in utero* gene transfer (Luton *et al.*, 2004). Using a fairly involved surgical procedure, the fetus was extruded from the uterus, and a catheter was inserted into the trachea. The lung fluid was aspirated and replaced with approximately 3 ml of BGTC/DOPE-DNA complex (300 μg DNA), with the trachea ligatured to avoid leakage. The fetus was then returned to the uterus. Three days later the lung was examined histologically. Focal epithelial lesions and underlying mesenchymal lesions were observed, with epithelial ulceration and cell shedding present in the most severe lesions (Luton *et al.*, 2004). Taken together, these results indicate that the complex-induced inflammatory response observed in mice also occurs in larger animals.

B. Studies in humans

Intrapulmonary delivery of cationic liposome-DNA complexes has also been attempted in humans. Phase I clinical trials were conducted in patients with cystic fibrosis (CF), in which complex was administered into the nose or lung. No signs of inflammation or toxicity were observed after DC-Chol/DOPE complexes were administered to the nasal septum using a syringe (0.4 mg of plasmid DNA in 8 ml applied over a period of 4 h) (Hyde *et al.*, 2000). Neither were adverse clinical events seen after aerosol delivery of cationic lipid GL-67/DOPE/DMPE-PEG5000 to the lungs of normal volunteers (Chadwick *et al.*, 1997). From these results this particular lipid formulation was considered to be safe for use in subsequent trials in CF patients. In one such trial, 8 CF patients received a complex of GL-67/DOPE/DMPE-PEG5000 and plasmid DNA, and 8 patients received lipid alone (Alton *et al.*, 1999). Although no adverse events were reported from patients that received lipid alone, seven of the eight patients that received complex reported mild flu-like symptoms— myalgia, headache, and fever—at 6 h after administration, which resolved by 36 h. In a second trial, complex was again aerosolized into the lung of CF patients. Four of eight patients developed fever, myalgia, and arthralgia within 6 h of administering the vector, and elevated levels of IL-6 were detected in the serum (Ruiz *et al.*, 2001). Again the symptoms resolved within 24–48 h, and there were no apparent long-term effects.

These results reaffirm the greater inflammatory character of cationic lipid-DNA complexes compared to liposomes alone. CF patients are perhaps more susceptible to these effects compared to normal individuals, since the CF lung contains a thickened, dehydrated mucous that leads to bacterial colonization and chronic inflammation. While the symptoms arising from administering complex in these trials were not serious, a better understanding of the underlying cause, as well as improved formulations and DNA vectors, will be required before conducting future clinical trials.

III. BIODISTRIBUTION OF CATIONIC LIPID-DNA COMPLEXES AFTER INTRAVENOUS DELIVERY

A second major route of delivery is via the bloodstream. IV injection is the favored route for those diseases requiring gene transfer to visceral organs, but carries greater potential for significant toxicity due to this broader distribution of complex. It is perhaps useful to briefly follow the fate of cationic lipid-DNA complexes after IV injection to provide some clues to the basis of the subsequent inflammatory response.

A. Interactions in the blood

Cationic lipid-DNA complexes interact with serum proteins, erythrocytes, monocytes, and other blood components upon IV injection. Some of the complex-binding proteins that have been identified include apolipoproteins, fibrinogen, and several others (Tandia *et al.*, 2003). Complexes aggregate immediately upon exposure to serum, and also interact with red blood cells, causing extensive agglutination (Eliyahu *et al.*, 2002). Prolonged interaction with serum causes disintegration of the complex (Li *et al.*, 1999a). In addition, cationic lipids such as DOTMA/DOPE can induce fusion between erythrocytes (Sakurai *et al.*, 2002).

B. Uptake by the lung and lymph nodes

After IV injection, complexes travel to the heart and then to the lung, the first capillary bed encountered. Five min after injecting fluorescently labeled DOTIM-cholesterol-DNA complexes, punctate fluorescence is observed within alveolar capillaries, and a patchy coating appears on the surface of the endothelial cells (McLean *et al.*, 1997). There is also intense staining of intravascular leukocytes. At 4 h after injection, complexes are found in endosomes of the luminal endothelial cells.

The lymph nodes and Peyer's patch (lymph nodes in the intestinal wall) are second only to the lung in taking up complex, as measured by fluorescence intensity 4 h after injection (McLean *et al.*, 1997). Abundant fluorescence is observed in the high endothelial venules and within macrophages. Complexes likely interact with the B cells, T cells, plasma cells, and macrophages that reside in the nodes.

C. Redistribution of complex to the liver and other organs

A sizeable portion of the complex initially taken up by the lung is gradually redistributed to the liver over time. Mahato *et al.* (1998) followed the biodistribution of DOTMA:DOPE-DNA complexes, more specifically the plasmid DNA portion, by quantitative PCR. Within 15 min post-injection, 83% of the recovered dose of DNA was found in the lung, 11% in the liver, 2% in the spleen, and 2% in the kidney. At 24 h post-injection, the percent of the recovered DNA in the lung fell to 35% and increased in the liver to nearly 60%. Using either fluorescently or radioactively labeled liposomes, the complex was primarily in Kupffer cells, with small amounts in sinusoidal endothelial cells, and none in hepatocytes (Litzinger *et al.*, 1996; McLean *et al.*, 1997). In the spleen, diffuse fluorescence was observed in the marginal zone of the white pulp (McLean *et al.*, 1997).

IV. ACUTE TOXICITY AFTER INTRAVENOUS DELIVERY

Within 24 h after intravenously injecting complex into mice, the animals often appear lethargic and "scruffy" due to the cessation of normal grooming activity. These outward symptoms are indicative of several adverse physiologic changes resulting from introducing cationic lipid-DNA complexes into the circulation (Fig. 7.1). The response is highly dose-dependent; at lower doses the mice appear quite normal after injection, or recover within one or two days, but at higher doses there is significant morbidity and mortality. There are a few reports characterizing the acute effects on different organ systems, but a real understanding of the actual cause of death remains to be elucidated.

A. Effects in the blood

Complexes composed of cationic lipid GL-67 and plasmid DNA (66 μg DNA) induce a significant loss of lymphocytes and platelets from the blood of approximately 80–90% at 24 h post-injection (Tousignant *et al.*, 2000). This lymphopenia and thrombocytopenia persists for several days, but eventually returns to normal. IV injection of liposomes alone has been shown to cause transient sequestration of platelets in the liver and spleen, although the levels were found to return to normal within 60 min (Reinish *et al.*, 1988).

Cationic lipid-DNA complexes have also been shown to deplete and activate complement (Plank *et al.*, 1996). DOTAP/Chol-DNA complexes deplete complement by 36% 24 h after injection into mice (Barron *et al.*, 1998). GL-67/DOPE-DNA (132 μg DNA) complexes also induce a rapid decrease in serum complement activity of greater than 50% within 1 hour after injection

(Tousignant *et al.*, 2000). When mice that are depleted of complement using cobra venom factor (CVF) are subsequently injected with complex, the mice continue to exhibit lymphopenia and thrombocytopenia, suggesting that complement is unlikely to be involved in these changes in blood (Tousignant *et al.*, 2000). Rather, complement proteins are likely opsonizing the complexes, as they would with any foreign microorganism. The anaphylotoxins C3a and C5a are then generated, which recruit and activate inflammatory cells, contributing to the acute toxicity that is observed.

B. Effects in the lung and liver

In the lung, DOTAP-Chol-DNA complexes or DOTAP-Chol-protamine-DNA complexes induce apoptosis in approximately 10% of the cells at 12 h after IV injection, likely due to the effects of TNF-α on the vascular endothelium (Li *et al.*, 1999b). There is also accumulation and activation of NK cells that release large amounts of IFN-γ (Dow *et al.*, 1999). This inflammatory response, however, does not lead to substantial necrosis or noticeable changes in the lung architecture, in sharp contrast to what is observed after instilling complexes into the lung lumen.

   In the liver, DOTAP-Chol-DNA complexes and GL-67-DOPE-DNA complexes (50–66 μg DNA dose) have been shown to induce acute inflammation and widespread, multifocal necrosis (Loisel *et al.*, 2001). This is at least in part due to the uptake of complex into the Kupffer cells, stimulating the secretion of TNF-α and other cytokines. The elevated levels of the serum transaminases ALT and AST correlate with the hepatocellular lesions that are observed.

V. ROLE OF IMMUNOSTIMULATORY CPG MOTIFS IN CATIONIC LIPID-DNA TOXICITY

The pleiotropic nature of the response to cationic lipid-DNA complexes suggests that many components of the host immune system are being activated, but how so is less clear. The DNA in the complex, in particular the unmethylated CpG motifs within the DNA sequence, appear to play a prominent, but not exclusive role in this activation.

A. Background

Microbial DNA contains a much higher frequency of unmethylated CpGs compared to mammalian DNA. These unmethylated CpGs are recognized by Toll-like receptor 9 (TLR9), a member of a family of receptors that have critical roles in innate immunity by recognizing different pathogen-associated molecular patterns that are features of various microorganisms (Akira *et al.*, 2001). In humans, TLR9 is expressed in B cells and plasmacytoid dendritic cells (DCs), whereas in mice, TLR9 is also expressed in myeloid cells (monocytes, macrophages and myeloid DCs) (Hornung *et al.*, 2002). TLR9 is located intracellularly and binds to CpG DNA in endocytic vesicles (Latz *et al.*, 2004). This binding initiates a signaling cascade that leads to the expression of NFkB and other transcription factors that upregulate proinflammatory gene expression. CpG DNA stimulates DCs to secrete high levels of IL-12 and IFN-α, and stimulates the maturation, differentiation and proliferation of NK cells, monocytes, macrophages and T cells (Krieg, 2002a).

B. Evidence from *in vivo* studies

The CpG dinucleotides within a plasmid DNA can be methylated readily *in vitro*. When cationic lipid complexes containing methylated plasmid DNA are delivered intravenously or intraluminally into the lung, they induce significantly lower levels of proinflammatory cytokines such as TNF-α, IFN-γ, IL-6, and KC compared to complexes containing unmethylated DNA (Li *et al.*, 1999b; McLachlan *et al.*, 2000; Yew *et al.*, 1999). In addition, complexes formed with CpG-containing oligonucleotides stimulate cytokine production, whereas complexes formed with non-CpG oligonucleotides or CpG-methylated oligonucleotides do not (Tan *et al.*, 1999). These data show that the unmethylated CpGs within plasmid DNA are largely responsible for the induction of the inflammatory cytokine response.

Further evidence has been provided by the use of plasmid DNA vectors that have been extensively reduced in CpG content. Compared to an unmodified vector, much lower levels of cytokines were present in the bronchoalveolar lavage fluid after instillation of complexes containing a plasmid DNA vector in which approximately 50% of its CpG motifs were eliminated (Yew *et al.*, 2000). Significantly lower cytokine levels were also detected in the serum after IV delivery, in this case using an even more CpG-depleted vector (80% of CpGs removed) (Yew *et al.*, 2002). This vector (pGZB) also reduced the acute loss of lymphocytes and platelets in the blood after IV injection of complex. The effect, however, was observable only at lower doses of complex (16.5 μg DNA), with no significant improvement in blood cell numbers at a 2-fold higher dose (33 μg DNA). Complexes containing the pGZB vector also exhibited reduced elevation of the liver transaminases ALT and AST compared to complexes containing an unmodified plasmid vector (Yew *et al.*, 2002).

Thus, the results from the CpG-reduced vectors confirm what was shown previously with CpG-methylated DNA, but with some notable differences. For example, methylation is not uniformly effective in reducing ALT and

AST levels, being effective with one complex but not another. Nor does methylation prevent the loss of lymphocytes or platelets after IV delivery (Tousignant *et al.*, 2000). So even though methylated CpGs do not stimulate TLR9 signaling, they are still recognized by the host immune system.

C. Possible consequences of CpG stimulation

Besides acute inflammation, there are other potential adverse effects that result from inappropriate immune activation by cationic lipid-DNA complexes. It should be stressed that these concerns arise from studies conducted principally in rodents, which may not predict the response in humans.

1. CpG and toxic shock

One concern with unwanted CpG stimulation is an increased risk of developing toxic shock. Mice treated with the hepatotoxic agent D-galactosamine become hypersensitive to both endotoxins (lipopolysaccharide, LPS) and to CpG DNA (Sparwasser *et al.*, 1997). Challenging these mice with LPS or bacterial DNA leads to lethality. In other studies, mice have been injected with bacterial DNA or CpG-containing oligonucleotides followed by a sublethal dose of LPS. The CpG DNA and LPS synergize to generate dramatically elevated levels of TNF-α and IL-6, and 75% of the mice die within 36 h (Cowdery *et al.*, 1996). The results suggest that administering cationic lipid-DNA complexes to individuals already compromised by infection may trigger a septic-like response, or conversely may greatly worsen the inflammatory response if a healthy individual that receives complex becomes exposed to a pathogen. However, the latter scenario is complicated by the finding that CpG DNA can actually increase resistance against microbial sepsis in some cases (Krieg, 2003; Weighardt *et al.*, 2000). So the concerns with regard to toxic shock remain quite speculative and uncertain at this point.

2. CpG and autoimmunity

Another concern is that repeated deposition of large quantities of stimulatory DNA into immune cells may enhance the risk of developing an immune response to either self-proteins or DNA, which may lead to autoimmune disease. TLR9 has been shown recently to have a role in initiating the production of autoantibodies to normal antibodies, which are known as rheumatoid factors (RF) (Krieg, 2002b). Complexes of immunoglobulin (Ig) and self-DNA were shown to activate RF[+] B cells through two signals: 1) the Ig portion of the complex binds to the B-cell receptor and 2) the unmethylated CpG motifs in the self-DNA activate TLR9. The results demonstrate that TLR9 signaling can activate B cells without T cell help (Leadbetter *et al.*, 2002). Other studies have shown that CpG oligonucleotides can induce the activation and expansion of autoreactive T cells, and in some animal models TLR9-mediated activation of antigen-presenting cells can break self-tolerance and induce autoimmune disease (Conant and Swanborg, 2004; Segal *et al.*, 2000; Waldner *et al.*, 2004).

However, while repeated immunization of normal mice with either plasmid DNA or bacterial DNA induces significant quantities of anti-DNA antibodies, the mice do not develop autoimmune disease (Gilkeson *et al.*, 1989; Mor *et al.*, 1997). Non-human primates given CpG oligonucleotides also do not develop autoimmunity, and human patients that have received DNA vaccines have not exhibited any significant adverse reactions (Epstein *et al.*, 2004). Thus, the degree of risk is probably quite small, and additional safety data from ongoing clinical trials will aid in assessing the potential for CpG-mediated autoimmunity in the context of gene therapy applications.

3. Immune organ damage with repetitive dosing

Yet another possible consequence of delivering immunostimulatory cationic lipid-DNA complexes is CpG-mediated damage from repeat dosing. When CpG-containing oligonucleotides are injected intraperitoneally once a day into mice, their spleens increase dramatically both in size and weight, as much as 5-fold by day 6–7 (Heikenwalder *et al.*, 2004). Splenic hemopoeisis is stimulated as evidenced by an increase in the number of granulocyte-macrophage and early erythroid progenitor cells. This effect is sequence dependent, as oligonucleotides lacking CpG motifs do not induce hemopoiesis (Sparwasser *et al.*, 1999). Repeated CpG dosing also leads to alteration and destruction of splenic, mesenteric and inguinal lymph nodes (Heikenwalder *et al.*, 2004).

However, the interval between injecting complex for most clinical applications will likely be several weeks to several months, rather than daily dosing. Also, the splenomegaly observed in mice may not pertain to humans, given the more restricted expression of human TLR9. Nonetheless, the results caution against protocols that involve frequent dosing.

VI. MODIFYING THE PLASMID DNA VECTOR TO DECREASE THE CPG RESPONSE

A. Methylation

Plasmid DNA can be methylated readily *in vitro* using the prokaryotic CpG methylase (Sss I methylase), which adds a methyl group to the cytosine residue at the C5 position within the sequence 5′-CG-3′. This modification can also occur *in vivo* if one uses a strain of *E. coli* that harbors a second plasmid expressing the Sss I methylase gene (Reyes-Sandoval and Ertl, 2004). As described above, methylated plasmid DNA induces significantly lower levels of proinflammatory cytokines compared to unmethylated DNA. Methylated DNA does not activate TLR9 signaling or stimulate spleen cells *in vitro*, and in fact, methylated DNA has been shown to inhibit at least some aspects of the inflammatory response both *in vitro* and *in vivo* (Chen *et al.*, 2001). This effect was observed when *E. coli* DNA and either calf thymus DNA, which is naturally predominantly methylated, or *in vitro* methylated plasmid DNA were applied together to spleen cells or co-injected intraperitoneally into mice. Significantly lower levels of IFN-γ or TNF-α were observed compared to the levels induced by *E.coli* DNA alone.

However, global CpG methylation of plasmid DNA often inhibits transgene expression quite significantly. Most promoters are negatively regulated by CpG methylation, although a few, such as the MMTV-LTR or SV40 early promoter, are relatively insensitive (Muiznieks and Doerfler, 1994). These latter promoters, however, possess relatively weak transcriptional activity and thus have little utility for most gene therapy applications. Furthermore, methylation of sequences lying outside of the promoter, such as the vector backbone or transgene, also can inhibit expression, likely through promoting the formation of a less active chromatin structure.

Lastly, methylated CpG DNA does not appear to be equivalent to DNA that does not contain CpG motifs, since methylated DNA still induces some of the adverse hematological changes (e.g., lymphopenia, thrombocytopenia) that were observed with unmethylated DNA complexes (Tousignant *et al.*, 2000).

B. Eliminating CpG motifs

Given the limitations of methylation, the most direct approach to reduce CpG-mediated responses is to eliminate CpGs from the plasmid vector. With this strategy, however, questions arise as to how many and which CpGs to remove. It was thought initially that eliminating only the most immunostimulatory CpG motifs, which are sequences consisting of 5'-purine-purine-CG-pyrimidine-pyrimidine-3', would be sufficient to see an effect. However, eliminating this small subset of all the CpGs in the vector did not noticeably alter the inflammatory response to cationic lipid-DNA complexes (Yew *et al.*, 1999). The choice of which CpGs to remove is also complicated by the finding that CpG motifs that are optimally stimulatory in mice are different than those in humans. Therefore, a vector that has been modified to be less immunostimulatory in rodents may still be stimulatory in humans. Lastly, CpGs in certain sequence contexts, usually consisting of direct repeats or clusters of C and G, have been shown to be suppressive, actually blocking the activity of neighboring immumostimulatory CpGs. Removing these suppressive CpGs has been shown to increase the Th1-type adjuvant effect of an antigen expressing plasmid vector (Krieg et al., 1998).

With these complications in mind, we nevertheless decided to eliminate as many CpGs from the plasmid vector as possible. Site-directed mutagenesis would be far too laborious to eliminate the greater than 200 CpGs present in a typical plasmid. Instead, plasmids were assembled using gene synthesis techniques that can generate any designed sequence. Within the regions encoding the transgene and the antibiotic resistance gene, the degeneracy of the genetic code can be exploited to eliminate CpGs. Except for methionine and tryptophan, all the amino acids are encoded by more than one codon, and thus through the appropriate choice of coding triplet all CpGs can be removed without altering the amino acid sequence. Within an intron, CpGs can be modified readily so long as functional splice donor, splice acceptor, and lariat branch point sites are preserved. Within the polyadenylation signal sequence, the few CpGs can be removed without consequence.

Within the promoter and enhancer sequence, removing CpGs has unpredictable effects on transcriptional activity. For example, removing CpGs from the promoter expressing the cellular ubiquitin B gene abolishes its activity, whereas removing CpGs from the promoter expressing the human cytomegalovirus (CMV) immediate early gene reduces its activity by less than 50% (Fig. 7.2A). The presence of multiple copies of several transcription factor binding sites may render the human CMV promoter less sensitive to alterations in its sequence. At present, the effect of removing CpGs is determined empirically, with some promoters requiring less modification than others. The murine CMV promoter, for example, has extensive regions of non-CpG sequence within its long enhancer region (Dorsch-Hasler et al., 1985), and a non-CpG version of this promoter has been incorporated recently into a CpG-free plasmid vector (http://www.invivogen.com/CpG/CpG_overview.htm, 2004).

C. CpG-free vectors

Until recently, the only CpGs that could not be eliminated from a given plasmid DNA vector were those that resided within the region involved in plasmid DNA replication (see Fig. 7.2B). Most vectors use a high-copy origin derived from pMB1, in which a 553 nucleotide RNA transcript, RNA II, serves to initiate replication. This RNA contains a high frequency of CpGs and forms complex stem-loop structures that are important for its function (Tomizawa and Itoh, 1982). Thus, while some CpGs can be modified, it is unlikely that this region can be made completely CpG-free. However, other plasmids, such as pSC101, P1, F, and R6K, use a protein rather than an RNA to initiate replication. The γ origin of replication from R6K has been incorporated into a

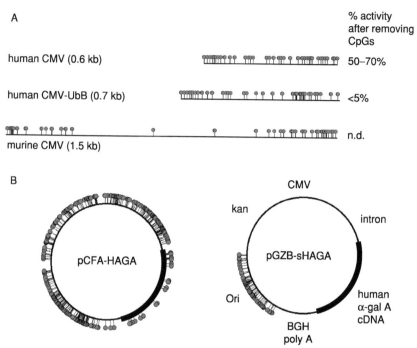

Figure 7.2. Eliminating CpGs from plasmid DNA vectors. A) Density and distribution of CpGs in different promoters. CMV, human CMV enhancer-promoter; CMV-Ub, human CMV enhancer-human ubiquitin B promoter hybrid; murine CMV, murine CMV enhancer-promoter; n.d., not determined. Shown is the approximate percent promoter activity (based on *in vivo* reporter gene expression studies) remaining after eliminating all CpGs in the sequence relative to the unmodified sequence. Each symbol represents one CpG (CpGs from both strands of the double-stranded DNA are depicted) B) Unmodified (pCFA) and CpG-depleted (pGZB) vectors expressing human α-galactosidase A (HAGA). CMV, human CMV enhancer-promoter; BGH poly A, bovine growth hormone polyadenylation signal; Ori, bacterial replication origin; kan, kanamycin resistance gene.

completely CpG-free plasmid and is commercially available. This plasmid is propagated in an *E. coli* strain in which a high-copy mutant pir gene, which expresses the replication initiator pi protein, has been integrated into the bacterial genome (Wu *et al.*, 1995).

D. Minimal plasmid vectors

The alternate approach to modifying individual CpGs is simply to excise those regions of the plasmid that are not required to express the transgene. This strategy both removes unwanted prokaryotic sequence and makes the vector smaller and potentially more efficient in transducing cells. Linear DNA molecules consisting of only the expression cassette (e.g., promoter-intron-transgene-polyadenylation signal) have been generated by either restriction endonuclease digestion or by PCR amplification. Such linear structures tend to yield lower levels of expression compared to supercoiled DNA, although Hofman *et al.* (2001) observed comparable levels of expression when a PCR amplified cassette (in a lipid-protamine-DNA complex) was delivered intravenously. Others have ligated short stem-loop-forming oligonucleotides to the ends of the linear cassettes resulting in a capped DNA molecule that is protected from exonuclease digestion (Schakowski *et al.*, 2001).

Minimal plasmids can also be generated by site-specific recombination. Darquet *et al.* (1999) exploited the bacteriophage λ integrase, which normally mediates integration between the attP site in the phage genome and the attB site in the *E. coli* chromosome. A plasmid harboring the attP and attB sites flanking an expression cassette was propagated in an *E. coli* strain that contained a thermoinducible λ integrase. Induction of the integrase generated two super-coiled DNA molecules, one consisting of the vector backbone, and the other, the minicircle containing the expression cassette. More recently, minicircles have also been generated successfully by placing the Streptomyces phage ØC31 integrase under the control of an arabinose inducible promoter, with minimal attP and attB sites in the plasmid vector (Chen *et al.*, 2003).

Such minimal vectors, either the linear or supercoiled versions, exhibit reduced inflammatory activity and increased persistence of expression (Chen *et al.*, 2003; Hofman *et al.*, 2001). It is now quite conceivable to incorporate CpG sequence modification within the minicircle vectors to generate a completely non-CpG minimal vector.

E. Limitations of CpG elimination

While much of the toxicity of cationic lipid-DNA complexes is CpG-mediated, a significant portion is likely not, as there appear to be TLR9-independent immune responses to prokaryotic DNA and lipid-DNA complexes. For example, methylating mycobacterial DNA does not inhibit its ability to stimulate IL-12 synthesis from monocytes and macrophages (Filion *et al.*, 2000). Secondly, methylated bacterial DNA is able to activate human neutrophils, and this activation is not inhibited by wortmannin or chloroquine, two drugs that inhibit TLR9 signaling (Trevani *et al.*, 2003). Moreover, immobilized bacterial DNA is still able to stimulate neutrophils, suggesting that uptake of the DNA into the cell and signaling through the internal TLR9 receptor is not required. The results imply that the host innate immune system is able to recognize prokaryotic DNA by an additional feature other than unmethylated CpG motifs.

Other investigators have observed TLR9 independent stimulation by cationic-lipid DNA complexes. Yasuda *et al.* (2002) found that complexes containing methylated plasmid DNA are able to induce TNF-α and IL-6 production in a macrophage cell line and from mouse peritoneal macrophages. Suzuki *et al.* (1999) showed that MHC class I expression, as well as expression of other genes required for antigen processing and presentation, are induced after transfecting thyroid cells with lipofectamine-DNA complexes. This effect was also shown in a variety of other cell lines, and appears to occur regardless of the methylation status or CpG content of the DNA.

Cationic lipid-DNA complexes have also been administered to transgenic mice that lack TLR9 (Zhao *et al.*, 2004). Splenocytes, macrophages and dendritic cells from these mice are completely non-responsive to CpG-containing DNA, and the mice are resistant to CpG-induced toxic shock (Hemmi *et al.*, 2000). While wild-type mice die within 12 h after being given D-galactosamine and CpG DNA, all the TLR9-/-mice survive, with no elevations of TNF-α, IL-6, and IL-12 in the serum. The absence of CpG signaling in these mice makes them useful in delineating the TLR9-independent acute toxic responses that remain after administering cationic lipid-DNA complexes.

As expected, TLR9-/-mice injected with cationic lipid-DNA complex exhibit a greatly reduced inflammatory response, with minimal to no induction of TNF-α, IL-1, IL-12, RANTES, and IFN-γ (Zhao *et al.*, 2004). Levels of the liver transaminases ALT and AST are significantly decreased compared to wild-type mice. Most importantly, the TLR9-/-mice are tolerant to doses of complex that are lethal in normal mice (Fig. 7.3). Unlike C57BL/6 mice, which have signs of ill effects from injecting cationic lipid GL-62-DNA complex even at lower doses, TLR9-/- mice behave and appear similar to naïve animals even at moderate doses (66 μg DNA).

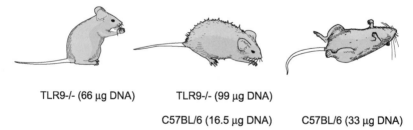

TLR9-/- (66 µg DNA)   TLR9-/- (99 µg DNA)

C57BL/6 (16.5 µg DNA)   C57BL/6 (33 µg DNA)

Figure 7.3. Increased tolerance and survival of TLR9 knockout mice after systemic delivery of cationic lipid DNA complex. Complexes were formed with cationic lipid GL-62 and pCF1-CAT plasmid DNA at doses ranging from 0.5:0.5 mM lipid:nucleotides of DNA (16.5 μg DNA) to 3:3 mM (99 μg DNA). Shown is the appearance of mice at different doses. Fifty percent of the mice that received a 1:1 mM dose (33 μg DNA) died within 24 h of receiving complex. Data from Zhao *et al.* (2004).

However, the lymphopenia and thrombocytopenia observed in normal mice also occurs in TLR9-/-mice, and at a high dose of complex (99 μg DNA), the majority of the TLR9-/-mice survive but appear scruffy, indicative of a CpG-independent inflammatory response (Zhao *et al.*, 2004). Additional studies, such as RNA expression analysis using microarrays, may be useful in elucidating the signaling pathways involved in these non-TLR9 mediated inflammatory responses.

VII. OTHER APPROACHES TO REDUCE THE TOXICITY OF CATIONIC LIPID-DNA COMPLEXES

A. Aerosol delivery

The route of delivering cationic lipid-DNA complexes has a major effect on the acute toxic response. Compared to systemic administration, intrapulmonary delivery is considered to be less problematic and has the advantage of being noninvasive and more amenable to repeat dosing. Delivery using an aerosolized form of the complex allows one to target the enormous surface area of the lung (approximately 100 m^2 in humans), which reduces the local concentration of complex on the lung epithelium when compared to using a bolus of liquid, and thereby reducing potential toxicity (Densmore, 2003; Eastman *et al.*, 1997). This theory has been borne out in studies showing that aerosolized cationic lipid-DNA complexes produce equivalent levels of expression but significantly reduced inflammation compared to instilling a bolus of the same amount of complex (Eastman *et al.*, 1997).

B. Immune suppression

A second approach to reduce complex-mediated toxicity is to dampen the host immune response. This can be achieved using glucocorticoids, which regulate the expression of many genes involved in the inflammatory response. Glucocorticoids are powerful immune modulators that inhibit the production of cytokines, prostaglandins, and nitric oxide, inhibit T-cell activation and proliferation, inhibit macrophage activation, and inhibit inflammatory cell migration by reducing expression of adhesion molecules (Janeway *et al.*, 1999). Dexamethasone, a synthetic glucocorticoid, has been shown to reduce the inflammatory response to IV delivery of cationic lipid-DNA complexes (Tan *et al.*, 1999). In addition, dexamethasone and other steroids such as β-estradiol and methylprednisolone increase transgene expression several fold, and also reduce the length of the refractory period that prevents effective re-administration of complex (Wiseman *et al.*, 2001).

While usually given prior to injection, dexamethasone has also been co-delivered with the complex. This approach has the advantage of targeting the glucocorticoid to the same sites within an organ or tissue that is receiving the lipid-DNA complex. Liu *et al.* (2004) encapsulated dexamethasone into DOTAP liposomes (at a weight/weight ratio of 1/10 dexamethasone to DOTAP) and then formed a complex with DNA. When injected IV, these DEX/DOTAP/DNA complexes induce significantly less TNF-α, IL-12, and IFN-γ compared to DOTAP/DNA complexes. Expression in the lung using the DEX/DOTAP/DNA complexes was similar to the expression using DOTAP/DNA complexes. Gruneich *et al.* (2004) synthesized a form of dexamethasone that can be used as a transfection reagent itself, while still retaining partial steroidal activity. Spermine was conjugated to dexamethasone to form a cationic steroid that can be complexed with DNA. When instilled into the lung, the dexamethasone-spermine-DNA complexes induced lower amounts of IFN-γ and an approximately 3–7-fold increase in expression compared to DC-Chol complexes.

Antibodies against cytokines or cell-surface molecules can also be used to inhibit the inflammatory response. A mixture of anti-IFN-γ and anti-TNF-α antibodies, when injected intravenously prior to injecting complex into mice, increase transgene expression 5-fold as a result of blocking the activity of these cytokines (Li *et al.*, 1999b). A mixture of two antibodies directed against LFA-1 and Mac-1a, two integrins present on the surface of neutrophils, has also been injected into mice just prior to instilling complex into the lung (Yew *et al.*, 1999). Reduced neutrophil influx into the lung and reduced levels of TNF-α, IFN-γ, and IL-12 were subsequently detected in the bronchoalveolar lavage fluid, and expression increased 4-fold.

C. Effect of free liposomes

A third approach to reduce toxicity involves first injecting free liposomes followed by either cationic lipid-DNA complexes or naked DNA. This idea originated from studies optimizing cationic lipid formulations such as DOTMA for systemic administration, which were found to require a high cationic lipid-to-DNA ratio for optimal transfection after IV delivery (Song and Liu, 1998). Complexes were formed at a lower cationic lipid/DNA ratio that is suboptimal, and free liposomes were preinjected one minute prior to injecting complex. Gene expression increased with increasing amounts of free liposomes, indicating that the determining factor was the total amount of cationic lipid and DNA injected. The proposed mechanism is that the free liposomes aggregate with serum proteins and become entrapped in the microvasculature of the lung. This in turn increases the retention time of the complex injected subsequently (Song *et al.*, 1998). Tan *et al.* (2001) then eliminated formulating complexes altogether by injecting naked DNA 2 to 20 min after injecting free cationic liposomes. This method decreased the levels of cytokines TNF-α, IL-12, and IFN-γ by 50–80%, reduced the loss of platelets and complement activity, and decreased elevations of the liver transaminases ALT and AST. Sequential injection, however, did not significantly alter the loss of lymphocytes and monocytes compared to complex injection. When the interval between the first and second injection occurred within 5 min, a 2–5-fold increase in transgene expression was observed relative to injecting complex alone. How transfection is occurring using this procedure is not known, but it is considered unlikely that the naked DNA is complexing with the cationic liposomes in the blood, since the liposomes would already be aggregated with serum proteins. Rather, it is postulated that the slowing of the pulmonary circulation may allow the naked DNA to associate with the endothelium and be taken up by endocytosis.

Elouahabi *et al.* (2003) have also shown that free liposomes reduce the inflammatory response when injected alone prior to injecting cationic lipid-DNA complexes. However, the kinetics and therefore the mechanism appear to be different than from previous studies. They injected the cationic lipid diC14-amidine prior to injecting complexes and observed an approximately 40-fold increase in expression compared to injecting complex alone. An interval of 30–60 min was required to see the maximal effect. It has been proposed that the diC14 liposomes are taken up by macrophages and inhibit their ability to respond to CpG DNA and release TNF-α (Elouahabi *et al.*, 2003).

D. Novel liposomes and formulations with reduced toxicity

Investigators continue to develop new liposomes, polycations and formulations to increase the efficiency of gene transfer. Some have been designed to reduce their interaction with plasma and prolong their survival in the circulation, while others aim to target the liver, brain, or tumors. Since many of these new complexes are reviewed elsewhere in this book, only a few examples in which some toxicity data have been reported will be described here.

An example of a novel formulation for pulmonary delivery has been described by Jenkins *et al.* (2000, 2003). They formed small (100 nm) complexes composed of Lipofectin, an integrin targeting peptide, and plasmid DNA. Using a relatively low dose (8 μg DNA), intratracheal instillation results in only mild changes in cytokine and cell profiles in the bronchoalveolar fluid, with no alteration in alveolar architecture even after multiple dosing. The ability of these complexes to express therapeutic levels of protein remains to be determined.

Other ternary complexes have been developed that have reduced interaction with serum proteins. Trubetskoy *et al.* (2003) added the synthetic polyanion polyacrylic acid (pAA) to either linear PEI-DNA or cationic lipid-DNA complexes. The addition of pAA increased expression and decreased toxicity compared to the binary complex. Linear PEI-pAA-DNA (50 μg DNA dose) complexes still cause substantial toxicity when injected IV, resulting in edema, congestion, and disintegration of alveoli in the lung, and necrosis in the liver. However, injecting a pAA "chaser" 30 min after injecting the ternary complex reduced toxicity in the lung and liver, as measured histologically (Fig. 7.4) and by the decrease in ALT levels. The addition of pAA is thought to electrostatically shield the complex from opsonizing serum proteins.

Polyethylene glycol (PEG) has also been used to reduce interactions with serum (Finsinger et al., 2000; Ogris et al., 1999). Zhang et al. (2003) encapsulated plasmid DNA within small (85 nm) neutral liposomes. The liposome surface was then pegylated with moderate molecular weight (2000 Da) PEG. To facilitate transfer across the blood–brain barrier, antibodies to the transferrin receptor were chemically conjugated to approximately 1–2% of the PEG. Rats were injected weekly for six weeks and sacrificed 3 days after the last dose, and although the timing was too late to measure the acute response, the

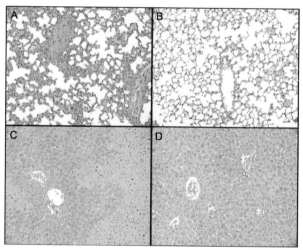

Figure 7.4. Haemotoxylin and eosin stained sections of lungs (A and B) and livers (C and D) from mice 24 h after injecting linear PEI/pAA/DNA (400 μg/50 μg/50 μg) complexes without pAA "chaser" (B and D) and linear PEI/pAA/DNA complexes (240 μg/40 μg/40 μg) with 1.5 mg of pAA "chaser" (A and C) injected 30 min after injecting complex. Magnification for all panels is 100X. Although equivalent levels of luciferase were expressed from the mice, significantly reduced toxicity was observed in the mice that received the pAA chaser. Reprinted with permission from Gene Therapy, Trubetskoy et al. (2003), copyright 2003, Macmillan Publishers Ltd. (See Color Insert.)

rats appeared normal based on serum chemistries and organ histology. This is not surprising given the low dose of DNA injected (5 μg/rat). What is unexpected is the widespread expression in the brain in both rodents and primates using these immunoliposome complexes, as well as efficacy in a rat model of Parkinson's disease (Zhang *et al.*, 2004).

VIII. SUMMARY

Although there are diseases, for example, cancer, where the inflammation induced by cationic lipid-DNA complexes is actually beneficial (Dow *et al.*, 1999; Rudginsky *et al.*, 2001), for most applications the inflammatory responses are dose-limiting, restricting the levels of expression that can be achieved and thus reducing efficacy. Much of the problem stems from the cationic lipid-mediated delivery of CpG-containing DNA into immune cells and their subsequent activation, although a portion of the response is independent of CpG signaling. New liposomes and formulations that avoid opsonization by complement and uptake into DCs and Kupffer cells will greatly improve the toxicity profile. Minimized plasmid vectors that are entirely devoid of CpGs, which is quite feasible, will eliminate the toxicity arising from CpG stimulation. Certainly most beneficial, however, will be complexes that have improved transduction efficiency and that require a lower dose of lipid and DNA to achieve efficacy. There are many novel complexes that show promise in these areas, and rigorous testing of their acute and long-term effects *in vivo* will be required to realize fully their potential for clinical use.

References

Akira, S., Takeda, K., and Kaisho, T. (2001). Toll-like receptors: Critical proteins linking innate and acquired immunity. *Nat. Immunol.* **2**, 675–680.

Alton, E. W., Stern, M., Farley, R., Jaffe, A., Chadwick, S. L., Phillips, J., Davies, J., Smith, S. N., Browning, J., Davies, M. G., Hodson, M. E., Durham, S. R., Li, D., Jeffery, P. K., Scallan, M., Balfour, R., Eastman, S. J., Cheng, S. H., Smith, A. E., Meeker, D., and Geddes, D. M. (1999). Cationic lipid-mediated CFTR gene transfer to the lungs and nose of patients with cystic fibrosis: A double-blind placebo-controlled trial. *Lancet* **353**, 947–954.

Audouy, S. A., de Leij, L. F., Hoekstra, D., and Molema, G. (2002). *In vivo* characteristics of cationic liposomes as delivery vectors for gene therapy. *Pharm. Res.* **19**, 1599–1605.

Barron, L. G., Meyer, K. B., and Szoka, F. C. Jr. (1998). Effects of complement depletion on the pharmacokinetics and gene delivery mediated by cationic lipid-DNA complexes. *Hum. Gene Ther.* **9**, 315–323.

Chadwick, S. L., Kingston, H. D., Stern, M., Cook, R. M., O'Connor, B. J., Lukasson, M., Balfour, R. P., Rosenberg, M., Cheng, S. H., Smith, A. E., Meeker, D. P., Geddes, D. M., and Alton, E. W. (1997). Safety of a single aerosol administration of escalating doses of the cationic lipid GL-67/DOPE/DMPE-PEG5000 formulation to the lungs of normal volunteers. *Gene Ther.* **4**, 937–942.

Chen, Y., Lenert, P., Weeratna, R., McCluskie, M., Wu, T., Davis, H. L., and Krieg, A. M. (2001). Identification of methylated CpG motifs as inhibitors of the immune stimulatory CpG motifs. *Gene Ther.* **8,** 1024–1032.

Chen, Z. Y., He, C. Y., Ehrhardt, A., and Kay, M. A. (2003). Minicircle DNA vectors devoid of bacterial DNA result in persistent and high-level transgene expression *in vivo. Mol. Ther.* **8,** 495–500.

Conant, S. B., and Swanborg, R. H. (2004). Autoreactive T cells persist in rats protected against experimental autoimmune encephalomyelitis and can be activated through stimulation of innate immunity. *J. Immunol.* **172,** 5322–5328.

Cowdery, J. S., Chace, J. H., Yi, A. K., and Krieg, A. M. (1996). Bacterial DNA induces NK cells to produce IFN-gamma *in vivo* and increases the toxicity of lipopolysaccharides. *J. Immunol.* **156,** 4570–4575.

Darquet, A. M., Rangara, R., Kreiss, P., Schwartz, B., Naimi, S., Delaere, P., Crouzet, J., and Scherman, D. (1999). Minicircle: An improved DNA molecule for *in vitro* and *in vivo* gene transfer. *Gene Ther.* **6,** 209–218.

Densmore, C. L. (2003). The re-emergence of aerosol gene delivery: A viable approach to lung cancer therapy. *Curr. Cancer Drug Targets* **3,** 275–286.

Dorsch-Hasler, K., Keil, G. M., Weber, F., Jasin, M., Schaffner, W., and Koszinowski, U. H. (1985). A long and complex enhancer activates transcription of the gene coding for the highly abundant immediate early mRNA in murine cytomegalovirus. *Proc. Natl. Acad. Sci. USA* **82,** 8325–8329.

Dow, S. W., Fradkin, L. G., Liggitt, D. H., Willson, A. P., Heath, T. D., and Potter, T. A. (1999). Lipid-DNA complexes induce potent activation of innate immune responses and antitumor activity when administered intravenously. *J. Immunol.* **163,** 1552–1561.

Eastman, S. J., Lukason, M. J., Tousignant, J. D., Murray, H., Lane, M. D., St George, J. A., Akita, G. Y., Cherry, M., Cheng, S. H., and Scheule, R. K. (1997). A concentrated and stable aerosol formulation of cationic lipid: DNA complexes giving high-level gene expression in mouse lung. *Hum. Gene Ther.* **8,** 765–773.

Eliyahu, H., Servel, N., Domb, A. J., and Barenholz, Y. (2002). Lipoplex-induced hemagglutination: Potential involvement in intravenous gene delivery. *Gene Ther.* **9,** 850–858.

Elouahabi, A., Flamand, V., Ozkan, S., Paulart, F., Vandenbranden, M., Goldman, M., and Ruysschaert, J. M. (2003). Free cationic liposomes inhibit the inflammatory response to cationic lipid-DNA complex injected intravenously and enhance its transfection efficiency. *Mol. Ther.* **7,** 81–88.

Emerson, M., Renwick, L., Tate, S., Rhind, S., Milne, E., Painter, H. A., Boyd, A. C., McLachlan, G., Griesenbach, U., Cheng, S. H., Gill, D. R., Hyde, S. C., Baker, A., Alton, E. W., Porteous, D. J., and Collie, D. D. (2003). Transfection efficiency and toxicity following delivery of naked plasmid DNA and cationic lipid-DNA complexes to ovine lung segments. *Mol. Ther.* **8,** 646–653.

Epstein, J. E., Charoenvit, Y., Kester, K. E., Wang, R., Newcomer, R., Fitzpatrick, S., Richie, T. L., Tornieporth, N., Heppner, D. G., Ockenhouse, C., Majam, V., Holland, C., Abot, E., Ganeshan, H., Berzins, M., Jones, T., Freydberg, C. N., Ng, J., Norman, J., Carucci, D. J., Cohen, J., and Hoffman, S. L. (2004). Safety, tolerability, and antibody responses in humans after sequential immunization with a PfCSP DNA vaccine followed by the recombinant protein vaccine RTS, S/AS02A. *Vaccine* **22,** 1592–1603.

Filion, M. C., Filion, B., Reader, S., Menard, S., and Phillips, N. C. (2000). Modulation of interleukin-12 synthesis by DNA lacking the CpG motif and present in a mycobacterial cell wall complex. *Cancer Immunol. Immunother.* **49,** 325–334.

Finsinger, D., Remy, J. S., Erbacher, P., Koch, C., and Plank, C. (2000). Protective copolymers for non-viral gene vectors: Synthesis, vector characterization and application in gene delivery. *Gene Ther.* **7,** 1183–1192.

Freimark, B. D., Blezinger, H. P., Florack, V. J., Nordstrom, J. L., Long, S. D., Deshpande, D. S., Nochumson, S., and Petrak, K. L. (1998). Cationic lipids enhance cytokine and cell influx levels in the lung following administration of plasmid: Cationic lipid complexes. *J. Immunol.* **160,** 4580–4586.

Gilkeson, G. S., Grudier, J. P., Karounos, D. G., and Pisetsky, D. S. (1989). Induction of anti-double stranded DNA antibodies in normal mice by immunization with bacterial DNA. *J. Immunol.* **142,** 1482–1486.

Griesenbach, U., Geddes, D. M., and Alton, E. W. (2004). Advances in cystic fibrosis gene therapy. *Curr. Opin. Pulm. Med.* **10,** 542–546.

Gruneich, J. A., Price, A., Zhu, J., and Diamond, S. L. (2004). Cationic corticosteroid for non-viral gene delivery. *Gene Ther.* **11,** 668–674.

Heikenwalder, M., Polymenidou, M., Junt, T., Sigurdson, C., Wagner, H., Akira, S., Zinkernagel, R., and Aguzzi, A. (2004). Lymphoid follicle destruction and immunosuppression after repeated CpG oligodeoxynucleotide administration. *Nature Med.* **10,** 187–192. Epub 2004 Jan 2025.

Hemmi, H., Takeuchi, O., Kawai, T., Kaisho, T., Sato, S., Sanjo, H., Matsumoto, M., Hoshino, K., Wagner, H., Takeda, K., and Akira, S. (2000). A Toll-like receptor recognizes bacterial DNA. *Nature* **408,** 740–745.

Hofman, C. R., Dileo, J. P., Li, Z., Li, S., and Huang, L. (2001). Efficient *in vivo* gene transfer by PCR amplified fragment with reduced inflammatory activity. *Gene Ther.* **8,** 71–74.

Hornung, V., Rothenfusser, S., Britsch, S., Krug, A., Jahrsdorfer, B., Giese, T., Endres, S., and Hartmann, G. (2002). Quantitative expression of toll-like receptor 1–10 mRNA in cellular subsets of human peripheral blood mononuclear cells and sensitivity to CpG oligodeoxynucleotides. *J. Immunol.* **168,** 4531–4537.

http://www.invivogen.com/CpG/CpG_overview.htm 2004. Company website. [Online.]

Hyde, S. C., Southern, K. W., Gileadi, U., Fitzjohn, E. M., Mofford, K. A., Waddell, B. E., Gooi, H. C., Goddard, C. A., Hannavy, K., Smyth, S. E., Egan, J. J., Sorgi, F. L., Huang, L., Cuthbert, A. W., Evans, M. J., Colledge, W. H., Higgins, C. F., Webb, A. K., and Gill, D. R. (2000). Repeat administration of DNA/liposomes to the nasal epithelium of patients with cystic fibrosis. *Gene Ther.* **7,** 1156–1165.

Janeway, C., Travers, P., Walport, M., and Capra, J. D. (1999). "Immunobiology: The Immune System in Health and Disease," 4th ed. Garland Publishing, New York.

Jenkins, R. G., Herrick, S. E., Meng, Q. H., Kinnon, C., Laurent, G. J., McAnulty, R. J., and Hart, S. L. (2000). An integrin-targeted non-viral vector for pulmonary gene therapy. *Gene Ther.* **7,** 393–400.

Jenkins, R. G., Meng, Q. H., Hodges, R. J., Lee, L. K., Bottoms, S. E., Laurent, G. J., Willis, D., Ayazi Shamlou, P., McAnulty, R. J., and Hart, S. L. (2003). Formation of LID vector complexes in water alters physicochemical properties and enhances pulmonary gene expression *in vivo. Gene Ther.* **10,** 1026–1034.

Krieg, A. M. (2003). CpG DNA: Trigger of sepsis, mediator of protection, or both? *Scand. J. Infect. Dis.* **35,** 653–659.

Krieg, A. M. (2002a). CpG motifs in bacterial DNA and their immune effects. *Annu. Rev. Immunol.* **20,** 709–760.

Krieg, A. M. (2002b). A role for Toll in autoimmunity. *Nat. Immunol.* **3,** 423–424.

Krieg, A. M., Wu, T., Weeratna, R., Efler, S. M., Love-Homan, L., Yang, L., Yi, A. K., Short, D., and Davis, H. L. (1998). Sequence motifs in adenoviral DNA block immune activation by stimulatory CpG motifs. *Proc. Natl. Acad. Sci. USA* **95,** 12631–12636.

Latz, E., Schoenemeyer, A., Visintin, A., Fitzgerald, K. A., Monks, B. G., Knetter, C. F., Lien, E., Nilsen, N. J., Espevik, T., and Golenbock, D. T. (2004). TLR9 signals after translocating from the ER to CpG DNA in the lysosome. *Nat. Immunol.* **5,** 190–198. Epub 2004 Jan 2011.

Leadbetter, E. A., Rifkin, I. R., Hohlbaum, A. M., Beaudette, B. C., Shlomchik, M. J., and Marshak-Rothstein, A. (2002). Chromatin-IgG complexes activate B cells by dual engagement of IgM and Toll-like receptors. *Nature* **416**, 603–607.

Li, S., Tseng, W. C., Stolz, D. B., Wu, S. P., Watkins, S. C., and Huang, L. (1999a). Dynamic changes in the characteristics of cationic lipidic vectors after exposure to mouse serum: Implications for intravenous lipofection. *Gene Ther.* **6**, 585–594.

Li, S., Wu, S. P., Whitmore, M., Loeffert, E. J., Wang, L., Watkins, S. C., Pitt, B. R., and Huang, L. (1999b). Effect of immune response on gene transfer to the lung via systemic administration of cationic lipidic vectors. *Am. J. Physiol.* **276**, L796–L804.

Li, X. L., Boyanapalli, M., Weihua, X., Kalvakolanu, D. V., and Hassel, B. A. (1998). Induction of interferon synthesis and activation of interferon-stimulated genes by liposomal transfection reagents. *J. Interferon Cytokine Res.* **18**, 947–952.

Litzinger, D. C., Brown, J. M., Wala, I., Kaufman, S. A., Van, G. Y., Farrell, C. L., and Collins, D. (1996). Fate of cationic liposomes and their complex with oligonucleotide *in vivo*. *Biochim. Biophys. Acta* **1281**, 139–149.

Liu, F., Shollenberger, L. M., and Huang, L. (2004). Non-immunostimulatory non-viral vectors. *Faseb J.* **18**, 1779–1781. Epub 2004 Sep 1713.

Liu, Y., Liggitt, D., Zhong, W., Tu, G., Gaensler, K., and Debs, R. (1995). Cationic liposome-mediated intravenous gene delivery. *J. Biol. Chem.* **270**, 24864–24870.

Loisel, S., Le Gall, C., Doucet, L., Ferec, C., and Floch, V. (2001). Contribution of plasmid DNA to hepatotoxicity after systemic administration of lipoplexes. *Hum. Gene Ther.* **12**, 685–696.

Luton, D., Oudrhiri, N., de Lagausie, P., Aissaoui, A., Hauchecorne, M., Julia, S., Oury, J. F., Aigrain, Y., Peuchmaur, M., Vigneron, J. P., Lehn, J. M., and Lehn, P. (2004). Gene transfection into fetal sheep airways in utero using guanidinium-cholesterol cationic lipids. *J. Gene Med.* **6**, 328–336.

Mahato, R. I., Anwer, K., Tagliaferri, F., Meaney, C., Leonard, P., Wadhwa, M. S., Logan, M., French, M., and Rolland, A. (1998). Biodistribution and gene expression of lipid/plasmid complexes after systemic administration. *Hum. Gene Ther.* **9**, 2083–2099.

McLachlan, G., Stevenson, B. J., Davidson, D. J., and Porteous, D. J. (2000). Bacterial DNA is implicated in the inflammatory response to delivery of DNA/DOTAP to mouse lungs. *Gene Ther.* **7**, 384–392.

McLean, J. W., Fox, E. A., Baluk, P., Bolton, P. B., Haskell, A., Pearlman, R., Thurston, G., Umemoto, E. Y., and McDonald, D. M. (1997). Organ-specific endothelial cell uptake of cationic liposome-DNA complexes in mice. *Am. J. Physiol.* **273**, H387–H404.

Mor, G., Singla, M., Steinberg, A. D., Hoffman, S. L., Okuda, K., and Klinman, D. M. (1997). Do DNA vaccines induce autoimmune disease? *Hum. Gene Ther.* **8**, 293–300.

Muiznieks, I., and Doerfler, W. (1994). The impact of 5'-CG-3' methylation on the activity of different eukaryotic promoters: A comparative study. *FEBS Lett.* **344**, 251–254.

Niidome, T., and Huang, L. (2002). Gene Therapy Progress and Prospects: Non-viral vectors. *Gene Ther.* **9**, 1647–1652.

Ogris, M., Brunner, S., Schuller, S., Kircheis, R., and Wagner, E. (1999). PEGylated DNA/transferrin-PEI complexes: Reduced interaction with blood components, extended circulation in blood and potential for systemic gene delivery. *Gene Ther.* **6**, 595–605.

Plank, C., Mechtler, K., Szoka, F. C. Jr., and Wagner, E. (1996). Activation of the complement system by synthetic DNA complexes: A potential barrier for intravenous gene delivery. *Hum. Gene Ther.* **7**, 1437–1446.

Reinish, L. W., Bally, M. B., Loughrey, H. C., and Cullis, P. R. (1988). Interactions of liposomes and platelets. *Thromb. Haemost.* **60**, 518–523.

Reyes-Sandoval, A., and Ertl, H. C. (2004). CpG methylation of a plasmid vector results in extended transgene product expression by circumventing induction of immune responses. *Mol. Ther.* **9**, 249–261.

Rudginsky, S., Siders, W., Ingram, L., Marshall, J., Scheule, R., and Kaplan, J. (2001). Antitumor activity of cationic lipid complexed with immunostimulatory DNA. *Mol. Ther.* **4**, 347–355.

Ruiz, F. E., Clancy, J. P., Perricone, M. A., Bebok, Z., Hong, J. S., Cheng, S. H., Meeker, D. P., Young, K. R., Schoumacher, R. A., Weatherly, M. R., Wing, L., Morris, J. E., Sindel, L., Rosenberg, M., van Ginkel, F. W., McGhee, J. R., Kelly, D., Lyrene, R. K., and Sorscher, E. J. (2001). A clinical inflammatory syndrome attributable to aerosolized lipid-DNA administration in cystic fibrosis. *Hum. Gene Ther.* **12**, 751–761.

Sakurai, F., Terada, T., Yasuda, K., Yamashita, F., Takakura, Y., and Hashida, M. (2002). The role of tissue macrophages in the induction of proinflammatory cytokine production following intravenous injection of lipoplexes. *Gene Ther.* **9**, 1120–1126.

Schakowski, F., Gorschluter, M., Junghans, C., Schroff, M., Buttgereit, P., Ziske, C., Schottker, B., Konig-Merediz, S. A., Sauerbruch, T., Wittig, B., and Schmidt-Wolf, I. G. (2001). A novel minimal-size vector (MIDGE) improves transgene expression in colon carcinoma cells and avoids transfection of undesired DNA. *Mol. Ther.* **3**, 793–800.

Scheule, R. K., St George, J. A., Bagley, R. G., Marshall, J., Kaplan, J. M., Akita, G. Y., Wang, K. X., Lee, E. R., Harris, D. J., Jiang, C., Yew, N. S., Smith, A. E., and Cheng, S. H. (1997). Basis of pulmonary toxicity associated with cationic lipid-mediated gene transfer to the mammalian lung. *Hum. Gene Ther.* **8**, 689–707.

Segal, B. M., Chang, J. T., and Shevach, E. M. (2000). CpG oligonucleotides are potent adjuvants for the activation of autoreactive encephalitogenic T cells *in vivo*. *J. Immunol.* **164**, 5683–5688.

Song, Y. K., and Liu, D. (1998). Free liposomes enhance the transfection activity of DNA/lipid complexes *in vivo* by intravenous administration. *Biochim. Biophys. Acta* **1372**, 141–150.

Song, Y. K., Liu, F., and Liu, D. (1998). Enhanced gene expression in mouse lung by prolonging the retention time of intravenously injected plasmid DNA. *Gene Ther.* **5**, 1531–1537.

Sparwasser, T., Hultner, L., Koch, E. S., Luz, A., Lipford, G. B., and Wagner, H. (1999). Immunostimulatory CpG-oligodeoxynucleotides cause extramedullary murine hemopoiesis. *J. Immunol.* **162**, 2368–2374.

Sparwasser, T., Miethke, T., Lipford, G., Erdmann, A., Hacker, H., Heeg, K., and Wagner, H. (1997). Macrophages sense pathogens via DNA motifs: Induction of tumor necrosis factor-alpha-mediated shock. *Eur. J. Immunol.* **27**, 1671–1679.

Stecenko, A. A., and Brigham, K. L. (2003). Gene therapy progress and prospects: Alpha-1 antitrypsin. *Gene Ther.* **10**, 95–99.

Suzuki, K., Mori, A., Ishii, K. J., Saito, J., Singer, D. S., Klinman, D. M., Krause, P. R., and Kohn, L. D. (1999). Activation of target-tissue immune-recognition molecules by double-stranded polynucleotides. *Proc. Natl. Acad. Sci. USA* **96**, 2285–2290.

Tan, Y., and Huang, L. (2002). Overcoming the inflammatory toxicity of cationic gene vectors. *J. Drug Target* **10**, 153–160.

Tan, Y., Li, S., Pitt, B. R., and Huang, L. (1999). The inhibitory role of CpG immunostimulatory motifs in cationic lipid vector-mediated transgene expression *in vivo*. *Hum. Gene Ther.* **10**, 2153–2161.

Tan, Y., Liu, F., Li, Z., Li, S., and Huang, L. (2001). Sequential injection of cationic liposome and plasmid DNA effectively transfects the lung with minimal inflammatory toxicity. *Mol. Ther.* **3**, 673–682.

Tandia, B. M., Vandenbranden, M., Wattiez, R., Lakhdar, Z., Ruysschaert, J. M., and Elouahabi, A. (2003). Identification of human plasma proteins that bind to cationic lipid/DNA complex and analysis of their effects on transfection efficiency: Implications for intravenous gene transfer. *Mol. Ther.* **8**, 264–273.

Tomizawa, J. I., and Itoh, T. (1982). The importance of RNA secondary structure in ColE1 primer formation. *Cell* **31**, 575–583.

Tousignant, J. D., Gates, A. L., Ingram, L. A., Johnson, C. L., Nietupski, J. B., Cheng, S. H., Eastman, S. J., and Scheule, R. K. (2000). Comprehensive analysis of the acute toxicities induced by systemic administration of cationic lipid:Plasmid DNA complexes in mice. *Hum. Gene Ther.* **11,** 2493–2513.

Trevani, A. S., Chorny, A., Salamone, G., Vermeulen, M., Gamberale, R., Schettini, J., Raiden, S., and Geffner, J. (2003). Bacterial DNA activates human neutrophils by a CpG-independent pathway. *Eur. J. Immunol.* **33,** 3164–3174.

Trubetskoy, V. S., Wong, S. C., Subbotin, V., Budker, V. G., Loomis, A., Hagstrom, J. E., and Wolff, J. A. (2003). Recharging cationic DNA complexes with highly charged polyanions for *in vitro* and *in vivo* gene delivery. *Gene Ther.* **10,** 261–271.

Uyechi, L. S., Gagne, L., Thurston, G., and Szoka, F. C. Jr. (2001). Mechanism of lipoplex gene delivery in mouse lung: Binding and internalization of fluorescent lipid and DNA components. *Gene Ther.* **8,** 828–836.

Waldner, H., Collins, M., and Kuchroo, V. K. (2004). Activation of antigen-presenting cells by microbial products breaks self tolerance and induces autoimmune disease. *J. Clin. Invest.* **113,** 990–997.

Weighardt, H., Feterowski, C., Veit, M., Rump, M., Wagner, H., and Holzmann, B. (2000). Increased resistance against acute polymicrobial sepsis in mice challenged with immunostimulatory CpG oligodeoxynucleotides is related to an enhanced innate effector cell response. *J. Immunol.* **165,** 4537–4543.

Wiseman, J. W., Goddard, C. A., and Colledge, W. H. (2001). Steroid hormone enhancement of gene delivery to a human airway epithelial cell line *in vitro* and mouse airways *in vivo. Gene Ther.* **8,** 1562–1571.

Wu, F., Levchenko, I., and Filutowicz, M. (1995). A DNA segment conferring stable maintenance on R6K gamma-origin core replicons. *J. Bacteriol.* **177,** 6338–6345.

Yasuda, K., Ogawa, Y., Kishimoto, M., Takagi, T., Hashida, M., and Takakura, Y. (2002). Plasmid DNA activates murine macrophages to induce inflammatory cytokines in a CpG motif-independent manner by complex formation with cationic liposomes. *Biochem. Biophys. Res. Commun.* **293,** 344–348.

Yew, N. S., Wang, K. X., Przybylska, M., Bagley, R. G., Stedman, M., Marshall, J., Scheule, R. K., and Cheng, S. H. (1999). Contribution of plasmid DNA to inflammation in the lung after administration of cationic lipid:pDNA complexes. *Hum. Gene Ther.* **10,** 223–234.

Yew, N. S., Zhao, H., Przybylska, M., Wu, I. H., Tousignant, J. D., Scheule, R. K., and Cheng, S. H. (2002). CpG-depleted plasmid DNA vectors with enhanced safety and long-term gene expression *in vivo. Mol. Ther.* **5,** 731–738.

Yew, N. S., Zhao, H., Wu, I. H., Song, A., Tousignant, J. D., Przybylska, M., and Cheng, S. H. (2000). Reduced inflammatory response to plasmid DNA vectors by elimination and inhibition of immunostimulatory CpG motifs. *Mol. Ther.* **1,** 255–262.

Zabner, J., Fasbender, A. J., Moninger, T., Poellinger, K. A., and Welsh, M. J. (1995). Cellular and molecular barriers to gene transfer by a cationic lipid. *J. Biol. Chem.* **270,** 18997–19007.

Zhang, Y., Schlachetzki, F., Zhang, Y. F., Boado, R. J., and Pardridge, W. M. (2004). Normalization of striatal tyrosine hydroxylase and reversal of motor impairment in experimental parkinsonism with intravenous non-viral gene therapy and a brain-specific promoter. *Hum. Gene Ther.* **15,** 339–350.

Zhang, Y. F., Boado, R. J., and Pardridge, W. M. (2003). Absence of toxicity of chronic weekly intravenous gene therapy with pegylated immunoliposomes. *Pharm. Res.* **20,** 1779–1785.

Zhao, H., Hemmi, H., Akira, S., Cheng, S. H., Scheule, R. K., and Yew, N. S. (2004). Contribution of toll-like receptor 9 signaling to the acute inflammatory response to non-viral vectors. *Mol. Ther.* **9,** 241–248.

Section 3

CATIONIC POLYMERS

8

Polyethylenimine (PEI)

Barbara Demeneix and Jean-Paul Behr

Evolution des Régulations Endocriniennes, Muséum National d'Histoire
Naturelle, 7, rue Cuvier, 75231 Paris
Chimie Génétique, Faculté de Pharmacie, route du Rhin
67401 Illkirch, France

ABSTRACT

Since the first edition of this book in 1999 the field of gene therapy has been the
arena both for major advances that justified the early hopes placed in the
concept, and for ever-present impatience with the slowness of overall progress.
On the positive side, gene therapy obtained its first brilliant success, though not
where most efforts were invested and not with a synthetic vector (Cavazzana-
Calvo et al., 2000). Yet the search for efficient molecules is still very active, in
part because the negative consequences of using viral vectors somewhat shadow
the brilliant picture (Hacein-Bey-Abina et al., 2003).

Advances in Genetics, Vol. 53
Copyright 2005, Elsevier Inc. All rights reserved.

0065-2660/05 $35.00
DOI: 10.1016/S0065-2660(05)53008-6

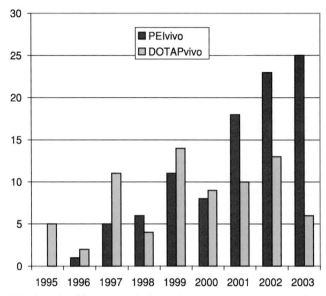

Figure 8.1. Number of publications with keywords «DOTAP» or «PEI» and «*in vivo* gene delivery» (source Science Citation Index; no s.d.)

Clinical trial reports using the first generation of non-viral vectors, that is, cationic lipids, emphasize safety more than efficacy. The next generation, namely cationic polymers, is coming to maturity. This is illustrated in Fig. 8.1 by a literature survey that compares the most used « open source » vector in each category, the lipid DOTAP and polymer PEI.

This "coming of age" is also highlighted by the number of planned clinical trials using PEI, trials we shall discuss in the last paragraph of this chapter. Advances have not only been made toward therapy, but also toward a better understanding of the mechanisms underlying gene delivery. Finally, PEI has become one of the most popular reagents for transfection of cells in culture, on a fast-growing market that is boosted by human and animal genome sequencing. © 2005, Elsevier Inc.

I. MECHANISM OF GENE DELIVERY

The chemical structure of PEI (polyethylenimine) is given in Fig. 8.2. It can be synthesized from ethyleneimine (aziridine) or oxazoline monomers, leading to branched or linear polymeric backbones, respectively. The molecule is a simple repetition of the 43 Da CH_2-CH_2-NH ethyleneimine motif.

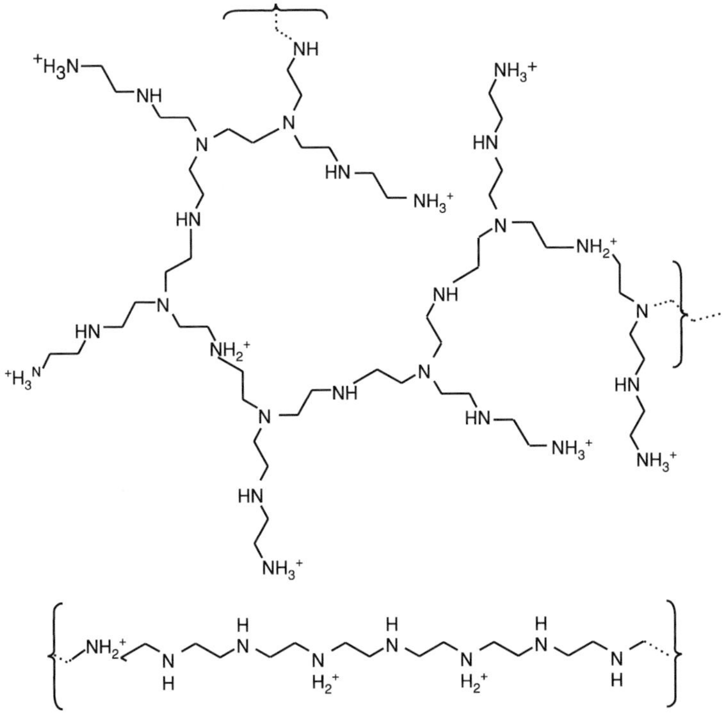

Figure 8.2. Chemical structures of branched (b-PEI, upper) and linear (l-PEI, lower) polyethylenimines.

Gene delivery is an extremely complex process. While a gene-delivering virus is made of several proteins whose unique sequences can store immense amounts of information refined through evolution, PEI made it alone, with a mere 43 Daltons and within a decade. To understand its success we need to analyze how multiple, sometimes conflicting (think of DNA binding/release), properties are stored in such a simple molecule.

Initial complex formation between DNA and PEI occurs by counter-condensation of the oppositely charged polymers, a process that relieves intra-molecular repulsions and releases many chloride and sodium ions. Due to the kinetic control of the complex formation process, complexes are multi-molecular, hence the polydispersity of the particles. Depending on the medium used for formulation, they will eventually aggregate further (see next paragraph).

The interior of the particles are electrically neutral; however, efficient transfection requires particles with a cationic surface for cell binding (Behr *et al.*, 1989), hence the crucial importance of charge ratio (N/P, for the atoms bearing the positive and negative charges on each polymer). An integrated mechanism for cell entry was described only recently (Kopatz *et al.*, 2004). In fact, it is quite banal and widely exploited for particle engulfment by animals (e.g., chylomicrons) and pathogens (bacteria and viruses). It uses syndecans as receptors and takes advantage of extracellular matrix catabolism by adherent cells. Syndecans are ubiquitous transmembrane adhesion molecules. Their polyanionic heparan sulfate moieties are bound at the distal end of their ectodomain, thus facilitating extracellular matrix binding, but also interaction with large cationic particles. After initial binding of the PEI/DNA complexes to their receptor, gradual electrostatic zippering of the plasma membrane around the particle is sustained by lateral diffusion of many syndecan molecules that cluster into cholesterol-rich rafts. Clustering, in turn, triggers PKC activation and linker protein-mediated actin binding to the cytoplasmic tail of syndecan. Resulting tension fibers and a growing network of cortical actin can then pull the particle into the cell, as schematically depicted in Fig. 8.3.

The end point of this catabolic process is the lysosomal compartment. Endosome trafficking from the cell surface to the lysosome occurs with acidification and is fast, which does not leave much time for DNA complexes to escape. After having acted successively as a DNA packaging matrix and as a ligand for cell-surface binding, PEI now exploits its *noncharged* nitrogen atoms to prolong the endosome trafficking time. Indeed, due to the close proximity of the nitrogen atoms within the structure (fig. 8.2), many amines are not protonated at physiological pH. PEI can thus act as a buffering « proton sponge » (Boussif *et al.*, 1995) that will delay acidification and fusion with the lysosome. At the same time osmotic swelling and finally rupture of some of the endosomes will allow escape of PEI/DNA complexes into the cytosol. Although questioned (Godbey *et al.*, 2000), this hypothesis is receiving convergent experimental support (Kichler *et al.*, 2001; Merdan *et al.*, 2002; Sonawane *et al.*, 2003).

The cytoplasm is a crowded and still dangerous place for DNA. Indeed, extended plasmid DNA molecules are unable to diffuse because of the cytoskeleton (Lukacs *et al.*, 2000) and cytosolic nucleases degrade it quickly (Lechardeur *et al.*, 1999; Pollard *et al.*, 2001). As long as it remains condensed with PEI, however, DNA is protected and expected to diffuse better. Competitive exchange of DNA with abundant cytosolic polyanions, such as actin or RNA, is probably slow and incomplete. From there, the most probable scenario is that a PEI/DNA complex gets access to the nucleus during mitosis (Pollard *et al.*, 1998). Even a still complex could have a chance to bind to chromosomes that move through the cytoplasm during anaphase. Finally, PEI would make a mass-action choice in favor of chromosomal DNA and the free plasmid would be enrolled into nucleosomes. Unfortunately, the next cell division would expel the nuclear plasmid: mitosis thus is the yin and the yang of transfection.

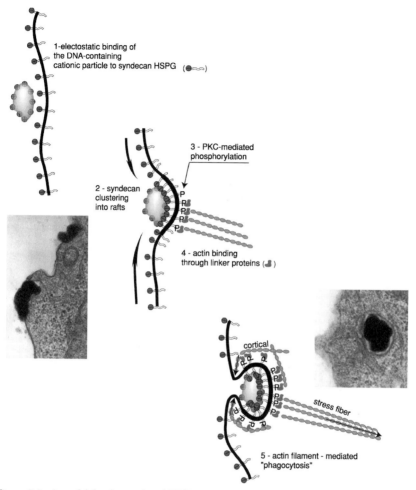

1-electostatic binding of
the DNA-containing
cationic particle to syndecan HSPG

2 - syndecan
clustering
into rafts

3 - PKC-mediated
phosphorylation

4 - actin binding
through linker proteins

cortical

stress fiber

5 - actin filament - mediated
"phagocytosis"

Figure 8.3. A model for the uptake of PEI/DNA complexes by adherent cells. Insets show TEM pictures of the complexes. (See Color Insert.)

Mitosis is not an absolute requirement for cell transfection since post-mitotic cells such as neurons can be well transfected with this vector (Horbinski *et al.*, 2001). In this context, PEI, and especially linear PEI, are much more efficient than the cationic lipid Lipofectamine at transfecting G1-arrested cells (Brunner *et al.*, 2002). This may have its origin in the high stability of the PEI/DNA complexes that may help nuclear barrier crossing, while cationic lipids release DNA in the cytoplasm by competitive exchange with phosphatidylserine (Zelphati *et al.*, 1996).

II. TRANSFECTION OF CELLS IN CULTURE

Cell transfection is a powerful technique used to study gene regulation and protein function. On a larger scale, it is used for protein and virus production. Genome sequencing has bolstered these areas of research and development. However, since non-adherent cell types lack syndecans (see above), transfection is efficient only when applied to adherent cells. Transfection of adherent cells in culture is straightforward because complexes can reach the cells easily and because cells are metabolically active and usually undergoing mitosis. Cells are active mainly in the presence of serum. However, serum albumin binds to the complexes and interferes with transfection. The best compromise is obtained when transfection is performed in the presence of serum, yet with large complexes that can fall onto the cells by sedimentation in order to find their receptor (subsequent cell entry is not size-restricted). Large cationic PEI/DNA complexes are formed by aggregation when Van der Waals attraction becomes stronger than coulombic repulsion; repulsion is weakened by shielding with salt. PEI/DNA complexes are therefore formed and allowed to grow for some time in 150 mM NaCl.

Over the years, there has been increasing experimental evidence that linear PEI (l-PEI) is a better vector than branched PEI (Wightman *et al.*, 2001): both its molecular weight distribution and purification are easier to control and transfection with l-PEI is less dependent on mitosis (see previous paragraph). The latter property is especially interesting when transfecting primary post-mitotic cells.

The other issue besides efficiency is cytotoxicity. Any transfection method incurs toxicity because the barriers met late in the transfection process (such as intracellular diffusion and crossing of the nuclear membrane) are so difficult to overcome that a large number of complexes need to be in the cell to ensure that some make it to the nucleus and get transcribed. The presence of a large number of complexes in the cell cannot be innocuous. However, it should be noted that cell detachment, a consequence of syndecan over-recruitment by the complexes, is not necessarily a sign of toxicity. Toxicity is reduced by the presence of serum and by removal of the lower molecular weight molecules present in the polymer mixture. In these conditions, linear PEI is among the top two reagents, irrespective of the cell line transfected (Fig. 8.4). PEI is the active species used in a number of gene and oligonucleotide transfection reagents such as ExGen 500, jet-PEI, Escort-V, TransIT-TKO, GeneTools and PolyMag.

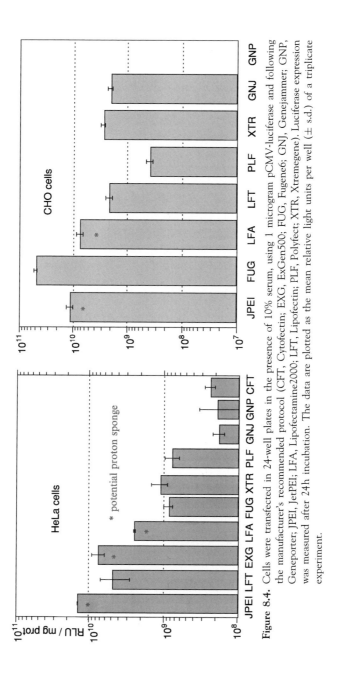

Figure 8.4. Cells were transfected in 24-well plates in the presence of 10% serum, using 1 microgram pCMV-luciferase and following the manufacturer's recommended protocol (CFT, Cytofectin; EXG, ExGen500; FUG, Fugene6; GNJ, Genejammer; GNP, Geneporter; JPEI, JetPEI; LFA, Lipofectin; LFT, Lipofectamine2000; PLF, Polyfect; XTR, Xtremegene). Luciferase expression was measured after 24h incubation. The data are plotted as the mean relative light units per well ($\pm$ s.d.) of a triplicate experiment.

III. *IN VIVO* GENE DELIVERY

As already mentioned in the introduction, PEI is by far the most-used non-viral vector for gene and oligonucleotide delivery in animals. Here again, linear PEI surpasses the branched isomer, as in the *in vivo* context its capacity for better transfection of slow-dividing cells (see above) provides a clear advantage. Moreover, perhaps as a consequence of a different cationic charge distribution, linear l-PEI/DNA complexes have a lesser tendency to aggregate erythrocytes than b-PEI (Kircheis *et al.*, 2001), hence a decreased risk of microembolies following intravenous injection (Abdallah *et al.*, 1996).

PEI has been tested successfully following various routes of delivery for transfection of a number of organs. Among them, gene delivery to the brain after intraventricular injection and lung transfection following systemic or tracheal delivery can lead to remarkably high transfection levels.

A. Use of b-PEI and l-PEI for transfection of the central nervous system

The initial description of use of PEI for intracerebral injection and gene delivery was in the brain of the newborn mouse (Boussif *et al.*, 1995). This demonstration was followed closely by the observation that the same methodology could be applied to the adult mouse brain (Abdallah *et al.*, 1996). In both instances branched PEI with a mean molecular weight of 25 kD was used.

Two key improvements were then made on these promising starts. First, using the newly synthesized 22 kD l-PEI formulated in 5% glucose, it was seen that small and stable virus like particles could be produced with diameters of about 50nm (Goula *et al.*, 1998b); and second, that intraventricular injection of these complexes was particularly efficient for transfecting large numbers of cells close to the ventricle throughout the brain, following one single injection (Goula *et al.*, 1998b). Almost simultaneous with this demonstration of efficient transfection of cells close to the ventricle came the demonstration by a number of groups that this brain area contains the neuronal stem cell population. This led us to closely define the cell types transfected by l-PEI following brain intraventricular transfection. A series of experimental approaches including triple immunochemistry combined with electron microscopy showed that the l-PEI-based transfection methodology was preferentially targeting the neuronal stem cells and their progenitors (Lemkine *et al.*, 2002).

The low toxicity and relatively high effiency of l-PEI when used for intrathecal injection (into the brain tissue as opposed to the ventricles) of defined brain areas and/or at defined developmental stages clearly provided a major experimental advantage for functional studies on protein function and gene regulation, providing information that could only otherwise be obtained by raising and complex crossing of mutant mice lines. l-PEI has been used to define the specific roles of different thyroid hormone receptor isoforms in regulating hypothalamic genes (Dupre et al., 2004; Guissouma et al., 2002). The methodology has also been used to deliver antisense coding plasmids so as to define the roles of the dopamine transporter in the substansia nigra (Martres et al., 1998), as well as the serotonin transporter in the raphe nucleus (Fabre et al., 2000).

Other, more applied, applications have included approaches related to therapeutic methods such as the demonstration of NGF-mediated protection of cholinergic neurons in the basal forebrain (Wu et al., 2004) and the demonstration of retrograde transport shown by the expression of both the luciferase and beta-galactosidase gene expression in hypoglossal neurons following injections into tongue muscle innervated by these neurons (Wang et al., 2001).

B. Transfection of the lung by tracheal or systemic delivery

Given the encouraging data coming out of the work done on the brain it was a logical step, strongly supported by charities connected to the cystic fibrosis patient groups, to see whether the methodology could be applied to the lung. The first experiments on this target organ used airway delivery by instillation and were applied to rabbits (Ferrari et al., 1997). Low levels of transfection were observed, but the process clearly required refining and improvement. Three to five years later a number of reports showed that these desired improvements could be achieved. First, experiments in mouse models showed that the order of efficiency for different vectors was l-PEI>b-PEI > DOTAP,GL67 in mice and rats (Bragonzi et al., 2000; Uduehi et al., 2001). Later confirmation came that when using l-PEI, expression lasting >7 days could be obtained in lung epithelial cells (Koping-Hoggard et al., 2001). Functional data followed with the demonstrations that the instillation methodology could be used to restore chloride channel function using l-PEI for CFTR gene delivery (Wiseman et al., 2003). Further improvements came with the use of aerosol delivery as opposed to instillation. Aerosol delivery to the airways showed PEI to provide 10 to 100 times higher transfection efficiencies than cationic lipids (Densmore et al., 2000) and led to the idea that aerosol delivery could be exploited to treat or vaccinate against lung metastases (Densmore 2003).

Another means of targeting lung tissue is to use systemic delivery. This was first shown to be possible with l-PEI formulated in 5% glucose by Goula and colleagues in 1998 (Goula et al., 1998a) and resulted in transfection of >5% of the airway cells! This, in turn, raised the question of the mechanism involved. The most plausible explanation (Goula et al., 2000; Zou et al., 2000) is that the rapid injection leads to a transient glucose bolus which avoids mixing with blood components and may cause endothelium retraction. Loose endothelial intercellular junctions (and possibly transcytosis) allow the PEI/DNA particles to reach the airway cells from the basolateral side where syndecans are concentrated. Fast turnover of these cells may also contribute to their transfection.

C. Tumor targeting and other *in vivo* applications with l-PEI

Systemic delivery of l-PEI complexes has also been shown to result in passive tumor targeting by leakage, especially if surface-shielded particles are used that abrogate the preferential uptake of charged particles by the lung and liver (Kircheis *et al.*, 1999). Indeed, better pharmacokinetics and targeting will rely on sophisticated derivatization of PEI (see chapters by Kabanov, Kim and Wagner).

In the past five years, over a hundred publications have appeared dealing with *in vivo* gene delivery with PEI for numerous functions as diverse as intramuscular vaccination (Oh *et al.*, 2003), delivery to the intestine (Cryan *et al.*, 2003), wound healing of pig skin (Berlanga *et al.*, 2000) or adventitial gene delivery to rabbit carotid artery (Turunen *et al.*, 1999).

The success in these diverse fields has promoted a number of preclinical studies moving towards therapy, with the large majority aiming for cancer-related targets. Studies particularly worthy of mention include the demonstration of a suicide gene therapy model of hepatocellular carcinoma (Iwai *et al.*, 2002) and the demonstration of increased apoptosis and decreased proliferative index in murine models of pancreatic tumors (Aoki *et al.*, 2001; Vernejoul *et al.*, 2002), a cancer that is almost always fatal in humans. In particular, Vernejoul and colleagues showed that l-PEI-based delivery of the somatostatin receptor 2 gene restored apoptosis with efficiencies that were in the same order of magnitude as adenoviral vectors (Vernejoul *et al.*, 2002).

Ovarian carcinoma is often characterized by resistance to chemotherapy. One therapeutic goal would be to restore chemosensitivity by gene delivery and this has been achieved in a number different ways with l-PEI (Poulain *et al.*, 2000; Ziller *et al.*, 2004), including facilitation of suicide gene therapy .

Inactivation of p53, a tumor suppressor gene, has been observed in a number of cancers, including a majority of small cell and non-small cell lung cancers and over 50% of head and neck cancers, engendering very poor prognoses. Thus, functional p53 delivery is a promising therapeutic approach for models of head-and-neck cancer and a proof of principle has been demonstrated with unsubstituted l-PEI or glucosylated PEI, the latter being the more effective (Dolivet *et al.*, 2002; Merlin *et al.*, 2001) and improved on with photochemical internalization (Ndoye *et al.*, 2004). The high levels of gene expression obtained in the normal lung after aerosol delivery of PEI complexes prompted Densmore and co-workers to adapt the delivery method for potential strategies for treating different forms of lung cancer (Densmore 2003; Densmore *et al.*, 2000, 2001, 2003; Gautam *et al.*, 2000, 2003; Jia *et al.*, 2003). Indeed, repeated aerosol delivery of b-PEI-complexed interleukin 12 significantly decreased the number of osteosarcoma-derived lung metastases in a mouse model over 5–6 weeks (Jia et al., 2003).

No doubt the most significant data on the potential use of PEI-based gene delivery for cancer therapy is the recently published report of local delivery of H19-driven diphteria toxin/l-PEI complexes for human bladder carcinoma (Ohana et al., 2004). The H19 enhancer sequence was chosen to drive the production of the gene encoding the diptheria toxin A chain as the H19 gene is significantly expressed in 85% of bladder cancers, and thus, use of its promoter should ensure high levels of expression of the gene of choice; in this case, the diptheria toxin gene. In this study, following successful use of l-PEI delivery of the therapeutic approach in a rat model, two male human patients with recurrent bladder cancer were treated. Both patients showed >85% tumor remission. This is, thus, an elegant demonstration of 'patient orientated DNA-based therapy.'

Finally, another therapeutic approach in which l-PEI-based delivery is finding useful applications is in AIDS immunotherapy trials. One such trial based on the use of l-PEI delivery of a plasmid encoding a replication and integration defective human immunodeficiency virus to genetically modify dendritic cells should be undertaken soon (Lisziewicz et al., 2001, 2004). This method exploits the possibility of obtaining expression of viral components in the absence of integration.

In conclusion, the early nineties saw great enthusiasm for gene therapy approaches and many vectors, particularly viral vectors and cationic lipids, took only a few years to go from the bench to the clinic. Unfortunately, the results obtained during this period have proven rather disappointing with cationic lipids, although simpler and safer than with viruses.

This disappointment adversely affected the adoption of PEI for toxicity studies and clinical trials. Indeed, when PEI appeared on the gene transfer scene in 1995 (Boussif et al., 1995), clinicians were rather hesitant to take on new non-viral vectors. Hence the much longer time lag for PEI from the first promising in vitro and in vivo demonstrations in the mid-90s to the first clinic trials that have started appearing this year (Ohana et al., 2004).

However, despite this delay the results suggest that PEI was certainly worth waiting for and that, overall, PEI is probably the most promising second-generation non-viral vector. Current studies suggest a quiet confidence that its use carries hope for a number of therapeutic settings.

References

Abdallah, B., Hassan, A., et al. (1996). A powerful non-viral vector for in vivo gene transfer into the adult mammalian brain: Polyethylenimine. Hum. Gene Ther. 7(16), 1947–1954.

Aoki, K., Furuhata, S., et al. (2001). Polyethylenimine-mediated gene transfer into pancreatic tumor dissemination in the murine peritoneal cavity. Gene Ther. 8(7), 508–514.

Behr, J. P., Demeneix, B., *et al.* (1989). Efficient gene transfer into mammalian primary endocrine cells with lipopolyamine-coated DNA. *Proc. Natl. Acad. Sci. USA* **86**(18), 6982–6986.

Berlanga, J., Saez, V., *et al.* (2000). Transfection of intact and wounded skin with a DNA/poly-ethylenimine complex. *Biotecnologia Aplicada* **17**(4), 235–240.

Boussif, O., Lezoualc'h, F., *et al.* (1995). A versatile vector for gene and oligonucleotide transfer into cells in culture and *in vivo*: Polyethylenimine. *Proc. Natl. Acad. Sci. USA* **92**(16), 7297–7301.

Bragonzi, A., Dina, G., *et al.* (2000). Biodistribution and transgene expression with non-viral cationic vector/DNA complexes in the lungs. *Gene Ther.* **7**(20), 1753–1760.

Brunner, S., Furtbauer, E., *et al.* (2002). Overcoming the nuclear barrier: Cell cycle independent non-viral gene transfer with linear polyethylenimine or electroporation. *Mol. Ther.* **5**(1), 80–86.

Cavazzana-Calvo, M., Hacein-Bey, S., *et al.* (2000). Gene therapy of human severe combined immunodeficiency (SCID)-X1 disease. *Science* **288**(5466), 669–672.

Cryan, S. A., and O'Driscoll, C. M. (2003). Mechanistic studies on non-viral gene delivery to the intestine using *in vitro* differentiated cell culture models and an *in vivo* rat intestinal loop. *Pharm. Res.* **20**(4), 569–575.

Densmore, C. L. (2003). The re-emergence of aerosol gene delivery: A viable approach to lung cancer therapy. *Curr. Cancer Drug Targets* **3**(4), 275–286.

Densmore, C. L., Kleinerman, E. S., *et al.* (2001). Growth suppression of established human osteosarcoma lung metastases in mice by aerosol gene therapy with PEI-p53 complexes. *Cancer Gene Ther.* **8**(9), 619–627.

Densmore, C. L., Orson, F. M., *et al.* (2000). Aerosol delivery of robust polyethyleneimine-DNA complexes for gene therapy and genetic immunization. *Mol. Ther.* **1**(2), 180–188.

Dolivet, G., Merlin, J. L., *et al.* (2002). *In vivo* growth inhibitory effect of iterative wild-type p53 gene transfer in human head and neck carcinoma xenografts using glucosylated polyethylenimine non-viral vector. *Cancer Gene Ther.* **9**(8), 708–714.

Dupre, S. M., Guissouma, H., *et al.* (2004). Both thyroid hormone receptor (TR)beta 1 and TR beta 2 isoforms contribute to the regulation of hypothalamic thyrotropin-releasing hormone. *Endocrinology* **145**(5), 2337–2345.

Fabre, V., Boutrel, B., *et al.* (2000). Homeostatic regulation of serotonergic function by the serotonin transporter as revealed by non-viral gene transfer. *J. Neurosci.* **20**(13), 5065–5075.

Ferrari, S., Moro, E., *et al.* (1997). ExGen 500 is an efficient vector for gene delivery to lung epithelial cells *in vitro* and *in vivo*. *Gene Ther.* **4**(10), 1100–1106.

Gautam, A., Densmore, C. L., *et al.* (2000). Enhanced gene expression in mouse lung after PEI-DNA aerosol delivery. *Mol. Ther.* **2**(1), 63–70.

Gautam, A., Waldrep, J. C., *et al.* (2003). Aerosol gene therapy for metastatic lung cancer using PEI-p53 complexes. *Methods Mol. Med.* **75,** 607–618.

Godbey, W. T., Barry, M. A., *et al.* (2000). Poly(ethylenimine)-mediated transfection: A new paradigm for gene delivery. *J. Biomed. Mater. Res.* **51**(3), 321–328.

Goula, D., Becker, N., *et al.* (2000). Rapid crossing of the pulmonary endothelial barrier by polyethylenimine/DNA complexes. *Gene Ther.* **7**(6), 499–504.

Goula, D., Benoist, C., *et al.* (1998a). Polyethylenimine-based intravenous delivery of transgenes to mouse lung. *Gene Ther.* **5**(9), 1291–1295.

Goula, D., Remy, J. S., *et al.* (1998b). Size, diffusibility and transfection performance of linear PEI/DNA complexes in the mouse central nervous system. *Gene Ther.* **5**(5), 712–717.

Guissouma, H., Dupre, S. M., *et al.* (2002). Feedback on Hypothalamic TRH Transcription Is Dependent on Thyroid Hormone Receptor N Terminus. *Mol. Endocrinol.* **16**(7), 1652–1666.

Hacein-Bey-Abina, S., Von Kalle, C., *et al.* (2003). LMO2-associated clonal T cell proliferation in two patients after gene therapy for SCID-X1. *Science* **302**(5644), 415–419.

Horbinski, C., Stachowiak, M. K., et al. (2001). Polyethyleneimine-mediated transfection of cultured postmitotic neurons from rat sympathetic ganglia and adult human retina. *BMC Neurosci.* **2**(1), 2.

Iwai, M., Harada, Y., et al. (2002). Polyethylenimine-mediated suicide gene transfer induces a therapeutic effect for hepatocellular carcinoma *in vivo* by using an Epstein-Barr virus-based plasmid vector. *Biochem. Biophys. Res. Commun.* **291**(1), 48–54.

Jia, S. F., Worth, L. L., et al. (2003). Aerosol gene therapy with PEI: IL-12 eradicates osteosarcoma lung metastases. *Clin. Cancer Res.* **9**(9), 3462–3468.

Kichler, A., Leborgne, C., et al. (2001). Polyethylenimine-mediated gene delivery: A mechanistic study. *J. Gene Med.* **3**(2), 135–144.

Kircheis, R., Schuller, S., et al. (1999). Polycation-based DNA complexes for tumor-targeted gene delivery *in vivo*. *J. Gene Med.* **1**(2), 111–120.

Kircheis, R., Wightman, L., et al. (2001). Polyethylenimine/DNA complexes shielded by transferrin target gene expression to tumors after systemic application. *Gene Ther.* **8**(1), 28–40.

Kopatz, I., Remy, J. S., et al. (2004). A model for non-viral gene delivery: Through syndecan adhesion molecules and powered by actin. *J. Gene Med.* **6**(7), 769–776.

Koping-Hoggard, M., Tubulekas, I., et al. (2001). Chitosan as a non-viral gene delivery system. Structure-property relationships and characteristics compared with polyethylenimine *in vitro* and after lung administration *in vivo*. *Gene Ther.* **8**(14), 1108–1121.

Lechardeur, D., Sohn, K.-J., et al. (1999). Metabolic instability of plasmid DNA in the cytosol: A potential barrier to gene transfer. *Gene Ther.* **6**(4), 482–497.

Lemkine, G. F., Mantero, S., et al. (2002). Preferential transfection of adult mouse neural stem cells and their immediate progeny *in vivo* with polyethylenimine. *Mol. Cell Neurosci.* **19**(2), 165–174.

Lisziewicz, J., Gabrilovich, D. I., et al. (2001). Induction of potent human immunodeficiency virus type 1-specific T-cell-restricted immunity by genetically modified dendritic cells. *J. Virol.* **75**(16), 7621–7628.

Lisziewicz, J., Trocio, J., et al. (2005). Dermavir: A novel topical vaccine for HIV/AIDS. *J. Invest. Dermatol.* **124**, 160–169.

Lukacs, G. L., Haggie, P., et al. (2000). Size-dependent DNA mobility in cytoplasm and nucleus. *J. Biol. Chem.* **275**(3), 1625–1629.

Martres, M. P., Demeneix, B., et al. (1998). Up- and down-expression of the dopamine transporter by plasmid DNA transfer in the rat brain. *Eur. J. Neurosci.* **10**(12), 3607–3616.

Merdan, T., Kunath, K., et al. (2002). Intracellular processing of poly(ethylene imine)/ribozyme complexes can be observed in living cells by using confocal laser scanning microscopy and inhibitor experiments. *Pharm. Res.* **19**(2), 140–146.

Merlin, J. L., Dolivet, G., et al. (2001). Improvement of non-viral p53 gene transfer in human carcinoma cells using glucosylated polyethylenimine derivatives. *Cancer Gene Ther.* **8**(3), 203–210.

Ndoye, A., Merlin, J.-L., et al. (2004). Enhanced gene transfer and cell death following p53 gene transfer using photochemical internalisation of glucosylated PEI-DNA complexes. *J. Gene Med.* **6**, 884–894.

Oh, Y. K., Park, J. S., et al. (2003). Enhanced adjuvanticity of interleukin-2 plasmid DNA administered in polyethylenimine complexes. *Vaccine* **21**(21–22), 2837–2843.

Ohana, P., Gofrit, O., et al. (2004). Regulatory sequences of the H19 gene in DNA based therapy of bladder cancer. *Gene Ther. Mol. Biol.* **8**, 181–192.

Pollard, H., Remy, J. S., et al. (1998). Polyethylenimine but not cationic lipids promotes transgene delivery to the nucleus in mammalian cells. *J. Biol. Chem.* **273**(13), 7507–7511.

Pollard, H., Toumaniantz, G., et al. (2001). Ca2+-sensitive cytosolic nucleases prevent efficient delivery of injected plasmids to the nucleus. *J. Gene Med.* **3**(2), 153–164.

Poulain, L., Ziller, C., *et al.* (2000). Ovarian carcinoma cells are effectively transfected by poly-ethylenimine (PEI) derivatives. *Cancer Gene Ther.* **7**(4), 644–652.

Sonawane, N. D., Szoka, Jr., F. C., *et al.* (2003). Chloride accumulation and swelling in endosomes enhances DNA transfer by polyamine-DNA polyplexes. *J. Biol. Chem.* **278**(45), 44826–44831.

Turunen, M. P., Hiltunen, M. O., *et al.* (1999). Efficient adventitial gene delivery to rabbit carotid artery with cationic polymer-plasmid complexes. *Gene Ther.* **6**(1), 6–11.

Uduehi, A. N., Stammberger, U., *et al.* (2001). Efficiency of non-viral gene delivery systems to rat lungs. *Eur. J. Cardiothorac. Surg.* **20**(1), 159–163.

Vernejoul, F., Faure, P., *et al.* (2002). Antitumor effect of *in vivo* somatostatin receptor subtype 2 gene transfer in primary and metastatic pancreatic cancer models. *Cancer Res.* **62**(21), 6124–6131.

Wang, S., Ma, N., *et al.* (2001). Transgene expression in the brain stem effected by intramuscular injection of polyethylenimine/DNA complexes. *Mol. Ther.* **3**(5 Pt 1), 658–664.

Wightman, L., Kircheis, R., *et al.* (2001). Different behavior of branched and linear polyethyleni-mine for gene delivery *in vitro* and *in vivo*. *J. Gene Med.* **3**(4), 362–372.

Wiseman, J. W., Goddard, C. A., *et al.* (2003). A comparison of linear and branched polyethyleni-mine (PEI) with DCChol/DOPE liposomes for gene delivery to epithelial cells *in vitro* and *in vivo*. *Gene Ther.* **10**(19), 1654–1662.

Wu, K., Meyers, C. A., *et al.* (2004). Polyethylenimine-mediated NGF gene delivery protects transected septal cholinergic neurons. *Brain Res.* **1008**(2), 284–287.

Zelphati, O., and Szoka, F. C. Jr. (1996). Mechanism of oligonucleotide release from cationic liposomes. *Proc. Natl. Acad. Sci. USA* **93**(21), 11493–11498.

Ziller, C., Lincet, H., *et al.* (2004). The cyclin-dependent kinase inhibitor p21(cip1/waf1) enhances the cytotoxicity of ganciclovir in HSV-tk transfected ovarian carcinoma cells. *Cancer Lett.* **212**(1), 43–52.

Zou, S. M., Erbacher, P., *et al.* (2000). Systemic linear polyethylenimine (L-PEI)-mediated gene delivery in the mouse. *J. Gene Med.* **2**(2), 128–134.

9

Pluronic Block Copolymers for Gene Delivery

Alexander Kabanov,* Jian Zhu,*,[1] and Valery Alakhov[†]
*Department of Pharmaceutical Sciences and Center for Drug Delivery and Nanomedicine, College of Pharmacy, University of Nebraska Medical Center Omaha, Nebraska 68198
†Supratek Pharma Inc, 215 Boul. Bouchard, Suite 1315, Dorval Quebec H9S 1A9, Canada

[1]Present address: Department of Pharmacology and Molecular Sciences, The Johns Hopkins School of Medicine, Baltimore, Maryland 21205.

Advances in Genetics, Vol. 53
Copyright 2005, Elsevier Inc. All rights reserved.

0065-2660/05 $35.00
DOI: 10.1016/S0065-2660(05)53009-8

ABSTRACT

Amphiphilic block copolymers of poly(ethylene oxide) and poly(propylene oxide) called Pluronic or poloxamer are commercially available pharmaceutical excipients. They recently attracted considerable attention in gene delivery applications. First, they were shown to increase the transfection with adenovirus and lentivirus vectors. Second, they were shown to increase expression of genes delivered into cells using non-viral vectors. Third, the conjugates of Pluronic with polycations, were used as DNA-condensing agents to form polyplexes. Finally, it was demonstrated that they can increase regional expression of the naked DNA after its injection in the skeletal and cardiac muscles or tumor. Therefore, there is substantial evidence that Pluronic block copolymers can improve gene expression with different delivery routes and different types of vectors, including naked DNA. These results and possible mechanisms of Pluronic effects are discussed. At least in some cases, Pluronic can act as biological adjuvants by activating selected signaling pathways, such as NF-κB, and upregulating the transcription of the genes.   © 2005, Elsevier Inc.

I. INTRODUCTION

Cationic lipids and polycations have been developed and used extensively for the design of non-viral gene delivery systems. These studies go back to the end of 80s and early 90s when it was discovered that the mixing of a plasmid DNA with cationic lipids or polycations resulted in formation of polymer complexes that efficiently transfected cells (Behr *et al.*, 1989; Boussif *et al.*, 1995; Felgner *et al.*, 1987; Haensler and Szoka, 1993; Kabanov *et al.*, 1989b; Wagner *et al.*, 1990; Wu and Wu, 1987; Zhou *et al.*, 1991). Numerous cationic lipids and polycations of different structure including linear, branched, dendrimer, block and graft copolymer have been developed for gene delivery since then (Duzgunes *et al.*, 2003; Garnett, 1999; Kabanov, 1999; Kakizawa and Kataoka, 2002; Kircheis *et al.*, 2001b; Liu and Huang, 2002). Some representative polycation molecules are shown in Fig. 9.1. Both the cationic lipids and the polycations electrostatically bind to DNA resulting in formation of polyelectrolyte complexes called "lipoplexes" (lipid-based) or "polyplexes" (polycation-based) (Felgner *et al.*, 1997). The underlying characteristic of these technologies is that the cationic molecules 1) bind and condense the DNA; 2) protect the DNA from the degradation; and 3) enhance transport of the DNA into the cell, which results in increased transgene expression (Felgner *et al.*, 1994; Godbey *et al.*, 1999; Kabanov and Kabanov, 1995; Suh *et al.*, 2003; Uyechi *et al.*, 2001). However, the effects of the delivery vehicles on the transcription and translation of the delivered transgenes have not been reported before.

Architecture	Polycation	Structure
A. Linear	Linear PEI (Exgen 500)	$H_3C-CH_2-NH-[CH_2-CH_2-NH]_n-CH_2-CH_2-NH_3$
B. Randomly Branched	Branched PEI	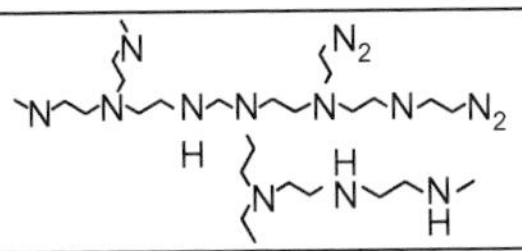
C. Dendrimer	Polyamidoamine (Superfect™)	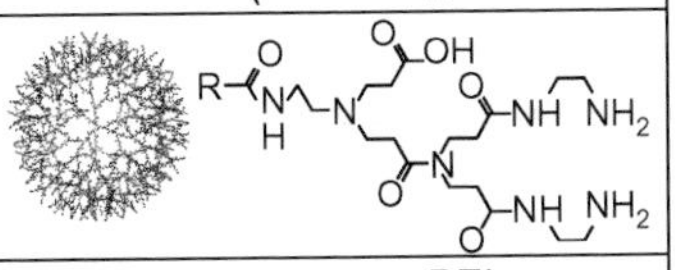
D. Graft/block copolymers	PEO-b-PLL P123-g-PEI(2K) Peptide-PEO-g-PEI	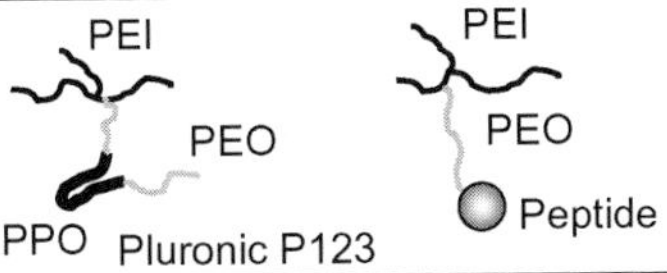
E. Dispersed Networks	**PEO-cross-PEI**	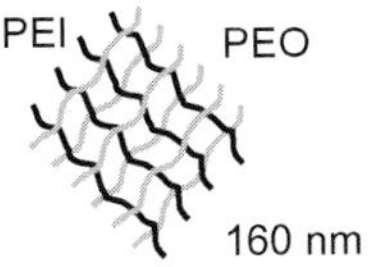

Figure 9.1. Representative polycations of different architecture developed for delivery of polynucleotides: (A) linear polyethyleneimine (PEI), ExGen 500 (Ferrari et al., 1997); (B) branched PEI (Boussif et al., 1995); (C) Polyamidoamine, Superfect™ (Haensler and Szoka, 1993; Kukowska-Latallo et al., 1996); (D) Block copolymer of poly(ethylene oxide) (PEO) and poly(L-lysine) (PLL) (Katayose and Kataoka, 1997; Wolfert and Seymour, 1996), PEI grafted with Pluronic P123 (Nguyen et al., 2000), PEI grafted with PEO carrying a peptide targeting group at the free end (Kursa et al., 2003; Ogris et al., 2003; Vinogradov et al., 1999a); (E) Nanoscale gels ("nanogels") from cross-linked PEI and PEO (Vinogradov et al., 1999b, 2004).

The cationic lipids and polycations are often inefficient when injected locally in the tissue, such as skeletal muscle or tumor sites, and display lower gene expression levels than that of the naked DNA. Few notable exceptions, including polyplexes formed by biodegradable cationic polymers, are believed to sustain release of the DNA at the site of injection as the polycation is hydrolyzed (Hosseinkhani et al., 2004; Wang et al., 2002). It is therefore noteworthy that a different group of molecules, nonionic water-soluble polymers that do not bind or condense DNA, displayed ability to increase transgene expression upon local administration of DNA in tissues. Among such agents are hydrophilic polymers, such as poly(N-vinyl pyrrolidone) and poly(vinyl alcohol) (Mumper et al., 1996, 1998) and Pluronic block copolymers (Lemieux et al., 2000). The evidence started to accumulate suggesting that these agents, in particular, Pluronic block copolymers, can increase gene expression in a way that does not fit the current non-viral gene delivery paradigm. This chapter considers the current advances in using Pluronic block copolymers in gene delivery.

II. PLURONIC BLOCK COPOLYMERS: THE STRUCTURE AND NOMENCLATURE

Pluronic block copolymers consist of hydrophilic ethylene oxide (EO) and hydrophobic propylene oxide (PO) blocks arranged in a basic A-B-A structure: $EO_x\text{-}PO_y\text{-}EO_x$. The structure formula of Pluronic block copolymers is presented in Fig. 9.2. Over 30 Pluronic molecules with different lengths of EO (N_{EO}) and PO (N_{PO}) blocks are available from BASF Corp. (Parsippany, NJ, USA) see Table 9.1. These molecules are characterized by different hydrophilic–lipophilic balance (HLB) and critical micelle concentration (CMC). Pluronic nomenclature includes a letter, F, P, or L, followed by a two- or three-digit numeric code. The letters stand for solid (F), paste (P) or liquid (L). The numeric code defines the structural parameters of the block copolymer. The last digit of this code approximates the weight content of EO block in tens of weight percent (for example, 80% wt. if the digit is 8 or 10% wt. if the digit is 1). The remaining first one or two digits encode the molecular mass of the central PO block. To decipher the code, one should multiply the corresponding number by 300 to obtain the approximate molecular mass in Da. Therefore, Pluronic nomenclature provides a convenient approach to estimate the characteristics of the block copolymer in the absence of reference literature.

The non-proprietary name "Poloxamer" is used for the block copolymers that are listed in the US Pharmacopoeia and for which the NF (National

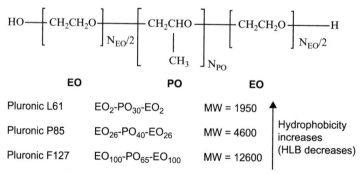

Figure 9.2. Pluronic block copolymers available from BASF Co. (Wyandotte, MI, USA), contain two hydrophilic EO blocks and a hydrophobic PO block.

Table 9.1. Physicochemical Characteristics of Pluronic Block Copolymers

Copolymer	MW[a]	Average no. of EO units (N_{EO})[b]	Average no. of PO units (N_{PO})[b]	HLB[c]	Cloud point in 1% aqueous solution, °C[c]	CMC, M[d]
L35	1900	21.59	16.38	19	73	$5.3 \cdot 10^{-3}$
L43	1850	12.61	22.33	12	42	$2.2 \cdot 10^{-3}$
L44	2200	20.00	22.76	16	65	$3.6 \cdot 10^{-3}$
L61	2000	4.55	31.03	3	24	$1.1 \cdot 10^{-4}$
L62	2500	11.36	34.48	7	32	$4.0 \cdot 10^{-4}$
L64	2900	26.36	30.00	15	58	$4.8 \cdot 10^{-4}$
F68	8400	152.73	28.97	29	>100	$4.8 \cdot 10^{-4}$
L81	2750	6.25	42.67	2	20	$2.3 \cdot 10^{-5}$
P84	4200	38.18	43.45	14	74	$7.1 \cdot 10^{-5}$
P85	4600	52.27	39.66	16	85	$6.5 \cdot 10^{-5}$
F87	7700	122.50	39.83	24	>100	$9.1 \cdot 10^{-5}$
F88	11400	207.27	39.31	28	>100	$2.5 \cdot 10^{-4}$
L92	3650	16.59	50.34	6	26	$8.8 \cdot 10^{-5}$
F98	13000	236.36	44.83	28	>100	$7.7 \cdot 10^{-5}$
L101	3800	8.64	58.97	1	15	$2.1 \cdot 10^{-6}$
P103	4950	33.75	59.74	9	86	$6.1 \cdot 10^{-6}$
P104	5900	53.64	61.03	13	81	$3.4 \cdot 10^{-6}$
P105	6500	73.86	56.03	15	91	$6.2 \cdot 10^{-6}$
F108	14600	265.45	50.34	27	>100	$2.2 \cdot 10^{-5}$
L121	4400	10.00	68.28	1	14	$1.0 \cdot 10^{-6}$
P123	5750	39.20	69.40	8	90	$4.4 \cdot 10^{-6}$
F127	12600	200.45	65.17	22	>100	$2.8 \cdot 10^{-6}$

[a]The average molecular weights provided by the manufacturer (BASF Co., Parsippany, NJ).
[b]The average numbers of EO and PO units were calculated using the average molecular weights.
[c]HLB values of the copolymers the cloud points were determined by the manufacturer.
[d]CMC values were determined previously using pyrene probe (Kozlov et al., 2000).

Formulary) grade is available. As of the year 2004, five block copolymers are listed in the US Pharmacopoeia: poloxamer 124 (Pluronic L44), poloxamer 188 (Pluronic F68), poloxamer 237 (Pluronic F87), poloxamer 338 (Pluronic F108), and poloxamer 407 (Pluronic F127). In addition to Pluronic nomenclature, several other proprietary names of the block copolymers, such as Synperonic, can also be found in the literature. Recently CytRx Corporation (Atlanta, GA, USA) has introduced a new name for its purified poloxamer copolymers, which includes the letter code CRL, for example, CRL-1605 (Newman et al., 1998). Supratek Pharma Inc. (Montreal, PQ, Canada) uses the code SP, for example, SP1049C or SP1017 for its various block copolymer formulations (Danson et al., 2004; Lemieux et al., 2000). For example, the GMP-manufactured SP1017 formulation represents the aqueous solution of the mixture of two copolymers, a hydrophobic Pluronic L61 (0.25% w./v.) and hydrophilic Pluronic F127 (2% w./v.).

Several other families of poly(ethylene oxide) PEO and poly(propylene oxide) (PPO) block copolymers available from BASF Corp. are worth mentioning. The first family, Pluronic R (or 'meroxapol'), possesses a B-A-B structure, which is the inverse of the Pluronic structure. The second family, Tetronic (or "polyoxamine"), contains a central ethylenediamine molecule with four PPO-PEO diblock branches attached to the nitrogen atoms. Finally, the order of the PEO and PPO blocks represented in Tetronic can be reversed, creating the third family, Tetronic-R. Currently, these copolymer families are not frequently used for the gene delivery applications reviewed in this chapter. However, at least some of these copolymers have properties similar to the properties of Pluronic and therefore can be used in similar applications (Houssami *et al.*, 1990; Moghimi and Hunter, 2000).

III. SYNTHESIS AND PURIFICATION OF PLURONIC BLOCK COPOLYMERS

Pluronic block copolymers are synthesized by sequential addition of PO and EO monomers in the presence of an alkaline catalyst (Schmolka, 1977). The reaction is initiated by polymerization of the PO block followed by the growth of EO chains at both ends of the PO block. Anionic polymerization usually produces polymers with a relatively low polydispersity index (M_w/M_n). However, the commercially available Pluronic preparations contain admixtures of the PO homopolymer as well as di- and triblock copolymers, exhibiting lower degrees of polymerization than expected. Chromatographic fractionation can be employed in procedures for the manufacture of highly purified block copolymers (Emanuele *et al.*, 1996a,b). This reduces the presence of admixtures, particularly, of the PPO homopolymer and of the block copolymers containing shorter PEO chains than expected.

IV. MICELLIZATION, SOLUBILIZATION, AND FORMATION OF GELS

A defining property of Pluronic is the ability of individual block copolymer molecules, termed *unimers*, to self-assemble into micelles in aqueous solutions above the CMC, a process called *micellization* (Fig. 9.3). The number of block copolymer unimers forming one micelle is referred to as the *aggregation number*. Usually, this number ranges from several to over a hundred. The micelles have a hydrophobic PO core and a hydrophilic EO shell and their shape can be spherical, rod-like or lamellar depending on the lengths of the PO and EO

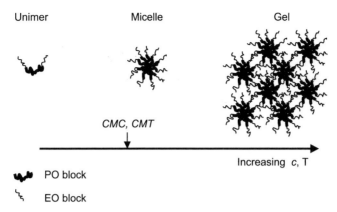

Figure 9.3. Self-assembly of Pluronic unimers into micelles occurs as the concentration of the block copolymer reaches the CMC and temperature reaches CMT. Further increase of the block copolymer concentration results in formation of Pluronic gel.

blocks (Nagarajan, 1999). The PO core within the micelles can serve as a "cargo hold" for the incorporation of various hydrophobic compounds. The process of transfer of water-insoluble compounds into the PO core in the micellar solution is referred to as *solubilization*.

Micelles are used as carriers for various drugs. The incorporation of drugs into the core of the Pluronic micelle results in increased solubility, increased metabolic stability and increased circulation time for the drug. A peptide or other biospecific molecule capable of promoting site-specific drug delivery can be attached to the surface of the EO corona, effectively targeting a micelle to a specific site in the body (Kabanov *et al.*, 1989a, 1992). The extensive recent reviews on the use of block copolymer micelles and related nanoparticles for drug delivery can be found in the literature (Adams *et al.*, 2003; Alakhov *et al.*, 2001; Allen *et al.*, 1999; Bittner and Mountfield, 2002; Duncan, 2003; Kabanov and Alakhov, 2000, 2002; Kakizawa and Kataoka, 2002; Kataoka *et al.*, 2001; Kwon, 2003; Kwon and Okano, 1999; Lavasanifar *et al.*, 2002; Lukyanov and Torchilin, 2004; Nishiyama and Kataoka, 2003; Otsuka *et al.*, 2003; Rosler *et al.*, 2001; Torchilin, 2001, 2002; van Nostrum, 2004).

The block copolymer unimers are present in the solution below, at and above the CMC. When the micelles are formed the unimers remain in equilibrium with the micelles, which serve as a source of the new unimers if the solution is diluted (e.g., in the body fluids). Due to the presence of the hydrophobic PO blocks, the Pluronic unimers can adsorb on surfaces, interact with the lipid membranes and even translocate inside the cells (Batrakova *et al.*,

2001b, 2003b). Upon binding with the cell membranes, Pluronic unimers induce structural transitions in the membrane accompanied by the decrease of the membrane microviscosity (Batrakova et al., 2001b), increase of the rate of the lipid "flip-flop" (Krylova et al., 2003) and enhancement of the transmembrane transport of ions and solutes (Erukova et al., 2000; Krylova and Pohl, 2004; Krylova et al., 2003).

As the concentration of the block copolymer increases, the micelles form tighter packed structures and, above a certain threshold concentration, the gels are formed (Fig. 9.3). These gels have microheterogeneous structure, are swollen, are generally biocompatible, and are capable of incorporating biomacromolecules, viruses and even living cells (Cellesi et al., 2002; Epperson et al., 2001; Johnston and Miller, 1989; Rill et al., 1998; Schmolka, 1977; Van Belle et al., 1998). The PO chains display lower critical solution temperature (LCST) behavior and, as a result, Pluronic micelles and gels are thermo-sensitive. While the PO chains are hydrated and soluble in water at low temperatures (e.g., 4 °C) they dehydrate and become insoluble as the temperature rises, resulting in the formation of the cores of the micelles. The temperature, at which micelles are formed is referred to as the *critical micelle temperature* or CMT (Linse and Malmsten, 1992). Similarly, the formation of the gels is observed only as the gelation temperature is reached (Schillen et al., 1993). This property is often used for entrapment of macromolecules, such as DNA into gels: the components are mixed at low temperature and then the temperature is elevated, which induces a sol-gel transition. Furthermore, the sterile filtration of Pluronic solutions is usually conducted at low temperatures when the viscosity of these solutions is low.

The structure of the block copolymers has profound effects on the micellization and gelation processes. As the length of the hydrophobic PO block increases, the formation of the micelles and gels becomes more favorable resulting in lower values of CMC, CMT, and gelation concentrations and temperatures (Alexandridis et al., 1994a,b; Linse, 1993). Conversely, an increase in the length of the hydrophilic EO block decreases the stability of the micelles. Extensive discussions of the properties of the Pluronic micelles and gels can be found elsewhere (Alexandridis and Lindman, 1999).

Intensive work has been performed on Pluronic-based gels as drug delivery and sustained release systems (Bochot et al., 1998; Carter et al., 1989; Chen Chow and Frank, 1981; Davidorf et al., 1990; Edsman et al., 1998; Fults and Johnston, 1990; Herbert et al., 1997; Hom et al., 1996; Johnston and Miller, 1989; Johnston et al., 1992; Lu et al., 1995; Morikawa et al., 1987; Paavola et al., 1995, 1998; Scherlund et al., 2000; Tobio et al., 1999; Yeh et al., 1996). Pluronic-based gels, such as Pluronic F127, were administered through various routes including intravenous (Johnston and Miller, 1989; Lu et al., 1995), intraperitoneal (Johnston et al., 1992; Pec et al., 1992), subcutaneous (Carter et al., 1989), rectal

(Barichello *et al.*, 1999), ocular (Bochot *et al.*, 1998; Davidorf *et al.*, 1990; Edsman *et al.*, 1998), topical (Lu *et al.*, 1995), intramuscular (Wang and Johnston, 1995) and nasal (Juhasz *et al.*, 1989). These studies examined formulations of both low molecular mass drugs and biomacromolecules, including proteins and polynucleo- tides. Recent studies exploited the well-known thermotropic behavior of Pluronic gels as a means for a controlled release of encapsulated drugs (Bromberg, 2001; Juhasz *et al.*, 1989).

V. BIOLOGICAL ACTIVITIES OF PLURONIC BLOCK COPOLYMERS

In addition to the use of Pluronic block copolymers as structural components of micellar and gel formulations, Pluronic molecules exhibit a remarkable set of biological activities. For example, water-in-oil, oil-in-water and water-in-oil-in-water emulsions formulated with selected Pluronic block copolymers display immunoadjuvant properties (Allison and Byars, 1986, 1990; Bomford, 1981; Byars and Allison, 1987; Hunter and Bennett, 1984, 1986; Hunter *et al.*, 1981, 1991, 1994; Ke *et al.*, 1997; Millet *et al.*, 1992; Takayama *et al.*, 1991). These studies reported significant enhancement of both cell-mediated and humoral immune response induced by addition of the block copolymer formulations with respect to a very broad spectrum of antigens. Selected block copolymers, such as Pluronic F127 have been found to significantly enhance the rate of wound and burn healing, and therefore have been included in cream formulations and skin substitutes for the treatment of burns and for other tissue engineering applications (Agren, 1998; Cao *et al.*, 1998; Follis *et al.*, 1996; Gear *et al.*, 1997; Nalbandian *et al.*, 1987; Rodeheaver *et al.*, 1976, 1980; Schmolka, 1972). Several Pluronic-based formulations were shown to effectively prevent postoperative adhesions or reduce adhesion area after surgery (Leach and Henry, 1990; Vlahos *et al.*, 2001). Furthermore, Pluronic block copolymers can enhance sealing of cell membranes permeabilized by ionizing radiation and electropora- tion, thus preventing cellular necrosis, which can be helpful for improving drug and gene delivery in skeletal muscle (Hannig *et al.*, 2000; Lee *et al.*, 1992, 1993, 1999). Recent studies have demonstrated that select Pluronic block copolymers can interact with multidrug-resistant (MDR) cancer cells resulting in chemo- sensitization of these cells (Alakhov *et al.*, 1996; Batrakova *et al.*, 1996, 1999, 2001a, 2003a; Venne *et al.*, 1996). These interactions involve alteration of the structure of cell membranes by block copolymer molecules as well as energy depletion in MDR cells leading to the inhibition of various energy-dependent drug resistance mechanisms, in particular, P-glycoprotein (P-gp) drug efflux systems (Kabanov *et al.*, 2002a). A formulation of doxorubicin with the mixture of Pluronic L61 and F127 (SP1049C), which is effective against multiple drug- resistant tumors, was evaluated in a Phase I human clinical trial (Danson *et al.*,

2004). This formulation currently undergoes Phase II trials against adenocarcinoma of esophagus and soft tissue sarcoma (both at Manchester Christy Hospital, UK) (Valle et al., 2004). The ability of Pluronic block copolymers to inhibit P-gp and other drug efflux transporters is a basis of the current studies using these copolymers to increase delivery of drugs across biological barriers, such as blood–brain barrier and intestinal epithelium (Kabanov and Alakhov, 2002; Kabanov et al., 2003). These examples show that the Pluronic block copolymers are valuable biological adjuvants exhibiting useful biological properties, which are of considerable importance for various therapeutic applications.

VI. SUMMARY OF APPLICATIONS OF PLURONIC IN GENE DELIVERY

Pluronic block copolymers attracted considerable attention as promising polymer excipients for gene delivery applications. Table 9.2 briefly summarizes some of these studies. First, Pluronic block copolymers were shown to increase the transfection in vitro and in vivo with adenovirus and lentivirus vectors (Dishart et al., 2003; Feldman et al., 1997; Maillard et al., 2000; March et al., 1995; Van Belle et al., 1998). Second, they were shown to increase expression of genes delivered into cells using non-viral vectors (Astafieva et al., 1996; Cho et al., 2000, 2001; Liu et al., 1996). Third, the conjugates of Pluronic with polycations, were used as DNA-condensing agents to form polyplexes (Gebhart et al., 2002; Jeon et al., 2003; Nguyen et al., 2000). Finally, it was demonstrated that they can increase regional expression of the naked DNA after its injection in the skeletal and cardiac muscles or tumor (Gebhart, 2003; Hartikka et al., 2001; Lemieux et al., 2000; Liaw et al., 2001; Lu et al., 2003; Pitard et al., 2002; Riera et al., 2004). Therefore, there is substantial evidence that Pluronic block copolymers can improve gene expression with different delivery routes and different types of vectors, including naked DNA. These studies are discussed below.

VII. EFFECTS OF PLURONIC BLOCK COPOLYMERS IN VIRAL GENE DELIVERY

March et al. was first to show that poloxamer 407 (Pluronic F127), at concentrations as high as 15% wt., increased the rate and frequency of transduction of vascular smooth muscle cell (SMC) with the first generation of recombinant adenovirus vectors (March et al., 1995). This study demonstrated that the pharmacokinetics of adenoviral-mediated gene delivery to vascular SMC can be modulated by the block copolymer. It was hypothesized that the gel of poloxamer 407 formed under the conditions of the experiment, served as the delivery reservoir for the virus and improved gene delivery by maintaining high

Table 9.2. Effects of Pluronic Block Copolymers on Expression of DNA[a]

Vector	Delivery route/cells or tissue transfected	Promoter: transgene	Pluronic Type[b]	Conc., %[c]
Adenovirus	In vitro (vascular smooth muscle cells) (March et al., 1995); I.A. (rabbit) (Maillard et al., 2000); I.A. (rats) (Feldman et al., 1997)	RSV:β-gal	F127	15–20
	I.A. (rabbit) (Van Belle et al., 1998)	CMV:Luc	F127	15–20
Lentivirus	In vitro (vascular endothelial cells, smooth muscle cells) (Dishart et al., 2003)	CMV:GFP	F127	5
DNA/PEVP	In vitro (NIH 3T3 cells) (Astafieva et al., 1996)	RSV:CAT	P85	0.1
DNA/PLL conjugates	In vitro (HepG2 and Hela cells) (Cho et al., 2000, 2001)	CMV: β-gal, p53	F127	n/a
DNA/PEI-P123	In vitro (Cos-7 and PC-3 cells) (Gebhart et al., 2002; Nguyen et al., 2000)	CMV:Luc	P123	0.03
Plasmid DNA	I.M. (mouse) (Lemieux et al., 2000)	CMV:Luc, β-gal, EPO	SP1017	0.01–0.05
	I.M. (mouse) (Hartikka et al., 2001)	CMV:Luc	F68	4
	Ocular drops (mouse) (Liaw et al., 2001)	CMV:β-gal	F68	0.01–0.3
	I.M. (mouse) and I.C. (rats) (Pitard et al., 2002)	CMV:Luc	L64	0.03–0.5
	I.M. (rats) (Riera et al., 2004)	CMV:β-gal, hHGF	SP1017	0.01
	I.M and I.T. (mouse, PC-3 tumors) (Gebhart, 2003)	CMV:Luc	SP1017, L61, P123, F127	0.1

[a]Abbreviations used: β-gal, β-galactosidase; CAT, chloramphenicol acetyltransferase; CMV, cytomegalovirus; EPO, erythropoietin; GFP, green fluorescent protein; hHGF, human hepatocyte growth factor; I.A., intra-arterial; I.C., intracardiac; I.M., intramuscular; I.T., intratumoral; Luc, luciferase; RSV, Rous sarcoma virus; PEI, polyethyleneimine; PEI-P123, graft copolymer of PEI and PBC P123; PEVP, poly(N-thyl-4-vinylpyridinium) bromide; PL, polylysine.

[b]Alternative polymer names: F127, poloxamer 407; L64, poloxamer 184; F68, poloxamer 188; SP1017, mixture of L61 and F127 (1:8).

[c]Optimal or only studied concentration of Pluronic in formulation.

pericellular concentrations of the vector (March et al., 1995). Subsequently, based on this earlier in vitro work, the study by Feldman et al. has shown that poloxamer 407 also facilitates adenovirus-mediated arterial transfection in vivo (Feldman et al., 1997). Gene transfer was performed in balloon-injured rat carotid arteries using adenoviral vectors diluted in poloxamer 407. The block copolymer-based formulation displayed higher transfection efficiency compared to the vector administered in the phosphate-buffered saline. Moreover, in the presence of poloxamer 407, it was possible to reduce the incubation time of adenoviral vectors without compromising the transfection efficiency. Notably, the site specificity of the arterial gene transfer was not altered by administration of poloxamer 407 and no specific tissue toxicity was observed even at high concentrations (20% wt.) of the copolymer administered. Similar results were obtained using the gels of poloxamer 407 for percutaneous adenovirus-mediated gene transfer both in nonstented rabbit iliac arteries and in conjunction with stent implantation (Maillard et al., 2000; Van Belle et al., 1998). The authors suggest that the use of block copolymer gels for percutaneous gene delivery is applicable to conventional stents and could present an attractive method by which to achieve local biological effects in a stent environment. Finally, in a recent study, Dishart et al. have evaluated gene delivery to endothelial cells and smooth muscle cells using alternate AAV serotypes and a third-generation vesicular stomatis virus glycoprotein-pseudotyped lentiviral system. As in the previously described studies, using adenoviral vectors, the transduction levels with the lentiviral system were enhanced with the poloxamer 407 (Dishart et al., 2003). Overall, these works suggest that poloxamer 407-based gels have potential as delivery systems for virus-mediated gene transfers.

VIII. PLURONIC BLOCK COPOLYMERS AS ADJUVANTS FOR NON-VIRAL VECTORS

Pluronic was also shown to increase expression of genes delivered into cells using polycation- and lipid-based non-viral vectors (Astafieva et al., 1996; Cho et al., 2000, 2001; Liu et al., 1996). The uptake of, and cell transfection with, the polyplex formed between plasmid DNA and poly(N-ethyl-4-vinylpyridinium) (PEVP) were significantly enhanced in several cell lines when Pluronic P85 was administered to the cells simultaneously with the polyplex (Astafieva et al., 1996). In this case the block copolymer was administered at the concentration of 1% wt. in a form of a liquid micellar solution in the serum-free medium. The suggested role of P85 was to intensify endocytic uptake of the complex into eukaryotic cells as well as enhance the liberation of the macromolecules from the endocytic compartments in the cytoplasm. Indeed, increased uptake of the DNA into the cells was observed in the presence of Pluronic P85 solution compared to the polyplexes alone. The authors speculated that Pluronic may interact with the hydrophobic sites of the polyplexes forming mixed Pluronic/DNA/PEVP nanoparticles. However, the formation of the triple component nanoparticles was not shown in this work.

Subsequently, a study by Kuo et al. has shown that hydrophilic Pluronic block copolymers applied at a concentration of 1 to 3% wt. reduced the serum-mediated inhibition of gene transfer with DNA/polyethylenimine (PEI) polyplexes in NIH/3T3 cells (Kuo, 2003). The efficiency of the block copolymers decreased as their hydrophobicity elevated: F68 > F127 > P105 > P94 > L122 > L61. Although the detailed mechanisms for such effects were not fully understood, it was suggested that the block copolymers incorporated in the particles. As a result the hydrophilic EO chains of the copolymers provided for the steric stabilization of the particle dispersions and prevented aggregation of the particles in the presence of serum. Similarly, poloxamer 407 was also shown to increase several fold the expression level of a reporter gene, β-galactosidase in HepG2 cells transfected with polyplexes formed by asialo-orosomucoid–polylysine conjugates and plasmid DNA (Cho et al., 2000). This conjugate-based system was designed to selectively target asialoglycoprotein receptor expressed at the surface of the hepatic cells. Thus Pluronic block copolymers can increase delivery of the DNA in the cells using both non-targeted and receptor-targeted polyplexes.

In another study Liu et al. has shown that emulsions composed of Castor oil, dioleoylphosphatidylethanolamine (DOPE) and 3 beta [N-(N', N'-dimethylaminoethane) carbamoyl] cholesterol (DC-Chol) were stabilized by adding nonionic surfactants, such as Pluronic F127, Pluronic F68, Tween and Brij 72 in the presence of serum (Liu et al., 1996). The particle size of the emulsions formulated with the plasmid DNA decreased when the surfactants were added. The resulting emulsions containing the surfactants displayed higher transfection efficiency in BL-6 cells in the presence of 20% fetal bovine serum than the surfactant-free emulsions (Liu et al., 1996) This study suggested that the particle size of the DNA/emulsion complexes and their ability to transfect cells was dependent on the structure of the hydrophilic head groups and the concentration of non-ionic surfactant in the formulation. Overall, these works suggest that Pluronic block copolymers are useful pharmaceutical excipients that can be applied to improve formulation and activity of various non-viral vectors.

IX. PLURONIC-POLYCATION GRAFT/BLOCK COPOLYMERS FOR GENE DELIVERY

The studies on the first generation of the polyplex systems for DNA delivery based on homopolymer or dendritic polycations have produced quite effective *in vitro* transfection agents such as linear PEI (Ferrari et al., 1997). However, the size and dispersion stability of these polyplexes is often quite sensitive to the nature of the buffer and to the presence of serum proteins. Consequently, such polyplexes have displayed significant limitations when used *in vivo* including toxicities, low transfection activities and poor pharmacokinetic and biodistribution characteristics in the body. These drawbacks are primarily the result of poor solubility and aggregation of polyplexes in the presence of cells and serum components, for example, within the capillary beds of the lung, which leads to severe pulmonary embolism (Mahato *et al.*, 1997). Overall, the current literature strongly suggests that both lipoplex and polyplex particles bearing a strong positive charge display poor pharmacokinetics and biodistribution and are preferentially taken up by the liver nonparenchymal cells mainly via Kupffer cell phagocytosis (Mahato *et al.*, 1995; Verbaan *et al.*, 2001). The positive charge is also responsible for the immune recognition of such particles and their subsequent rapid clearance from the circulation (Hwang and Davis, 2001).

Thus, we and others developed a second generation of polyplexes using "cationic copolymers" to overcome such problems (Choi *et al.*, 1998; Kabanov *et al.*, 1995; Kataoka *et al.*, 1996; Katayose and Kataoka, 1997; Maruyama *et al.*, 1997; Vinogradov *et al.*, 1998; Wolfert *et al.*, 1996). These molecules have graft or block copolymer architecture comprising nonionic water-soluble polymer, such as PEO or poly(N-(2-hydroxypropyl)methacrylamide) (PHPMA) linked to a polycation, such as polyspermine (Kabanov *et al.*, 1995), poly(L-lysine) (Choi *et al.*, 1998; Kataoka *et al.*, 1996; Katayose and Kataoka, 1997; Wolfert *et al.*, 1996), poly[2-(diethylamino)ethyl methacrylate] (PDEAEMA) (Asayama *et al.*, 1997), 2-(trimethylammonio)ethyl methacrylate (PTMAEMA) (Oupicky *et al.*, 1999, 2000), PEI (Kircheis *et al.*, 2001a; Nguyen *et al.*, 2000; Ogris *et al.*, 2000; Vinogradov *et al.*, 1998) and others. During interaction with the DNA such cationic copolymers form practically uncharged hydrophilic particles called *block ionomer complexes* or *polyion complex micelles* that contain a core of neutralized polycation and DNA chains and a shell of water-soluble nonionic chains. These systems have shown substantial promise for the delivery of short DNA oligonucleotides *in vitro* and *in vivo* (Harada *et al.*, 2001; Kabanov *et al.*, 1995; Roy *et al.*, 1999). However, they are much less efficient when the plasmid DNA is delivered. The complexes formed by the plasmid DNA with the cationic copolymer are much larger than those formed by the oligonucleotides and interaction of these complexes with the membranes is hindered by the nonionic chains displayed at the surface of these complexes (Bronich *et al.*, 2001; Harper *et al.*, 1991; Lasic *et al.*, 1991; Nguyen *et al.*, 2000).

As one approach to overcome this problem Nguyen *et al.* proposed to use Pluronic block copolymers as substitutes for nonionic EO chains in the block ionomer complexes (Nguyen *et al.*, 2000). The idea behind this approach was that hydrophobic PO chains of the Pluronic in the polyplex could facilitate membrane interactions and possibly act as a synthetic fusogen moiety to improve the transport of the polyplex into the cell interior. At the same time, the EO chains of the Pluronic molecules provide for the polyplex solubility in aqueous dispersions. To test this hypothesis the graft/block copolymers of Pluronic and PEI have been synthesized. One such polyplex system, based on a Pluronic P123-g-PEI(2K) reacted with plasmid DNA resulting in formation of stable 100 to 160 nm particles, which did not aggregate in the presence of the serum and displayed improved gene delivery characteristics, relative to PEI alone, both *in vitro* and *in vivo*. In order to obtain such particles, the free Pluronic P123 was added as the third component, along with Pluronic P123-g-PEI(2K) copolymer and DNA. It was suggested that the resulting polyplex represented mixed complexes, in which the free Pluronic molecules were bound through the hydrophobic interactions with the PO chains of the Pluronic P123-g-PEI(2K) conjugates (Fig. 9.4). The resulting particles are stabilized in the aqueous dispersions by hydrophilic EO chains very much like the Pluronic micelles. It was shown that addition of the free Pluronic P123 in such polyplexes was essential to prevent aggregation of the particles, accomplish efficient transport of the plasmid into cells and result in high levels of gene expression (Gebhart *et al.*, 2002). Furthermore, the free Pluronic P123 in this formulation was replaced with a more hydrophilic Pluronic F127 practically without the loss of

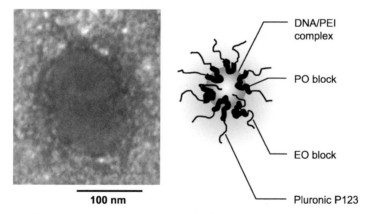

DNA/PEI
complex

PO block

EO block

100 nm

Pluronic P123

Figure 9.4. Transmission electron micrograph of polyplex particle formed in the mixture of Pluronic P123, Pluronic P123-g-PEI(2K) conjugate and plasmid DNA (left) and schematic representation of such particle (right): The sphere represents the polyion complex core of condensed DNA and PEI chains. The Pluronic P123 chains that are grafted to the PEI within the core are not incorporated into the core, but form a micelle-like structure around the core, orienting the lipophilic PO blocks near the core and the hydrophilic EO blocks forming the exterior shell. Additional non-modified Pluronic P123 chains fill in to optimize the stability of the micelle-like structure. Based on data reported in (Gebhart *et al.*, 2002).

the polyplex transfection activity. The polyplexes formed were practically electroneutral and displayed little toxicity *in vitro* and *in vivo* (Gebhart and Kabanov, 2001; Nguyen et al., 2000). Using these polyplexes, administered intravenously, efficient gene transfer into the liver cells was achieved as shown based on expression of a reporter gene, lucliferase (Nguyen et al., 2000) and ICAM-1 molecules in ICAM-1 deficient mouse (Ochietti et al., 2002). The reporter gene expression was observed in spleen, heart, lungs and liver 24 hours after intravenous injection of this polyplex in mice, and the distribution of gene expression between these organs was more uniform compared to unmodified branched PEI (25 kDa), which is mainly expressed in the lungs. In a recent study a similar type of graft/block copolymer molecule, Pluronic F127 grafted to PLL, was proposed for gene delivery (Jeon et al., 2003). As in the study discussed above, this copolymer formed compact complex with plasmid DNA that displayed relatively high transfection activity *in vitro*. Overall, the Pluronic-grafted polycations represent a promising type of gene delivery molecules, which, in our view, will attract further attention.

X. EFFECTS OF PLURONIC BLOCK COPOLYMERS ON GENE TRANSFER USING NAKED DNA

One promising approach in gene therapy involves the direct injection of Pluronic block copolymers in combination with a plasmid DNA in skeletal and cardiac muscles (Gebhart, 2003; Hartikka et al., 2001; Lemieux et al., 2000; Liaw et al., 2001; Lu et al., 2003; Pitard et al., 2002; Riera et al., 2004). It is well known that plasmid DNA can be injected into skeletal muscle and generate therapeutically meaningful levels of gene expression (Wolff et al., 1990). This was demonstrated using therapeutic genes encoding systemically acting secreted proteins (Tokui et al., 1997; Tripathy et al., 1996) as well as locally acting proteins (Braun et al., 2000; Laham et al., 2000; Lathi et al., 2001). However, the potential of this technique for gene therapy is limited, due to poor transduction efficiency and high variability of gene expression observed with the naked DNA (Danko et al., 1994; Wolff et al., 1992). Moreover, the commercially available cationic carriers, such as cationic dendrimers and lipids that are known to improve gene expression *in vitro*, commonly inhibit intramuscular gene expression.

It was recently reported that a mixed Pluronic formulation, termed SP1017, composed of Pluronics L61 and F127, increased the level and duration of expression of reporter and therapeutic genes injected in the skeletal muscle (Lemieux et al., 2000). (This mixture has the same composition as the Pluronic mixture with doxorubicin, SP1049C, that has been evaluated in clinical trials (Danson et al., 2004)). The dose-dependency study suggested that maximal stimulation of gene expression in skeletal muscle with this formulation was observed at a relatively low concentration of the block copolymer (0.01% wt.), which provides for at least a 500-fold safety margin in animals. Compared to poly(vinyl pyrrolidone), another non-ionic carrier that is under development for non-viral gene therapy (Mumper et al., 1996; Rolland and Mumper, 1998), SP1017 appears to be more efficient and requires less DNA to achieve similar levels of the transgene expression (Lemieux et al., 2000). The high levels of transgene expression observed with SP1017 sustained for a few weeks as opposed to the expression of the naked DNA alone which faded after several days.

The capacity of SP1017 to enhance plasmid-driven expression of secreted therapeutic genes was illustrated using the murine erythropoietin (EPO) gene (Lemieux et al., 2000). The hematocrits of mice injected with pCMV-mEPO formulated with SP1017 increased in comparison with non-treated and naked DNA-treated mice (Lemieux et al., 2000). The resulting hematocrits levels were comparable to those achievable with adeno-associated virus (Zhou et al., 1998). The increase in hematocrits of pCMV-mEPO/SP1017-treated mice correlated with an increase in mEPO levels in both the serum and in the muscle, as detected by ELISA on day 7 post-injection. Furthermore, the effect of the pCMV-mEPO and SP1017 formulation on the hematocrits of mice was sustained for a significant period of time, from 21 to 50 days post-injection. SP1017 was also formulated with an Flt3-ligand plasmid (pNGVL-hFlex) to augment the immune response to a p17 HIV plasmid vaccine (Sang et al., 2003). The intramuscular injection SP1017-formulated pNVGL-hFlex increased the number of dendritic cells in draining lymph nodes and at the site of injection, thereby resulting in a significant increase of antigen-specific T cell response. Overall the efficacy of the Pluronic formulations in gene delivery in the muscle was comparable to that achieved using the electrotransfer of the plasmid DNA (Pitard et al., 2002). Furthermore, the SP1017 formulation applied in combination with electrotransfer resulted in further increase in gene expression in the muscle (Riera et al., 2004).

More recently it was shown that the formulations containing single Pluronic copolymers with plasmid DNA also enhanced the gene transfer in the muscle. In particular, Pluronic L64 (PE6400) improved the level of transfection efficiency over the naked DNA in the cardiac and skeletal muscle (Pitard et al., 2002). Moreover, Pluronic P85 has been identified as a very potent agent that significantly enhanced and prolonged the transgene expression in the skeletal muscle with the efficiency higher than that of the SP1017 formulation (paper in preparation). In the case of this copolymer, the levels of the reporter gene luciferase, found in the muscle at the peak of the gene expression, were about 20-times higher than the luciferase levels after administration of the naked DNA. Furthermore, the high levels of gene expression were observed for over one month and diminished only after 40 days. The area under the gene expression curve for the Pluronic P85-formulated plasmid DNA was about ten times greater compared to the expression of the naked DNA, a very substantial increase in the time-average exposure to the delivered transgene. Moreover, Pluronic P85 enhanced gene expression of a plasmid DNA for a broad range of DNA doses (5 to 50 μg per muscle per mouse). The Pluronic P85-formulated plasmid DNA was further examined in the antitumor DNA vaccination experiment suggesting significant enhancement in the protection of wild type p53 DNA vaccine against the tumor challenge in mice (data not shown).

XI. POSSIBLE MECHANISMS OF PLURONIC EFFECTS ON GENE DELIVERY

Mechanism(s) by which Pluronic acted in the applications described above has not been established. In brief, the following speculations were made: (1) At high concentrations (e.g., 15–20%), Pluronic forms a gel that may act as a local reservoir for adenovirus release (Feldman *et al.*, 1997); (2) Interactions between Pluronic and membrane facilitate cellular uptake of adenovirus (Cho *et al.*, 2000), polyplex (Astafieva *et al.*, 1996; Gebhart *et al.*, 2002) or naked DNA (Lemieux *et al.*, 2000); (3) Pluronic promotes membrane resealing and decreases trauma after electroporation (Hartikka *et al.*, 2001; Lee *et al.*, 1993); (4) Pluronic enhances DNA distribution through the muscle (Lemieux *et al.*, 2000); and (5) Pluronic increases transport of free DNA from the cytoplasm in the nucleus of the muscle cells (Pitard *et al.*, 2002). Overall, the mechanisms, by which Pluronic enhanced the transgene expression in these studies, are different from those of the cationic lipids or polycations. Notably, unlike these cationic molecules, neither Pluronic P85 nor SP1017 bind with and condense the plasmid DNA. This conclusion was based on several experiments including agarose electrophoresis, ethidium bromide displacement of the plasmid DNA formulated with Pluronic, and measurements of the heat effects of mixing DNA and Pluronic by microcalorimetry technique (unpublished).

At the same time, the block copolymers enhanced distribution of gene expression through the muscle around the injection site (Lemieux *et al.*, 2000; Pitard *et al.*, 2002). This may be due to enhanced extravasation of the DNA in the muscle tissue. The *in vitro* studies, however, suggested that SP1017 did not improve the *in vitro* transfection of either the myoblast or the myofiber developmental stages of the murine muscle cell line, C2C12; perhaps because the transport of the naked DNA in these cell lines was negligible (Kabanov *et al.*, 2002b). Similarly, Pitard *et al.* concluded that Pluronic L64 formulated with the plasmid DNA was inefficient *in vitro* in established cell lines and in isolated cardiomyocytes (Pitard *et al.*, 2002). However, alternative *in vitro* models were developed to investigate the effects of Pluronic in gene delivery. In particular, cytoplasm microinjection of plasmid DNA formulated with Pluronic L64 increased the percentage of the cells expressing the reporter gene, β-galactosidase (Pitard et al., 2002). At the same time, no changes in the presence of Pluronic L64 were observed when the DNA was directly microinjected in the nucleus. This suggested that Pluronic might have a role in the transport of the DNA from the cytoplasm into the nucleus. Though the formulation of plasmid DNA with Pluronic did not appear to enhance DNA uptake by cells in vitro, recent studies from our laboratory suggest that Pluronic P85 and other block copolymers enhance expression of reporter genes under the control of CMV promoter in the stably transfected mouse fibroblasts and other cells (Kabanov et al., 2004). This underscores that Pluronic block copolymers may act as biological response modifying agents, perhaps by up-regulating the transcription of the genes through activating selected signal pathways. Moreover, while Pluronic enhanced gene delivery in various strains of mouse and in other species, there was no enhancement of gene expression in the athymic nude mouse (although the levels of the naked DNA expression in the nude mice were comparable to those in other animals) (in preparation). Since the athymic nude mouse lacks a normal thymus gland and has a defective immune system, this result indicated that some molecules or cells of the immune system participate in the activation of transgene expression by Pluronic.

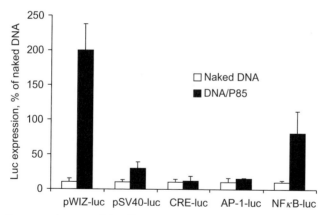

Figure 9.5. Promoter selectivity of the effect of Pluronic P85 on luciferase gene expression in the skeletal muscle. Groups of five Balb/c mice were injected intramuscularly with 0.3% P85-formulated (black bars) or non-formulated (white bars) plasmid DNA. Luciferase encoding plasmids were driven by CMV or SV40 promoters or CRE, AP-1 or NF-κB response elements. Luciferase activity was determined 24h after injections. Data are the percentage of increase of Pluronic P85-formulated DNA over the naked DNA, mean $\pm$ s.e.m (n = 10). *p < 0.05, n.s. not significant (p = 0.06). The experiment was conducted by Dr. Zhihui Yang in the University of Nebraska Medical Center (paper in preparation).

We further demonstrated that Pluronic effects on the transgene expression in the muscle were promoter-selective. In particular, the plasmids driven by CMV promoter and NF-κB response element were much more responsive to Pluronic than plasmids driven by SV-40 promoter or AP-1 and CRE response elements (Fig. 9.5). The promoter selectivity of Pluronic effect on transgene expression strongly suggested that Pluronic affected certain signal transduction pathways, in particular, NF-κB, which plays a central role in the regulation of the cellular defense and immunological responses. In the most recent unpublished study using the stably transfected fibroblast NIH 3T3 cell line as an *in vitro* model, we demonstrated that Pluronic P85 up-regulates the expression of an endogenous NF-κB-dependent gene, hsp68, to the same extent as it affects the reporter gene, luciferase, driven by CMV promoter in these cells. Future studies should identify molecular mechanisms of Pluronic action and possibly the genes upstream or downstream in NF-κB signal transduction pathway that are affected by the block copolymer.

XII. CONCLUSION

In conclusion, although the studies described in this chapter are relatively recent, the results obtained already suggest that Pluronic block copolymers are promising agents for gene therapy applications. First, it appears that at least some of these molecules can modify the biological response during gene therapy, resulting in an enhancement of the transgene expression as well as an enhancement of the therapeutic effect of the transgene. Second, Pluronic block copolymers are versatile molecules that can be used as structural elements in novel self-assembling gene delivery systems that are superior to currently known vectors. Used as nonionic surfactants to stabilize non-viral vectors in dispersion in the presence of the serum, Pluronic block copolymers have also shown some promise as formulation agents. Finally, these block copolymers can be formulated in a form of gels that have shown substantial improvement of *in vivo* delivery of the viral vectors. Based on these studies, the use of Pluronic block copolymers in gene delivery is a very exciting area of research in which new and important developments are expected in the near future.

XIII. MAIN ABBREVIATIONS

EO	ethylene oxide
CMC	critical micelle concentration
CMT	critical micelle temperature
DC-Chol	3 beta [N-(N′,N′-dimethylaminoethane) carbamoyl] cholesterol

DOPE   dioleoylphosphatidylethanolamine
HLB   hydrophilic–lipophilic balance
LCST   lower critical solution temperature
MDR   multidrug resistant
M_n number-average molecular mass
M_w weight-average molecular mass
P-gp   P-glycoprotein
PDEAEMA   poly[2-(diethylamino)ethyl methacrylate]
PEI   polyethyleneimine
PEVP   poly(N-ethyl-4-vinylpyridinium)
PHPMA   poly(N-(2-hydroxypropyl)methacrylamide)
PTMAEMA   2-(trimethylammonio)ethyl methacrylate
PEO   poly(ethylene oxide)
PO   propylene oxide
PPO   poly(propylene oxide)
SMC   smooth muscle cell

Acknowledgments

AVK acknowledges support from the National Science Foundation (BES-9907281) and the Nebraska Research Initiative Gene Therapy program. AVK and VYA are been co-founders, shareholders of Supratek Pharma Inc. (Montreal, PQ, Canada). AVK is an advisor, and VYA is an employee of this Company.

References

Adams, M. L., Lavasanifar, A., and Kwon, G. S. (2003). Amphiphilic block copolymers for drug delivery. *J. Pharm. Sci.* **92**, 1343–1355.

Agren, M. S. (1998). An amorphous hydrogel enhances epithelialisation of wounds. *Acta. Derm. Venereol.* **78**, 119–122.

Alakhov, V., Klinski, E., Lemieux, P., Pietrzynski, G., and Kabanov, A. (2001). Block copolymeric biotransport carriers as versatile vehicles for drug delivery. *Expert Opin. Biol. Ther.* **1**, 583–602.

Alakhov, V. Y., Moskaleva, E. Y., Batrakova, E. V., and Kabanov, A. V. (1996). Hypersensitization of multidrug resistant human ovarian carcinoma cells by pluronic P85 block copolymer. *Bioconjug. Chem.* **7**, 209–216.

Alexandridis, P., Athanassiou, V., Fukuda, S., and Hatton, T. A. (1994a). Surface activity of poly (ethylene oxide)-block-poly(propylene oxide)-block-poly(ethylene oxide) copolymers. *Langmuir* **10**, 2604–2612.

Alexandridis, P., Holzwarth, J. F., and Hatton, T. A. (1994b). Micellization of poly(ethylene oxide)-poly(propylene oxide)-poly(ethylene oxide) triblock copolymers in aqueous solutions: thermodynamics of copolymer association. *Macromolecules* **27**, 2414–2425.

Alexandridis, P., and Lindman, B. (eds.) (1999). Amphiphilic Block Copolymers. Self-Assembly and Applications (Amsterdam, Lausanne, New York, Oxford, Shannon, Singapore, Tokyo, Elsevier).

Allen, C., Maysinger, D., and Eisenberg, A. (1999). Nano-engineering block copolymer aggregates for drug delivery. Coll Surfaces, B. *Biointerfaces* **16,** 3–27.

Allison, A. C., and Byars, N. E. (1986). An adjuvant formulation that selectively elicits the formation of antibodies of protective isotypes and of cell-mediated immunity. *J. Immunol. Methods* **95,** 157–168.

Allison, A. C., and Byars, N. E. (1990). Adjuvant formulations and their mode of action. *Semin. Immunol.* **2,** 369–374.

Asayama, S., Maruyama, A., Cho, C. S., and Akaike, T. (1997). Design of comb-type polyamine copolymers for a novel pH-sensitive DNA carrier. *Bioconjug. Chem.* **8,** 833–838.

Astafieva, I., Maksimova, I., Lukanidin, E., Alakhov, V., and Kabanov, A. (1996). Enhancement of the polycation-mediated DNA uptake and cell transfection with Pluronic P85 block copolymer. *FEBS Lett.* **389,** 278–280.

Barichello, J. M., Morishita, M., Takayama, K., Chiba, Y., Tokiwa, S., and Nagai, T. (1999). Enhanced rectal absorption of insulin-loaded Pluronic F-127 gels containing unsaturated fatty acids. *Int. J. Pharm.* **183,** 125–132.

Batrakova, E. V., Dorodnych, T. Y., Klinskii, E. Y., Kliushnenkova, E. N., Shemchukova, O. B., Goncharova, O. N., Arjakov, S. A., Alakhov, V. Y., and Kabanov, A. V. (1996). Anthracycline antibiotics non-covalently incorporated into the block copolymer micelles: *In vivo* evaluation of anti-cancer activity. *Br. J. Cancer* **74,** 1545–1552.

Batrakova, E. V., Lee, S., Li, S., Venne, A., Alakhov, V., and Kabanov, A. (1999). Fundamental relationships between the composition of pluronic block copolymers and their hypersensitization effect in MDR cancer cells. *Pharm. Res.* **16,** 1373–1379.

Batrakova, E. V., Li, S., Alakhov, V. Y., Elmquist, W. F., Miller, D. W., and Kabanov, A. V. (2003a). Sensitization of cells overexpressing multidrug-resistant proteins by pluronic P85. *Pharm. Res.* **20,** 1581–1590.

Batrakova, E. V., Li, S., Alakhov, V. Y., Miller, D. W., and Kabanov, A. V. (2003b). Optimal structure requirements for Pluronic block copolymers in modifying P-glycoprotein drug efflux transporter activity in bovine brain microvessel endothelial cells. *J. Pharmacol. Exp. Ther.* **304,** 845–854.

Batrakova, E. V., Li, S., Elmquist, W. F., Miller, D. W., Alakhov, V. Y., and Kabanov, A. V. (2001a). Mechanism of sensitization of MDR cancer cells by Pluronic block copolymers: Selective energy depletion. *Br. J. Cancer* **85,** 1987–1997.

Batrakova, E. V., Li, S., Vinogradov, S. V., Alakhov, V. Y., Miller, D. W., and Kabanov, A. V. (2001b). Mechanism of pluronic effect on P-glycoprotein efflux system in blood-brain barrier: Contributions of energy depletion and membrane fluidization. *J. Pharmacol. Exp. Ther.* **299,** 483–493.

Behr, J. P., Demeneix, B., Loeffler, J. P., and Perez-Mutul, J. (1989). Efficient gene transfer into mammalian primary endocrine cells with lipopolyamine-coated DNA. *Proc. Natl. Acad. Sci. USA* **86,** 6982–6986.

Bittner, B., and Mountfield, R. J. (2002). Intravenous administration of poorly soluble new drug entities in early drug discovery: The potential impact of formulation on pharmacokinetic parameters. *Curr. Opin. Drug. Discov. Devel.* **5,** 59–71.

Bochot, A., Mashhour, B., Puisieux, F., Couvreur, P., and Fattal, E. (1998). Comparison of the ocular distribution of a model oligonucleotide after topical instillation in rabbits of conventional and new dosage forms. *J. Drug. Target* **6,** 309–313.

Bomford, R. (1981). The adjuvant activity of fatty acid esters. The role of acyl chain length and degree of saturation. *Immunology* **44,** 187–192.

Boussif, O., Lezoualc'h, F., Zanta, M. A., Mergny, M. D., Scherman, D., Demeneix, B., and Behr, J. P. (1995). A versatile vector for gene and oligonucleotide transfer into cells in culture and *in vivo*: Polyethylenimine. *Proc. Natl. Acad. Sci. USA* **92,** 7297–7301.

Braun, S., Thioudellet, C., Rodriguez, P., Ali-Hadji, D., Perraud, F., Accart, N., Balloul, J. M., Halluard, C., Acres, B., Cavallini, B., and Pavirani, A. (2000). Immune rejection of human dystrophin following intramuscular injections of naked DNA in mdx mice. *Gene Ther.* **7**, 1447–1457.

Bromberg, L. E. (2001). Enhanced nasal retention of hydrophobically modified polyelectrolytes. *J. Pharm. Pharmacol.* **53**, 109–114.

Bronich, T. K., Kabanov, A. V., and Marky, L. A. (2001). A thermodynamic characterization of the interaction of a cationic copolymer with DNA. *J. Phys. Chem.* **105**, 6042–6050.

Byars, N. E., and Allison, A. C. (1987). Adjuvant formulation for use in vaccines to elicit both cell-mediated and humoral immunity. *Vaccine* **5**, 223–228.

Cao, Y. L., Lach, E., Kim, T. H., Rodriguez, A., Arevalo, C. A., and Vacanti, C. A. (1998). Tissue-engineered nipple reconstruction. *Plast. Reconstr. Surg.* **102**, 2293–2298.

Carter, K. C., Gallagher, G., Baillie, A. J., and Alexander, J. (1989). The induction of protective immunity to Leishmania major in the BALB/c mouse by interleukin 4 treatment. *Eur. J. Immunol.* **19**, 779–782.

Cellesi, F., Tirelli, N., and Hubbell, J. A. (2002). Materials for cell encapsulation via a new tandem approach combining reverse thermal gelation and covalent cross-linking. *Macromol. Chem. Phys.* **203**, 1466–1472.

Chen Chow, P., and Frank, S. G. (1981). Comparison of lidocaine release from Pluronic F-127 gels and other formulations. *Acta Pharm. Suec.* **18**, 239–244.

Cho, C. W., Cho, Y. S., Kang, B. T., Hwang, J. S., Park, S. N., and Yoon, D. Y. (2001). Improvement of gene transfer to cervical cancer cell lines using non-viral agents. *Cancer Lett.* **162**, 75–85.

Cho, C. W., Cho, Y. S., Lee, H. K., Yeom, Y. I., Park, S. N., and Yoon, D. Y. (2000). Improvement of receptor-mediated gene delivery to HepG2 cells using an amphiphilic gelling agent. *Biotechnol. Appl. Biochem.* **32**, 21–26.

Choi, Y. H., Liu, F., Kim, J.-S., Choi, Y. K., Park, J. S., and Kim, S. W. (1998). Polyethylene glycol-grafted poly-L-lysine as polymeric gene carrier. *J. Contr. Release* **54**, 39–48.

Danko, I., Fritz, J. D., Jiao, S., Hogan, K., Latendresse, J. S., and Wolff, J. A. (1994). Pharmacological enhancement of *in vivo* foreign gene expression in muscle. *Gene Ther.* **1**, 114–121.

Danson, S., Ferry, D., Alakhov, V., Margison, J., Kerr, D., Jowle, D., Brampton, M., Halbert, G., and Ranson, M. (2004). Phase I dose escalation and pharmacokinetic study of pluronic polymer-bound doxorubicin (SP1049C) in patients with advanced cancer. *Br. J. Cancer* **90**, 2085–2091.

Davidorf, F. H., Chambers, R. B., Kwon, O. W., Doyle, W., Gresak, P., and Frank, S. G. (1990). Ocular toxicity of vitreal pluronic polyol F-127. *Retina* **10**, 297–300.

Dishart, K. L., Denby, L., George, S. J., Nicklin, S. A., Yendluri, S., Tuerk, M. J., Kelley, M. P., Donahue, B. A., Newby, A. C., Harding, T., and Baker, A. H. (2003). Third-generation lentivirus vectors efficiently transduce and phenotypically modify vascular cells: Implications for gene therapy. *J. Mol. Cell. Cardiol.* **35**, 739–748.

Duncan, R. (2003). The dawning era of polymer therapeutics. *Nat. Rev. Drug. Discov.* **2**, 347–360.

Duzgunes, N., De Ilarduya, C. T., Simoes, S., Zhdanov, R. I., Konopka, K., and Pedroso de Lima, M. C. (2003). Cationic liposomes for gene delivery: Novel cationic lipids and enhancement by proteins and peptides. *Curr. Med. Chem.* **10**, 1213–1220.

Edsman, K., Carlfors, J., and Petersson, R. (1998). Rheological evaluation of poloxamer as an in situ gel for ophthalmic use. *Eur. J. Pharm. Sci.* **6**, 105–112.

Emanuele, M. R., Balasubramanian, M., and Alludeen, H. S. (1996a). Polyoxypropylene/polyoxyethylene copolymers with improved biological activity (US, CytRx Corporation (Norcross, GA)).

Emanuele, M. R., Hunter, R. L., and Culbreth, P. H. (1996b). Polyoxypropylene/polyoxyethylene copolymers with improved biological activity (US, CYTRX Corporation (Norcross, GA)).

Epperson, J. D., Dodge, J., Rill, R. L., and Greenbaum, N. L. (2001). Analysis of oligonucleotides and unincorporated nucleotides from *in vitro* transcription by capillary electrophoresis in Pluronic F127 gels. *Electrophoresis* **22**, 771–778.

Erukova, V. Y., Krylova, O. O., Antonenko, Y. N., and Melik-Nubarov, N. S. (2000). Effect of ethylene oxide and propylene oxide block copolymers on the permeability of bilayer lipid membranes to small solutes including doxorubicin. *Biochim. Biophys. Acta* **1468**, 73–86.

Feldman, L. J., Pastore, C. J., Aubailly, N., Kearney, M., Chen, D., Perricaudet, M., Steg, P. G., and Isner, J. M. (1997). Improved efficiency of arterial gene transfer by use of poloxamer 407 as a vehicle for adenoviral vectors. *Gene Ther.* **4**, 189–198.

Felgner, J., Kumar, R., Sridhar, C., Wheller, C., Tsai, Y., Border, R., Ramsey, P., Martin, M., and Felgner, P. (1994). Enhanced gene delivery and mechanism studies with a novel series of cationic lipid formulations. *J. Biol. Chem.* **269**, 2550–2561.

Felgner, P. L., Barenholz, Y., Behr, J. P., Cheng, S. H., Cullis, P., Huang, L., Jessee, J. A., Seymour, L., Szoka, F., Thierry, A. R., *et al.* (1997). Nomenclature for synthetic gene delivery systems. *Hum. Gene Ther.* **8**, 511–512.

Felgner, P. L., Gadek, T. R., Holm, M., Roman, R., Chan, H. W., Wenz, M., Northrop, J. P., Ringold, G. M., and Danielsen, M. (1987). Lipofection: A highly efficient, lipid-mediated DNA-transfection procedure. *Proc. Natl. Acad. Sci. USA* **84**, 7413–7417.

Ferrari, S., Moro, E., Pettenazzo, A., Behr, J. P., Zacchello, F., and Scarpa, M. (1997). ExGen 500 is an efficient vector for gene delivery to lung epithelial cells *in vitro* and *in vivo*. *Gene Ther.* **4**, 1100–1106.

Follis, F., Jenson, B., Blisard, K., Hall, E., Wong, R., Kessler, R., Temes, T., and Wernly, J. (1996). Role of poloxamer 188 during recovery from ischemic spinal cord injury: A preliminary study. *J. Invest. Surg.* **9**, 149–156.

Fults, K. A., and Johnston, T. P. (1990). Sustained-release of urease from a poloxamer gel matrix. *J. Parenter. Sci. Technol.* **44**, 58–65.

Garnett, M. C. (1999). Gene-delivery systems using cationic polymers. *Crit. Rev. Ther. Drug. Carrier Syst.* **16**, 147–207.

Gear, A. J., Hellewell, T. B., Wright, H. R., Mazzarese, P. M., Arnold, P. B., Rodeheaver, G. T., and Edlich, R. F. (1997). A new silver sulfadiazine water soluble gel. *Burns* **23**, 387–391.

Gebhart, C. L., Alakhov, V. Yu., and Kabanov, A. V. (2003). Pluronic block copolymers enhance local transgene expression in skeletal muscle and solid tumor. Paper presented at: Controlled Release Society 30th Annual Meeting Proceedings (Glasgow, UK).

Gebhart, C. L., and Kabanov, A. V. (2001). Evaluation of polyplexes as gene transfer agents. *J. Contr. Release* **73**, 401–416.

Gebhart, C. L., Sriadibhatla, S., Vinogradov, S., Lemieux, P., Alakhov, V., and Kabanov, A. V. (2002). Design and formulation of polyplexes based on pluronic-polyethyleneimine conjugates for gene transfer. *Bioconjug. Chem.* **13**, 937–944.

Godbey, W., Wu, K., and Mikos, A. (1999). Tracking the intracellular path of poly(ethylenimine)/ DNA complexes for gene delivery. *Proc. Natl. Acad. Sci. USA* **96**, 5177–5181.

Haensler, J., and Szoka, F. C., Jr. (1993). Polyamidoamine cascade polymers mediate efficient transfection of cells in culture. *Bioconjug. Chem.* **4**, 372–379.

Hannig, J., Zhang, D., Canaday, D. J., Beckett, M. A., Astumian, R. D., Weichselbaum, R. R., and Lee, R. C. (2000). Surfactant sealing of membranes permeabilized by ionizing radiation. *Radiat. Res.* **154**, 171–177.

Harada, A., Togawa, H., and Kataoka, K. (2001). Physicochemical properties and nuclease resistance of antisense-oligodeoxynucleotides entrapped in the core of polyion complex micelles composed of poly(ethylene glycol)-poly(L-Lysine) block copolymers. *Eur. J. Pharm. Sci.* **13**, 35–42.

Harper, G. R., Davies, M. C., Davis, S. S., Tadros, T. F., Taylor, D. C., Irving, M. P., and Waters, J. A. (1991). Steric stabilization of microspheres with grafted polyethylene oxide reduces phagocytosis by rat Kupffer cells in vitro. Biomaterials 12, 695–700.

Hartikka, J., Sukhu, L., Buchner, C., Hazard, D., Bozoukova, V., Margalith, M., Nishioka, W. K., Wheeler, C. J., Manthorp, M., and Sawdey, M. (2001). Electroporation-facilitated delivery of plasmid DNA in skeletal muscle: Plasmid dependence of muscle damage and effect of poloxamer 188. Mol. Ther. 4, 407–415.

Herbert, J. M., Guy, A. F., Lamarche, I., Mares, A. M., Savi, P., and Dol, F. (1997). Intimal hyperplasia following vascular injury is not inhibited by an antisense thrombin receptor oligodeoxynucleotide. J. Cell. Physiol. 170, 106–114.

Hom, D. B., Medhi, K., Assefa, G., Juhn, S. K., and Johnston, T. P. (1996). Vascular effects of sustained-release fibroblast growth factors. Ann. Otol. Rhinol. Laryngol. 105, 109–116.

Hosseinkhani, H., Azzam, T., Tabata, Y., and Domb, A. J. (2004). Dextran-spermine polycation: An efficient non-viral vector for in vitro and in vivo gene transfection. Gene Ther. 11, 194–203.

Houssami, R., Check, I. J., and Hunter, R. L. (1990). Immunoendocrine modulation and stimulation of hematopoiesis with the ionophore copolymer T150R1. Proc. Soc. Exp. Biol. Med. 194, 274–282.

Hunter, R., Olsen, M., and Buynitzky, S. (1991). Adjuvant activity of non-ionic block copolymers. IV. Effect of molecular weight and formulation on titre and isotype of antibody. Vaccine 9, 250–256.

Hunter, R., Strickland, F., and Kezdy, F. (1981). The adjuvant activity of nonionic block polymer surfactants. I. The role of hydrophile-lipophile balance. J. Immunol. 127, 1244–1250.

Hunter, R. L., and Bennett, B. (1984). The adjuvant activity of nonionic block polymer surfactants. II. Antibody formation and inflammation related to the structure of triblock and octablock copolymers. J. Immunol. 133, 3167–3175.

Hunter, R. L., and Bennett, B. (1986). The adjuvant activity of nonionic block polymer surfactants. III. Characterization of selected biologically active surfaces. Scand. J. Immunol. 23, 287–300.

Hunter, R. L., McNicholl, J., and Lal, A. A. (1994). Mechanisms of action of nonionic block copolymer adjuvants. AIDS Res. Hum. Retroviruses 10, S95–S98.

Hwang, S. J., and Davis, M. E. (2001). Cationic polymers for gene delivery: Designs for overcoming barriers to systemic administration. Curr. Opin. Mol. Ther. 3, 183–191.

Jeon, E., Kim, H. D., and Kim, J. S. (2003). Pluronic-grafted poly-(L)-lysine as a new synthetic gene carrier. J. Biomed. Mater. Res. 66A, 854–859.

Johnston, T. P., and Miller, S. C. (1989). Inulin disposition following intramuscular administration of an inulin/poloxamer gel matrix. J. Parenter. Sci. Technol. 43, 279–286.

Johnston, T. P., Punjabi, M. A., and Froelich, C. J. (1992). Sustained delivery of interleukin-2 from a poloxamer 407 gel matrix following intraperitoneal injection in mice. Pharm. Res. 9, 425–434.

Juhasz, J., Lenaerts, V., Raymond, P., and Ong, H. (1989). Diffusion of rat atrial natriuretic factor in thermoreversible poloxamer gels. Biomaterials 10, 265–268.

Kabanov, A. V. (1999). Taking polycation gene delivery systems from in vitro to in vivo. Pharm. Sci. Tech. Today 2, 365–372.

Kabanov, A. V., and Alakhov, V. Y. (2000). Micelles of amphilphilic block copolymers as vehicles for drug delivery. In "Amphiphilic Block Copolymers: Self-Assembly and Applications" (P. Alexandridis and B. Lindman, eds.), pp. 347–376. Elsevier, Amsterdam, Lausanne, New York, Oxford, Shannon, Singapore, Tokyo.

Kabanov, A. V., and Alakhov, V. Y. (2002). Pluronic block copolymers in drug delivery: From micellar nanocontainers to biological response modifiers. Crit. Rev. Ther. Drug Carrier Syst. 19, 1–72.

Kabanov, A. V., Batrakova, E. V., and Alakhov, V. Y. (2002a). Pluronic block copolymers for overcoming drug resistance in cancer. Adv. Drug Deliv. Rev. 54, 759–779.

Kabanov, A. V., Batrakova, E. V., Melik-Nubarov, N. S., Fedoseev, N. A., Dorodnich, T. Y., Alakhov, V. Y., Chekhonin, V. P., Nazarova, I. R., and Kabanov, V. A. (1992). A new class of drug carriers: Micelles of poly(oxyethylene)-poly(oxypropylene) block copolymers as microcontainers for drug targeting from blood in brain. *J. Contr. Release*. **22,** 141–157.

Kabanov, A. V., Batrakova, E. V., and Miller, D. W. (2003). Pluronic((R)) block copolymers as modulators of drug efflux transporter activity in the blood-brain barrier. *Adv. Drug Deliv. Rev.* **55,** 151–164.

Kabanov, A. V., Chekhonin, V. P., Alakhov, V. Y., Batrakova, E. V., Lebedev, A. S., Melik-Nubarov, N. S., Arzhakov, S. A., Levashov, A. V., Morozov, G. V., Severin, E. S., and Kabanov, V. A. (1989a). The neuroleptic activity of haloperidol increases after its solubilization in surfactant micelles. Micelles as microcontainers for drug targeting. *FEBS Lett.* **258,** 343–345.

Kabanov, A. V., and Kabanov, V. A. (1995). DNA complexes with polycations for the delivery of genetic material into cells. *Bioconjug. Chem.* **6,** 7–20.

Kabanov, A. V., Kiselev, V. I., Chikindas, M. L., Astafieva, I. V., Glukhov, A. I., Gordeev, S. A., Izumrudov, V. A., Zezin, A. B., Levashov, A. V., Severin, E. S., and Kabanov, V. A. (1989b). [Increase in the transforming activity of plasmid DNA by means of its inclusion in the interpolyelectrolytic complex with a carbon-chain cation]. *Doklady Biochemistry* **306,** 133–136 (Dokl Akad Nauk SSSR).

Kabanov, A. V., Lemieux, P., Vinogradov, S., and Alakhov, V. (2002b). Pluronic block copolymers: Novel functional molecules for gene therapy. *Adv. Drug Deliv. Rev.* **54,** 223–233.

Kabanov, A. V., Sriadibhatla, S., Gebhart, C. L., Yang, Z., and Alakhov, V. Y. (2004). Effect of Pluronic block copolymers on gene expression. *Polym. Prepr.* **45,** 392–393.

Kabanov, A. V., Vinogradov, S. V., Suzdaltseva, Y. G., and Alakhov, V. Y. (1995). Water-soluble block polycations as carriers for oligonucleotide delivery. *Bioconjug. Chem.* **6,** 639–643.

Kakizawa, Y., and Kataoka, K. (2002). Block copolymer micelles for delivery of gene and related compounds. *Adv. Drug Deliv. Rev.* **54,** 203–222.

Kataoka, K., Harada, A., and Nagasaki, Y. (2001). Block copolymer micelles for drug delivery: Design, characterization and biological significance. *Adv. Drug Deliv. Rev.* **47,** 113–131.

Kataoka, K., Togawa, H., Harda, A., Yasugi, K., Matsumoto, T., and Katayose, S. (1996). Spontaneous formation of polyion complex micelles with narrow distribution from antisense oligonucleotide and cationic copolymer in physiological saline. *Macromolecules* **29,** 8556–8557.

Katayose, S., and Kataoka, K. (1997). Water-soluble polyion complex associates of DNA and poly (ethylene glycol)-poly(L-lysine) block copolymer. *Bioconj. Chem.* **8,** 702–707.

Ke, Y., McGraw, C. L., Hunter, R. L., and Kapp, J. A. (1997). Nonionic triblock copolymers facilitate delivery of exogenous proteins into the MHC class I and class II processing pathways. *Cell Immunol.* **176,** 113–121.

Kircheis, R., Wightman, L., Schreiber, A., Robitza, B., Rossler, V., Kursa, M., and Wagner, E. (2001a). Polyethylenimine/DNA complexes shielded by transferrin target gene expression to tumors after systemic application. *Gene Ther.* **8,** 28–40.

Kircheis, R., Wightman, L., and Wagner, E. (2001b). Design and gene delivery activity of modified polyethylenimines. *Adv. Drug Deliv. Rev.* **53,** 341–358.

Kozlov, M. Y., Melik-Nubarov, N. S., Batrakova, E. V., and Kabanov, A. V. (2000). Relationship between pluronic block copolymer structure, critical micellization concentration and partitioning coefficients of low molecular mass solutes. *Macromolecules* **33,** 3305–3313.

Krylova, O. O., Melik-Nubarov, N. S., Badun, G. A., Ksenofontov, A. L., Menger, F. M., and Yaroslavov, A. A. (2003). Pluronic L61 accelerates flip-flop and transbilayer doxorubicin permeation. *Chemistry* **9,** 3930–3936.

Krylova, O. O., and Pohl, P. (2004). Ionophoric activity of pluronic block copolymers. *Biochemistry* **43,** 3696–3703.

Kukowska-Latallo, J. F., Bielinska, A. U., Johnson, J., Spindler, R., Tomalia, D. A., and Baker, J. R. Jr. (1996). Efficient transfer of genetic material into mammalian cells using Starburst polyamidoamine dendrimers. *Proc. Natl. Acad. Sci. USA* **93**, 4897–4902.

Kuo, J. H. (2003). Effect of Pluronic-block copolymers on the reduction of serum-mediated inhibition of gene transfer of polyethyleneimine-DNA complexes. *Biotechnol. Appl. Biochem.* **37**, 267–271.

Kursa, M., Walker, G. F., Roessler, V., Ogris, M., Roedl, W., Kircheis, R., and Wagner, E. (2003). Novel shielded transferrin-polyethylene glycol-polyethylenimine/DNA complexes for systemic tumor-targeted gene transfer. *Bioconjug. Chem.* **14**, 222–231.

Kwon, G. S. (2003). Polymeric micelles for delivery of poorly water-soluble compounds. *Crit. Rev. Ther. Drug Carrier Syst.* **20**, 357–403.

Kwon, G. S., and Okano, T. (1999). Soluble self-assembled block copolymers for drug delivery. *Pharm. Res.* **16**, 597–600.

Laham, R. J., Chronos, N. A., Pike, M., Leimbach, M. E., Udelson, J. E., Pearlman, J. D., Pettigrew, R. I., Whitehouse, M. J., Yoshizawa, C., and Simons, M. (2000). Intracoronary basic fibroblast growth factor (FGF-2) in patients with severe ischemic heart disease: Results of a phase I open-label dose escalation study. *J. Am. Coll. Cardiol.* **36**, 2132–2139.

Lasic, D. D., Martin, F. J., Gabizon, A., Huang, S. K., and Papahadjopoulos, D. (1991). Sterically stabilized liposomes: A hypothesis on the molecular origin of the extended circulation times. *Biochim. Biophys. Acta* **1070**, 187–192.

Lathi, K. G., Vale, P. R., Losordo, D. W., Cespedes, R. M., Symes, J. F., Esakof, D. D., Maysky, M., and Isner, J. M. (2001). Gene therapy with vascular endothelial growth factor for inoperable coronary artery disease: Anesthetic management and results. *Anesth. Analg.* **92**, 19–25.

Lavasanifar, A., Samuel, J., and Kwon, G. S. (2002). Poly(ethylene oxide)-block-poly(L-amino acid) micelles for drug delivery. *Adv. Drug Deliv. Rev.* **54**, 169–190.

Leach, R. E., and Henry, R. L. (1990). Reduction of postoperative adhesions in the rat uterine horn model with poloxamer 407. *Am. J. Obstet. Gynecol.* **162**, 1317–1319.

Lee, R. C., Canaday, D. J., and Hammer, S. M. (1993). Transient and stable ionic permeabilization of isolated skeletal muscle cells after electrical shock. *J. Burn. Care Rehabil.* **14**, 528–540.

Lee, R. C., Hannig, J., Matthews, K. L., Myerov, A., and Chen, C. T. (1999). Pharmaceutical therapies for sealing of permeabilized cell membranes in electrical injuries. *Ann. NY Acad. Sci.* **888**, 266–273.

Lee, R. C., River, L. P., Pan, F. S., Ji, L., and Wollmann, R. L. (1992). Surfactant-induced sealing of electropermeabilized skeletal muscle membranes *in vivo*. *Proc. Natl. Acad. Sci. USA* **89**, 4524–4528.

Lemieux, P., Guerin, N., Paradis, G., Proulx, R., Chistyakova, L., Kabanov, A., and Alakhov, V. (2000). A combination of poloxamers increases gene expression of plasmid DNA in skeletal muscle. *Gene Ther.* **7**, 986–991.

Liaw, J., Chang, S. F., and Hsiao, F. C. (2001). *In vivo* gene delivery into ocular tissues by eye drops of poly(ethylene oxide)-poly(propylene oxide)-poly(ethylene oxide) (PEO-PPO-PEO) polymeric micelles. *Gene Ther.* **8**, 999–1004.

Linse, P. (1993). Micellization of poly(ethylene oxide)-poly(propylene oxide) block copolymers in aqueous solution. *Macromolecules* **26**, 4437–4449.

Linse, P., and Malmsten, M. (1992). Temperature-dependent micellization in aqueous block copolymer solutions. *Macromolecules* **25**, 5434–5439.

Liu, F., and Huang, L. (2002). Development of non-viral vectors for systemic gene delivery. *J. Control Release* **78**, 259–266.

Liu, F., Yang, J., Huang, L., and Liu, D. (1996). Effect of non-ionic surfactants on the formation of DNA/emulsion complexes and emulsion-mediated gene transfer. *Pharm. Res.* **13**, 1642–1646.

Lu, G. W., Jun, H. W., Dzimianski, M. T., Qiu, H. C., and McCall, J. W. (1995). Pharmacokinetic studies of methotrexate in plasma and synovial fluid following i.v. bolus and topical routes of administration in dogs. *Pharm. Res.* **12,** 1474–1477.

Lu, Q. L., Bou-Gharios, G., and Partridge, T. A. (2003). Non-viral gene delivery in skeletal muscle: A protein factory. *Gene Ther.* **10,** 131–142.

Lukyanov, A. N., and Torchilin, V. P. (2004). Micelles from lipid derivatives of water-soluble polymers as delivery systems for poorly soluble drugs. *Adv. Drug Deliv. Rev.* **56,** 1273–1289.

Mahato, R., Kawabata, K., Nomura, T., Takakura, Y., and Hashida, M. (1995). Physiochemical and pharmacokinetics of plasmid DNA/cationic liposome complexes. *J. Pharm. Sci.* **84,** 1267–1271.

Mahato, R. I., Takakura, Y., and Hashida, M. (1997). Non-viral vectors for *in vivo* gene delivery: Physicochemical and pharmacokinetic considerations. *Crit. Rev. Ther. Drug Carrier Syst.* **14,** 133–172.

Maillard, L., Van Belle, E., Tio, F. O., Rivard, A., Kearney, M., Branellec, D., Steg, P. G., Isner, J. M., and Walsh, K. (2000). Effect of percutaneous adenovirus-mediated Gax gene delivery to the arterial wall in double-injured atheromatous stented rabbit iliac arteries. *Gene Ther.* **7,** 1353–1361.

March, K. L., Madison, J. E., and Trapnell, B. C. (1995). Pharmacokinetics of adenoviral vector-mediated gene delivery to vascular smooth muscle cells: Modulation by poloxamer 407 and implications for cardiovascular gene therapy. *Hum. Gene Ther.* **6,** 41–53.

Maruyama, A., Katoh, M., Ishihara, T., and Akaike, T. (1997). Comb-type polycations effectively stabilize DNA triplex. *Bioconjug. Chem.* **8,** 3–6.

Millet, P., Kalish, M. L., Collins, W. E., and Hunter, R. L. (1992). Effect of adjuvant formulations on the selection of B-cell epitopes expressed by a malaria peptide vaccine. *Vaccine* **10,** 547–550.

Moghimi, S. M., and Hunter, A. C. (2000). Poloxamers and poloxamines in nanoparticle engineering and experimental medicine. *Trends Biotechnol.* **18,** 412–420.

Morikawa, K., Okada, F., Hosokawa, M., and Kobayashi, H. (1987). Enhancement of therapeutic effects of recombinant interleukin 2 on a transplantable rat fibrosarcoma by the use of a sustained release vehicle, pluronic gel. *Cancer Res.* **47,** 37–41.

Mumper, R. J., Duguid, J. G., Anwer, K., Barron, M. K., Nitta, H., and Rolland, A. P. (1996). Polyvinyl derivatives as novel interactive polymers for controlled gene delivery to muscle. *Pharm. Res.* **13,** 701–709.

Mumper, R. J., Wang, J., Klakamp, S. L., Nitta, H., Anwer, K., Tagliaferri, F., and Rolland, A. P. (1998). Protective interactive noncondensing (PINC) polymers for enhanced plasmid distribution and expression in rat skeletal muscle. *J. Control Release* **52,** 191–203.

Nagarajan, R. (1999). Solubilization of hydrocarbons and resulting aggregate shape transitions in aqueous solutions of Pluronic (PEO-PPO-PEO) block copolymers. *Coll Surfaces, B. Biointerfaces* **16,** 55–72.

Nalbandian, R. M., Henry, R. L., Balko, K. W., Adams, D. V., and Neuman, N. R. (1987). Pluronic F-127 gel preparation as an artificial skin in the treatment of third-degree burns in pigs. *J. Biomed. Mater. Res.* **21,** 1135–1148.

Newman, M. J., Actor, J. K., Balusubramanian, M., and Jagannath, C. (1998). Use of nonionic block copolymers in vaccines and therapeutics. *Crit. Rev. Ther. Drug Carrier Syst.* **15,** 89–142.

Nguyen, H. K., Lemieux, P., Vinogradov, S. V., Gebhart, C. L., Guerin, N., Paradis, G., Bronich, T. K., Alakhov, V. Y., and Kabanov, A. V. (2000). Evaluation of polyether-polyethyleneimine graft copolymers as gene transfer agents. *Gene Ther.* **7,** 126–138.

Nishiyama, N., and Kataoka, K. (2003). Polymeric micelle drug carrier systems: PEG-PAsp(Dox) and second generation of micellar drugs. *Adv. Exp. Med. Biol.* **519,** 155–177.

Ochietti, B., Lemieux, P., Kabanov, A. V., Vinogradov, S., St-Pierre, Y., and Alakhov, V. (2002). Inducing neutrophil recruitment in the liver of ICAM-1-deficient mice using polyethyleneimine grafted with Pluronic P123 as an organ-specific carrier for transgenic ICAM-1. *Gene Ther.* **9,** 939–945.

Ogris, M., Wagner, E., and Steinlein, P. (2000). A versatile assay to study cellular uptake of gene transfer complexes by flow cytometry. *Biochim. Biophys. Acta* **1474,** 237–243.

Ogris, M., Walker, G., Blessing, T., Kircheis, R., Wolschek, M., and Wagner, E. (2003). Tumor-targeted gene therapy: Strategies for the preparation of ligand-polyethylene glycol-polyethylenimine/DNA complexes. *J. Control Release* **91,** 173–181.

Otsuka, H., Nagasaki, Y., and Kataoka, K. (2003). PEGylated nanoparticles for biological and pharmaceutical applications. *Adv. Drug Deliv. Rev.* **55,** 403–419.

Oupicky, D., Konak, C., and Ulbrich, K. (1999). DNA complexes with block and graft copolymers of N-(2-hydroxypropyl)methacrylamide and 2-(trimethylammonio)ethyl methacrylate. *J. Biomater. Sci. Polym. Ed.* **10,** 573–590.

Oupicky, D., Konak, C., Ulbrich, K., Wolfert, M. A., and Seymour, L. W. (2000). DNA delivery systems based on complexes of DNA with synthetic polycations and their copolymers. *J. Control Release* **65,** 149–171.

Paavola, A., Yliruusi, J., Kajimoto, Y., Kalso, E., Wahlstrom, T., and Rosenberg, P. (1995). Controlled release of lidocaine from injectable gels and efficacy in rat sciatic nerve block. *Pharm. Res.* **12,** 1997–2002.

Paavola, A., Yliruusi, J., and Rosenberg, P. (1998). Controlled release and dura mater permeability of lidocaine and ibuprofen from injectable poloxamer-based gels. *J. Control Release* **52,** 169–178.

Pec, E. A., Wout, Z. G., and Johnston, T. P. (1992). Biological activity of urease formulated in poloxamer 407 after intraperitoneal injection in the rat. *J. Pharm. Sci.* **81,** 626–630.

Pitard, B., Pollard, H., Agbulut, O., Lambert, O., Vilquin, J. T., Cherel, Y., Abadie, J., Samuel, J. L., Rigaud, J. L., Menoret, S., *et al.* (2002). A nonionic amphiphile agent promotes gene delivery *in vivo* to skeletal and cardiac muscles. *Hum. Gene Ther.* **13,** 1767–1775.

Riera, M., Chillon, M., Aran, J. M., Cruzado, J. M., Torras, J., Grinyo, J. M., and Fillat, C. (2004). Intramuscular SP1017-formulated DNA electrotransfer enhances transgene expression and distributes hHGF to different rat tissues. *J. Gene Med.* **6,** 111–118.

Rill, R. L., Locke, B. R., Liu, Y., and Van Winkle, D. H. (1998). Electrophoresis in lyotropic polymer liquid crystals. *Proc. Natl. Acad. Sci. USA* **95,** 1534–1539.

Rodeheaver, G., Turnbull, V., Edgerton, M. T., Kurtz, L., and Edlich, R. F. (1976). Pharmacokinetics of a new skin wound cleanser. *Am. J. Surg.* **132,** 67–74.

Rodeheaver, G. T., Kurtz, L., Kircher, B. J., and Edlich, R. F. (1980). Pluronic F-68: A promising new skin wound cleanser. *Ann. Emerg. Med.* **9,** 572–576.

Rolland, A. P., and Mumper, R. J. (1998). Plasmid delivery to muscle: Recent advances in polymer delivery systems. *Adv. Drug Deliv. Rev.* **30,** 151–172.

Rosler, A., Vandermeulen, G. W., and Klok, H. A. (2001). Advanced drug delivery devices via self-assembly of amphiphilic block copolymers. *Adv. Drug Deliv. Rev.* **53,** 95–108.

Roy, S., Zhang, K., Roth, T., Vinogradov, S., Kao, R. S., and Kabanov, A. (1999). Reduction of fibronectin expression by intravitreal administration of antisense oligonucleotides. *Nat. Biotechnol.* **17,** 476–479.

Sang, H., Pisarev, V. M., Munger, C., Robinson, S., Chavez, J., Hatcher, L., Parajuli, P., Guo, Y., and Talmadge, J. E. (2003). Regional, but not systemic recruitment/expansion of dendritic cells by a pluronic-formulated Flt3-ligand plasmid with vaccine adjuvant activity. *Vaccine* **21,** 3019–3029.

Scherlund, M., Brodin, A., and Malmsten, M. (2000). Micellization and gelation in block copolymer systems containing local anesthetics. *Int. J. Pharm.* **211,** 37–49.

Schillen, K., Glatter, O., and Brown, W. (1993). Characterization of a PEO-PPO-PEO block copolymer system. *Prog. Colloid. Polym. Sci.* **93,** 66–71.

Schmolka, I. R. (1972). Artificial skin. I. Preparation and properties of pluronic F-127 gels for treatment of burns. *J. Biomed. Mater. Res.* **6,** 571–582.

Schmolka, I. R. (1977). A review of block polymer surfactants. *J. Am. Oil. Chem. Soc.* **54**, 110–116.

Suh, J., Wirtz, D., and Hanes, J. (2003). Efficient active transport of gene nanocarriers to the cell nucleus. *Proc. Natl. Acad. Sci. USA* **100**, 3878–3882.

Takayama, K., Olsen, M., Datta, P., and Hunter, R. L. (1991). Adjuvant activity of non-ionic block copolymers. V. Modulation of antibody isotype by lipopolysaccharides, lipid A and precursors. *Vaccine* **9**, 257–265.

Tobio, M., Nolley, J., Guo, Y., McIver, J., and Alonso, M. J. (1999). A novel system based on a poloxamer/PLGA blend as a tetanus toxoid delivery vehicle. *Pharm. Res.* **16**, 682–688.

Tokui, M., Takei, I., Tashiro, F., Shimada, A., Kasuga, A., Ishii, M., Ishii, T., Takatsu, K., Saruta, T., and Miyazaki, J. (1997). Intramuscular injection of expression plasmid DNA is an effective means of long-term systemic delivery of interleukin-5. *Biochem. Biophys. Res. Commun.* **233**, 527–531.

Torchilin, V. P. (2001). Structure and design of polymeric surfactant-based drug delivery systems. *J. Control Release* **73**, 137–172.

Torchilin, V. P. (2002). PEG-based micelles as carriers of contrast agents for different imaging modalities. *Adv. Drug Deliv. Rev.* **54**, 235–252.

Tripathy, S. K., Svensson, E. C., Black, H. B., Goldwasser, E., Margalith, M., Hobart, P. M., and Leiden, J. M. (1996). Long-term expression of erythropoietin in the systemic circulation of mice after intramuscular injection of a plasmid DNA vector. *Proc. Natl. Acad. Sci. USA* **93**, 10876–10880.

Uyechi, L. S., Gagne, L., Thurston, G., and Szoka, F. C., Jr. (2001). Mechanism of lipoplex gene delivery in mouse lung: Binding and internalization of fluorescent lipid and DNA components. *Gene Ther.* **8**, 828–836.

Valle, J. W., Lawrance, J., Brewer, J., Clayton, A., Corrie, P., Alakhov, V., and Ranson, M. (2004). A phase II, window study of SP1049C as first-line therapy in inoperable metastatic adenocarcinoma of the oesophagus. Paper presented at: 2004 ASCO Annual Meeting.

Van Belle, E., Maillard, L., Rivard, A., Fabre, J. E., Couffinhal, T., Kearney, M., Branellec, D., Feldman, L. J., Walsh, K., and Isner, J. M. (1998). Effects of poloxamer 407 on transfection time and percutaneous adenovirus-mediated gene transfer in native and stented vessels. *Hum. Gene Ther.* **9**, 1013–1024.

van Nostrum, C. F. (2004). Polymeric micelles to deliver photosensitizers for photodynamic therapy. *Adv. Drug Deliv. Rev.* **56**, 9–16.

Venne, A., Li, S., Mandeville, R., Kabanov, A., and Alakhov, V. (1996). Hypersensitizing effect of pluronic L61 on cytotoxic activity, transport, and subcellular distribution of doxorubicin in multiple drug-resistant cells. *Cancer Res.* **56**, 3626–3629.

Verbaan, F. J., Oussoren, C., van Dam, I. M., Takakura, Y., Hashida, M., Crommelin, D. J., Hennink, W. E., and Storm, G. (2001). The fate of poly(2-dimethyl amino ethyl)methacrylate-based polyplexes after intravenous administration. *Int. J. Pharm.* **214**, 99–101.

Vinogradov, S., Batrakova, E., Li, S., and Kabanov, A. (1999a). Polyion complex micelles with protein-modified corona for receptor-mediated delivery of oligonucleotides into cells. *Bioconjug. Chem.* **10**, 851–860.

Vinogradov, S. V., Batrakova, E. V., and Kabanov, A. V. (1999b). Poly(ethylene glycol)-polyethyleneimine NanoGel particles: Novel drug delivery systems for antisense oligonucleotides. *Coll Surf B. Biointerfaces* **16**, 291–304.

Vinogradov, S. V., Batrakova, E. V., and Kabanov, A. V. (2004). Nanogels for oligonucleotide delivery to the brain. *Bioconjug. Chem.* **15**, 50–60.

Vinogradov, S. V., Bronich, T. K., and Kabanov, A. V. (1998). Self-assembly of polyamine-poly (ethylene glycol) copolymers with phosphorothioate oligonucleotides. *Bioconjug. Chem.* **9**, 805–812.

Vlahos, A., Yu, P., Lucas, C. E., and Ledgerwood, A. M. (2001). Effect of a composite membrane of chitosan and poloxamer gel on postoperative adhesive interactions. *Am. Surg.* **67**, 15–21.

Wagner, E., Zenke, M., Cotten, M., Beug, H., and Birnstiel, M. L. (1990). Transferrin-polycation conjugates as carriers for DNA uptake into cells. *Proc. Natl. Acad. Sci. USA* **87,** 3410–3414.

Wang, J., Zhang, P. C., Mao, H. Q., and Leong, K. W. (2002). Enhanced gene expression in mouse muscle by sustained release of plasmid DNA using PPE-EA as a carrier. *Gene Ther.* **9,** 1254–1261.

Wang, P.-L., and Johnston, T. P. (1995). Sustained-release interleukin-2 following intramuscular injection in rats. *Int. J. Pharm.* **113,** 73–81.

Wolfert, M. A., Schacht, E. H., Toncheva, V., Ulbrich, K., Nazarova, O., and Seymour, L. W. (1996). Characterization of vectors for gene therapy formed by self-assembly of DNA with synthetic block co-polymers. *Hum. Gene Ther.* **7,** 2123–2133.

Wolfert, M. A., and Seymour, L. W. (1996). Atomic force microscopic analysis of the influence of the molecular weight of poly(L)lysine on the size of polyelectrolyte complexes formed with DNA. *Gene Ther.* **3,** 269–273.

Wolff, J. A., Dowty, M. E., Jiao, S., Repetto, G., Berg, R. K., Ludtke, J. J., Williams, P., and Slautterback, D. B. (1992). Expression of naked plasmids by cultured myotubes and entry of plasmids into T tubules and caveolae of mammalian skeletal muscle. *J. Cell Sci.* **103**(Pt. 4), 1249–1259.

Wolff, J. A., Malone, R. W., Williams, P., Chong, W., Acsadi, G., Jani, A., and Felgner, P. L. (1990). Direct gene transfer into mouse muscle *in vivo. Science* **247,** 1465–1468.

Wu, G. Y., and Wu, C. H. (1987). Receptor-mediated *in vitro* gene transformation by a soluble DNA carrier system. *J. Biol. Chem.* **262,** 4429–4432.

Yeh, M. K., Davis, S. S., and Coombes, A. G. (1996). Improving protein delivery from microparticles using blends of poly(DL lactide co-glycolide) and poly(ethylene oxide)-poly(propylene oxide) copolymers. *Pharm. Res.* **13,** 1693–1698.

Zhou, S., Murphy, J. E., Escobedo, J. A., and Dwarki, V. J. (1998). Adeno-associated virus-mediated delivery of erythropoietin leads to sustained elevation of hematocrit in nonhuman primates. *Gene Ther.* **5,** 665–670.

Zhou, X. H., Klibanov, A. L., and Huang, L. (1991). Lipophilic polylysines mediate efficient DNA transfection in mammalian cells. *Biochim. Biophys. Acta* **1065,** 8–14.

10

Terplex Gene Delivery System

Sung Wan Kim

Department of Pharmaceutics and Pharmaceutical Chemistry
University of Utah, Salt Lake City, Utah 84112

ABSTRACT

Polymeric gene delivery systems have been developed to overcome problems caused by viral carriers. They are low cytotoxic, have no size limit, are convenient in handling, of low cost and reproducible. A Terplex gene delivery system consisting of plasmid DNA, low density lipoprotein and hydropholized poly-L-lysine was designed and characterized. The plasmid DNA, when formulated with stearyl PLL and LDL, forms a stable and hydrophobicity/charge-balanced Terplex system of optimal size for efficient cellular uptake. DNA is still intact after the Terplex formation. This information is expected to be utilized for the development of improved transfection vector for *in vivo* gene therapy. Terplex DNA complex showed significantly longer retention in the vascular space than naked DNA. This system was used in the augmentation of myocardial transfection at an infarction site with the VEGF gene. © 2005, Elsevier Inc.

Advances in Genetics, Vol. 53
0065-2660/05 $35.00
DOI: 10.1016/S0065-2660(05)53010-4

I. POLYMERIC GENE DELIVERY SYSTEMS

Although viral vectors are the most common methods in gene transfer, they have several disadvantages such as endogenous virus recombination, oncogenic effects and immunological reactions (Temin, 1990; Kabanov and Kabanov, 1995). Polymer carriers have advantages over viral carriers which include cationic liposomes, cationic polymers and their modifications (Han et al., 2002; Lee and Kim, 2002). The cationic polymer forms complex with plasmid by electrostatic interaction between positive charges of polycations and negatively charged phosphate groups of the DNA. Condensing plasmid with a polymeric carrier improves plasmid resistance against enzymatic degradation. Polymer carriers include poly-L-lysine (PLL), polyethylenimine (PEI) and their conjugates. PLL is cytotoxic and undergoes aggregation; therefore, PLL was conjugated with dextran (Maruyama et al., 1998), methylglycosylate (Boussif et al., 1999) and polyethylene glycol (PEG) (Choi et al., 1998). PEG can shield excess positive charges of the polymer/DNA complex and reduce interaction of complex with serum protein. PEG can increase solubility of polymer/DNA complex and decrease cytotoxicity of gene carriers.

PEI is commonly used for gene delivery which contains primary, secondary and tertiary amines. PEI condenses plasmid and transfects the complex in cells (Boussif et al., 1995; Dunlop et al., 1997).

PEI presence in endosome causes proton accumulation due to the proton buffering effect with resultant osmotic swelling of the endosome, thereby disrupting the endosome. PEI cytotoxicity and aggregation was reduced by its conjugation with various functional groups.

The polymeric carriers can be divided into three groups according to the administration routes: site-specific, local and systemic carriers. An example of a site-specific gene carrier is galactose-PEG-PEI. Galactose was conjugated to PEI or PLL to achieve specific targeting to hepatocytes using asialoglycoprotein receptors (Bissen et al., 1995; Sagara and Kim, 2002). PEI-PEG-gal condensed plasmid effectively and had higher transfection efficiency than PEI to Hep G2, while it had lower transfection efficiency than PEI to NIH3T3 cells. PEI-PEG-gal/plasmid complex was administered to mice intravenously via tail vein. It was found that the main organ of the expressed luciferase was the liver. The plasmid which contains Factor IX gene was complexed with PEI-PEG-gal and injected into mice via tail vein. The FIX level in the plasma reached 10 $\mu g/ml$ after three days of injection.

Artery wall-binding peptide (AWBP) was conjugated to PEG-PLL. The complex size of PLL-PEG-AWBP/plasmid was about 86 nm. Transfection of PLL-PEG-AWBP/plasmid complexes to bovine aorta endothelial cells and smooth muscle cells were 150–180 times higher than PLL and PEG-PLL which indicated gene transfer of AWBP-PEG-PLL/plasmid via receptor-mediated endocytosis pathway (Nah, et al., 2002).

For cancer treatment, RGD peptide-PEG-PEI was synthesized as a site-specific delivery carrier. $\alpha\nu\beta3$ and $\alpha\nu\beta5$ integrins are over-expressed in angiogenic endothelial cells within tumors, where they are minimally expressed in normal vascular endothelial cells (Westlin, et al., 2001). Therefore, angiogenic endothelial cell-targeted polymer gene carrier was designed by conjugating $\alpha\nu\beta3/\alpha\nu\beta5$ integrin-binding PGD-4C peptide (RGD peptide = ACDCRGDCGC) into the PEI via PEG spacer (Suh and Kim, 2002). Stimulation of human dermal microvascular endothelial cells (HDMEC) with VEGF was found to elevate $\alpha\nu\beta$ integrin level on the HDMEC surface. RGD-PEG-PEI enhanced the transfection efficiency to HDMED with VEGF. Nontargeting PEI-PEG-RAE showed decreased transfection.

A new gene carrier, water soluble lipopolymer (WSLP), was developed to achieve high transfection and reduce cytotoxicity by local delivery. Low molecular weight PEI (1800 Da) was conjugated to cholesterol. This WSLP condensed the plasmid to 70 nm. Low MW PEI did not show recognizable cytotoxicity but showed very low transfection. However, conjugation with cholesterol showed higher transfection than high MW PEI (25000 Da) (Han et al., 2001; Lee et al., 2001). Transfection of WSLP was enhanced by the cholesterol moiety WSLP and was applied to cancer gene therapy. P2CMV-MIL12/WSLP complex was injected into a tumor locally and growth was significantly delayed in tumor-bearing mice (Mahato et al., 2001). Combination therapy of this system with paclitaxol demonstrated complete eradication of the tumor (Janat et al., 2004).

A soluble biodegradable gene carrier, poly (α-(4-aminobutyl)-L-glycolic acid) (PAGA), was synthesized for systemic gene delivery (Lim et al., 2000). The polymer degradation was characterized by rapid initiation and gradual degradation to the monomer, L-oxylysine. However, the complex of PLGA/plasmid was slower than PAGA alone and this polymer did not show cytotoxicity. PAGA was evaluated as a gene carrier for cytokine gene delivery to prevent type I diabetes. The plasmids of interleukin-4 (IL-4) and IL-10 were complexed with PAGA. The complexes were injected into NOD mice intravenously via tail vein. The expression level of cytokines in plasma reached up to 200 pg/ml one day after the injection and slowly decayed for up to nine weeks. It was observed that autoimmune-induced insulitis and diabetes were effectively prevented by cytokine expression (Koh et al., 2000; Lee et al., 2002; Ko et al., 2001).

The biodegradable PEI was synthesized with low MW PEI (1800 Da) and difunctional PEG succinimidyl succinat(PEG-SS). This polymer reduces cytotoxicity and increased solubility of PEI/plasmid complex (Ahn et al., 2002).

II. DESIGN AND CHARACTERIZATION OF TERPLEX

The Terplex system is unique. The main driving force of complex formation
with plasmid is a hydrophobicity/charge balance between hydrophobized PLL
(HPLL), low density lipoproein (LDL) and plasmid DNA. H-PLL was synthesized by N-alkylation of PLL with alkylbromide. 41 mg of stearyl bromide in a
solution of 2 ml dioxane and 200 μl in NaOH was added to 100 mg of PLL-
hydrobromide in 2 ml DMSO and kept for 24 hours. The solution was then
poured into a large excess of diethyl ether. The precipitated polymer was
dissolved in DMSO and precipitated in diethyl ether. The formed stearyl PLL
(H-PLL) was dialyzed and lyophilized. Various Terplex formulations of different
weights of H-PLL/LDL/DNA were prepared in PBS for transfection with a fixed
amount of plasmid DNA as shown in Fig. 10.1 (Kim et al., 1997).

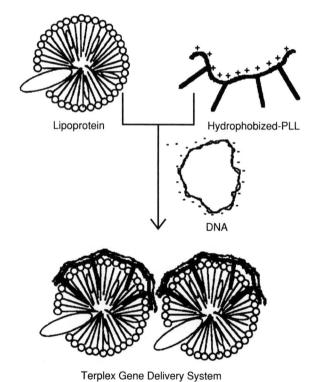

Terplex Gene Delivery System

Figure 10.1. Formation of the Terplex/DNA delivery system. Equal amount of LDL and H-PLL
were mixed in PBS solution at room temperature, to which varying amounts of
plasmid DNA were added to form a stable Terplex system.

Table 10.1. Pharmacokinetics and Biodistribution Studies

	Clearance (L/h)	Half-life (h)	Mean residual time (h)	Steady-state volume of distribution (L/kg)
Terplex/DNA	0.116	6.18	7.84	2.61
Naked DNA	0.132	1.96	2.54	0.95

Formulation of the Terplex system was verified by NMR, atomic force microscopy agarose gel electrophoresis and zeta potential (Kim *et al.*, 1998). The tertiary structure of plasmid DNA in the Terplex system was examined using a circular dichroism (DD) spectropolarimetry. The CD spectrum of Terplex DNA was similar to that of natural DNA (Kim *et al.*, 1998). The pharmacokinetic studies of Terplex/DNA and naked DNA in blood were carried out via rat intravenous injection. The Terplex/DNA was more slowly eliminated from the body and showed much higher steady state volume of distribution than naked DNA. The high value of this volume of distribution suggests that the plasmid DNA, especially the Terplex DNA, distributed into the intracellular space as well as interstitial and vascular space (Yu *et al.*, 2001) (Table 10.1). The stearyl groups have hydrophobic interactions with lipoprotein. LDL incorporation into the polymer gene carrier enhances transfection by augmentation of the LDL receptor-mediated endocytosis pathway, since LDL binds to various types of cells—vascular endothelial cells, smooth muscle cells, hepatocytes and macophages. The receptor-mediated endocytosis of LDL is mainly mediated by apolipoprotein B (APOB), which appears to be the sole protein component of LDL. The LDL receptor on the artery wall belongs to the C-type lectin family, showing high affinity to LDL.

III. *IN VITRO* VASCULAR CELL TRANSFECTION

It was demonstrated that Terplex/DNA is a stable complex with high gene transfection and lower cytotoxicity compared to liposome-mediated gene transfer in HepG2 and A7R5 cell lines (Kim *et al.*, 1997; Kim, *et al.*, 1998). To test whether the Terplex/DNA system was able to target artery wall cells, primary bovine or wall cells (endothelial cells and smooth muscle cells) were used as target cells. Two reporter genes were used to evaluate the Terplex/DNA system for gene transfection: the β galactosidase gene and luciferase gene (Yu *et al.*, 2001). The levels of luciferase activity from cell lysates of endothelial cell (EC) and smooth muscle cell (SMC) showed that the Terplex/DNA system produced significant high levels of luciferase activity in both ECs and SMCs compared to

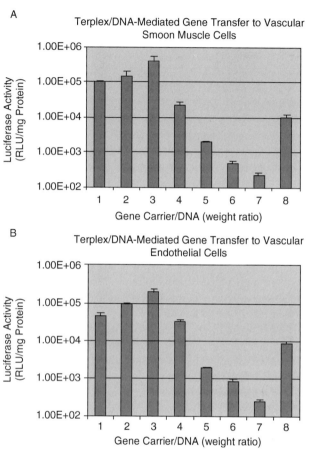

Figure 10.2. Chemiluminescent analysis. Terplex/DNA transfection efficiency assay in artery wall cell: bovine aorta endothelial cells (A) and bovine aorta endothelial smooth muscle cells (B). Transfection efficiency is expressed as luciferase activity per mg of protein. Columns 1 to 7: Terplex-mediated plasmid DNA (pLUC) transfection to artery wall cells with fixed DNA (1.5 μg) and PLL stearyl (1.5 μg), but various amounts of LDL (from 1.5 to 15 μg). The ratio of LDL:PLL-stearyl:DNA starts from 1:1:1 (column 1), and increased to 2:1:1 (column 2), 3:1:1 (column 3), 4:1:1 (column 4), and then up to 6:1:1 (column 5), 8:1:1 (column 6) and 10:1:1 (column 7). Column 8 is a control of Lipofectamine gene carrier with the same amount of plasmid DNA. Each experiment was performed triplicately.

the control gene carrier, lipofectamine. The highest levels of luciferase activities were obtained when the LDL:PLL-steryl:DNA ratio was 3:1:1 in Terplex DNA-mediated transfection. As shown in Fig. 10.2A and 2B, the luciferase activities from cell lysates of both SMCs and ECs transfected by the Terplex/DNA systems were high and 37 times greater in SMCs and 23 times greater in ECs when compared to lipofectamine-mediated gene transfer in the same cells. This was further supported by the data from cytochemical staining of both SMCs and ECs transfected with the β galactosidase gene. In addition, vascular endothelial growth factor (VEGF165) showed that Terplex/DNA produced more VEGF than that which was transfected by lipofectamine. The Terplex/DNA system showed a lower cytotoxicity in primary SMCs and ECs compared to lipofectamine; although there was similar cytotoxicity in primary artery wall cells when Terplex/DNA was increased to 6 μg per well. However, the level of Terplex/DNA concentration that caused cytotoxicity was far from its therapeutic range (less than 1 μg per well) (Fig. 10.3A and 3B). It was found that Terplex/DNA-mediated gene transfer to artery wall cells was LDL dose-dependent, saturable and could be inhibited by excess free LDL. The Terplex/DNA-mediated primary artery wall cell gene transfer is LDL receptor-specific and receptor-mediated endocytosis pathway. The Terplex/DNA system can be a powerful gene delivery tool to deliver therapeutic genes to the cardiovascular system.

IV. TERPLEX/DNA FOR MYOCARDIAL TRANSFECTION

Although the drawbacks of viral gene transfection are known, nonviral gene delivery has not been well accepted due to its low transfection rate (Kabanov et al., 1999). We delivered the Terplex/DNA system into myocardium. A series of experiments utilizing direct myocardium injection of this carrier/DNA were conducted (Affleck, et al., 2001). The apices of hearts of Sprague-Dawley male rats, 400–500 g were injected with 10 μg of luciferase cDNA either as naked plasmid DNA or complexed with Terplex. Rat heart with naked DNA showed a transition elevation in luciferase activity to Day 5, peaking 3 days post-injection. However, by Day 7, the luciferase activity among the naked DNA injection had fallen back to baseline.

Meanwhile, the Terplex/PCMV-Luc hearts demonstrated significantly elevated transfection efficiency on Day 3 and, further, the transfection efficacy in this group remained elevated to Day 30 (Fig. 10.4). Transgene expression in the rabbit model was evaluated—3, 5, and 30 days post-injection—for luciferase in the left ventricles of New Zealand white rabbit (Bull et al., 2003). The luciferase activity was significantly higher in the group injected with Terplex/PCMv-Luc vs. naked PCMV-Luc. At three days, rabbits were injected with volumes of 100 μl; luciferase activity was approximately 45000 RLU for the Terplex/PCMV-Luc animals vs. 1600 RLU for the naked PCMV-Luc animals. Thirty days after injections, luciferase activity was 677 $\pm$ 52 RLU for the naked PCMV-Luc hearts (Fig. 10.5). There were no side effects of Terplex/DNA

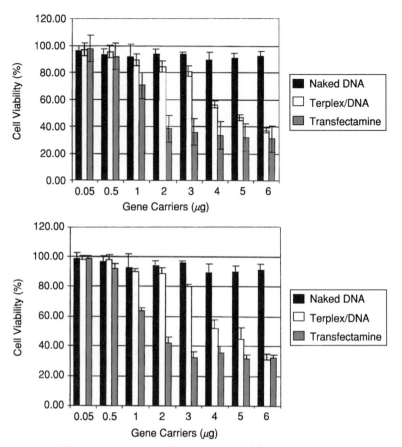

Figure 10.3. MTT Cytotoxicity Assay. The cytotoxicity of the Terplex/DNA on bovine aorta endothelial cells (A) and bovine aorta smooth muscle cells (B). Naked plasmid DNA was taken as a reference and transfectamine (Lipofectamine Plus) as a control gene carrier.

injection demonstrated by cardiac monitoring after the injection. Hearts appeared similar at the time of excision with minor ecchymosis in both groups.

We also used Terplex/DNA to transfect myocardium with plasmid vascular endothelial growth factor-165 (pVEGF). pVEGF was evaluated for preserving left ventricle function and structure after coronary ligation in a rabbit model. A plasmid-carrying VEGF-165 coding region, under the control of cytomegalovirus (CMV) early-promoter region, and chicken beta globulin intron, was constructed. The therapeutic gene was inserted into PCMV-lei based on the Mlu I and Bam HI restriction sites. Treated animals received 50 μg of LDL, and 50 μg of PCMV-VEGF in PBS. Myocardial infarction was

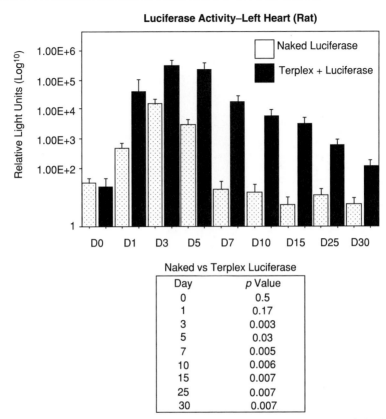

Figure 10.4. Luciferase activity Day 0 (D0) to Day 30 (D30) after injection of the firefly luciferase gene into the apex of the heart of Sprague-Dawley rats. Hearts were injected with either naked luciferase DNA or complexed with Terplex. Injectate volume was $150\,\mu l$ containing $10\,\mu g$ of luciferase plasmid under the influence of a CMV promoter. Luciferase activity is expressed as relative light units $\log^{10}$ (RLU = total light units/ml homogenate/mg protein). The naked luciferase group demonstrates a brief elevation in luciferase activity to Day 5, falling back to baseline after that time-point. Luciferase activity in the Terplex + Luciferase group is significantly elevated at every time-point after Day 3 and remains significantly elevated to Day 30.

induced after completion of injection. Ischemia was confirmed by visual inspection of the myocardium and by electrocardiographic changes on continuous EKG monitoring. It was observed that the ejection fraction in the VEGF-treated group and the control group were not different pre-operatively and on post-ligation Day 1. However, there was significant improvement in ejection fraction in the VEGF-treated group at Day 21 (Fig. 10.6). There was no improvement in ejection fraction between Day 1 and Day 21 in the control group.

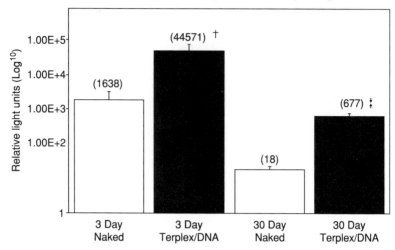

† $p = 0.002$ versus 3 Day Naked Luciferas
‡ $p = 0.002$ versus 30 Day Naked Luciferase

Figure 10.5. Luciferase activity 3 days and 30 days after injection of the firefly luciferase gene into the left ventricles of New Zealand white rabbits. Hearts were injected with naked luciferase DNA (pCMV-Luc) or complexed with the Terplex gene delivery system (Terplex + pCMV-Luc). Injectate volume was $1000 \mu l$. Luciferase activity is expressed as relative light units $\log^{10}$ (RLU = total light units/ml homogenate/mg protein). Luciferase activity at 3 days for the naked pCMV-Luc injected hearts was 1638 ± 567 RLU versus 44571 ± 8370 RLU for the Terplex + pCMV-Luc hearts (P = 0.002). At 30 days, luciferase activity for the naked pCMV-Luc injected hearts was 18 ± 3 RLU versus 677 ± 52 RLU for the Terplex + pCMV-Luc hearts (P = 0.002).

Representative M-mode short axis tracings were illustrated for VEGF-treated animals at Day 1 and Day 21 and a control animal on Day 1 and Day 21. Our findings indicate that supplementing endogenous VEGF protects left ventricular anatomy and enhances the ventricular function. We have demonstrated that Terplex/VEGF-165 gene therapy provides a protective effect on the structure and function of the left ventricle following acute myocardial ischemia and infarction (Bull *et al.*, 2003).

In conclusion, we have developed a new gene delivery system based on stearyl-PLL and LDL. This supermolecular gene carrier, the Terplex system, is unique because the main driving force for the Terplex formation is a balance between hydrophobic and electrostatic interactions between stearyl-PLL, LDL and DNA. The designed system was well characterized and demonstrated both *in vitro* and *in vivo* as an effective carrier of plasmid DNA with the ultimate goal of human application.

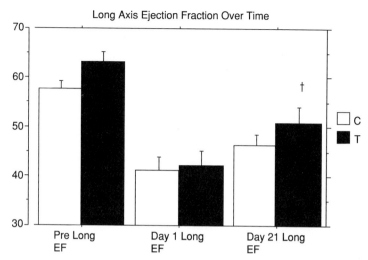

Figure 10.6. Long axis ejection fraction over time. Left ventricular function as measured by right parasternal long axis views. † p = 0.02 vs. Day 1 Long EF. Pre Long = pre-operative long axis ejection fraction. Day 1 Long = day 1 long axis ejection fraction. Day 21 Long = day 21 long axis ejection fraction. C = control, T = treatment.

Acknowledgment

This work was supported by NIH grant HL-65477.

References

Affleck, D. G., Yu, L., Bull, D. A., Bailey, S. H., and Kim, S. W. (2001). Augmentation of myocardial transfection using Terplex DNA: A novel gene delivery system. *Gene Ther.* **8,** 349–353.

Ahn, C. H., Chae, S. Y., Bae, Y. H., and Kim, S. W. (2002). Biodegradable polyethylenimine for plasma DNA delivery. *J. Control Rel.* **80,** 273–282.

Bissen, E. A., Beuting, D. M., Roefen, H. C. P. F., Van de Marel, G. A., Van Boom, J. H., and Van Berkel, T. J. C. (1995). Synthesis of cluster galactosides. *J. Med. Chem.* **38,** 1538–1546.

Boussif, O., Delair, T., Brua, C., Veron, L., Parirani, A., and Kolbe, H. V. (1999). Synthesis of polyallylamine derivatives and their use as gene transfer vectors *in vitro. Bioconjugate. Chem.* **10,** 877–883.

Boussif, O., and Zanta, M. A. (1996). Optimized galencies improve *in vitro* gene transfer with cationic molecules up to 100 fold. *Gene Ther.* **3,** 1074–1080.

Boussif, O., Lezoualc, H. F., Zanta, M. A., Mergny, M. D., Scheman, D., and Demeneix, B. (1995). A versatile vector for gene and oligonucleotide transfer in cell culture and *in vivo. Proc. Nat. Acad. Sci. USA* **92,** 7297–72301.

Bull, D. A., Bailey, S. H., Rentz, J. J., Zebrack, J. S., Lee, M., Litwin, S. E., and Kim, S. W. (2003). Effect of Terplex/VEGF165 gene therapy on left ventricular function and structure following myocardial infarction. *J. Control Rel.* 175–181.

Choi, Y. H., Liu, F., Kim, J., Choi, Y. K., Park, J. S., and Kim, S. W. (1998). Polyethylene glycol grafted poly-L-lysine as polymeric gene carriers. *J. Control Rel.* **54**, 39–48.

Dunlop, D. D., Maggi, A., Soria, M. R., and Monaco, L. (1997). Nanoscopic structure of DNA condensed for gene delivery. *Nucleic Acids Res.* **25**, 3095–3101.

Han, S., Mahato, R. I., Sung, Y. K., and Kim, S. W. (2002). Development of biomaterials for gene therapy. *Mol. Ther.* **2**, 302–317.

Han, S., Mahato, R. I., and Kim, S. W. (2001). Water soluble lipopolymer for gene delivery. *Bioconjugate. Chem.* **12**, 337–345.

Janat, J. W., Yockman, J., Furgeson, D., Lee, M., Kern, S., and Kim, S. W. (2004). Combination of local, nonviral IL-12 gene therapy and systemic paclitaxel treatment in metastatic breast cancer model. *Mol. Ther.* **9**, 829–836.

Kabanov, A. V., and Kabanov, V. A. (1995). DNA complexes with polycations for the delivery of generic materials into cells. *Bioconjugate. Chem.* **6**, 7–20.

Kim, J. S., Maruyama, A., Akaike, T., and Kim, S. W. (1997). *In vitro* gene expression on smooth muscle cells using terplex delivery system. *J. Control Rel.* **80**, 273–282.

Kim, J. S., Kim, B. I., Maruyama, A., Akaike, T., and Kim, S. W. (1998). A new non-viral delivery vector: Terplex system. *J. Control Rel.* **53**, 175–182.

Ko, K. S., Lee, M., Koh, J. J., and Kim, S. W. (2001). Combined administration of plasmid encoding IL-4 and IL-10 prevents the development of autoimmune diabetes in non-obese diabetic mice. *Mol. Ther.* **4**, 313–316.

Koh, J. J., Ko, K. S., Lee, M., Han, S., Park, J. S., and Kim, S. W. (2000). Degradable polymer carrier for the delivery of IC-10 plasmid DNA to prevent autoimmune insulitis of NOD mice. *Gene Ther.* **7**, 2099–2105.

Lee, M., and Kim, S. W. (2002). Polymeric gene carriers. *Pharmaceutical. News* **9**, 1407–1415.

Lee, M., Han, S., Ko, K. S., and Kim, S. W. (2001). Cell type specific and glucose responsive expression of IL-4 by using insulin promoter and water soluble lipopolymer. *J. Control Rel.* **75**, 421–429.

Lee, M., Koh, J. J., Han, S., Ko, K. S., and Kim, S. W. (2002). Prevention of autoimmune insulitis by delivery of IL-4 plasmid using biodegradable polymer carrier. *Pharm. Res.* **19**, 246–249.

Lim, Y. B., Han, S., Kong, H. U., Lee, Y., Park, J. S., Jeong, B., and Kim, S. W. (2000). *Pharm. Res.* **17**, 811–816.

Mahato, R. I., Lee, M., Han, S., Maheshwari, A., and Kim, S. W. (2001). Intratumoral delivery of P2CMVmIL-12 using water soluble lipopolymer. *Mol. Ther.* **4**, 130–138.

Maruyama, A., Watanabe, H., Ferdous, A., Katoh, M., Ishikara, T., and Meaike, T. (1998). Characterization of interpolyelectrolyte complex between DNA and polylysine comb type copolymer. *Bioconjugate. Chem.* **9**, 292–299.

Nah, J. W., Yu, L., Han, S., Ahn, C. H., and Kim, S. W. (2002). Artery wall binding peptide-PEG grafted poly-l-lysine based gene delivery to artery wall cells. *J. Contol Rel.* **78**, 273–284.

Sagara, K., and Kim, S. W. (2002). A new synthesis of galactose-PEG-PEI for gene delivery to hepatocytes. *J. Control Rel.* **79**, 271–281.

Suh, W. H., and Kim, S. W. (2002). RGD peptide-PEG grafted PEI as a tumor targeting gene carrier. *Mol. Ther.* **6**, 664–672.

Temin, H. M. (1990). Safety consideration in somatic gene therapy of human disease with retrovirus vectors. *Human Gene Ther.* **1**, 111–123.

Westlin, W. F. (2001). Integrins as targets of angiogenesis inhibition. *Cancer J.* **7**, S139–S143.

Yu, L., Suh, H., Koh, J. J., and Kim, S. W. (2001). Systemic administration of Terplex DNA: Pharmacokinetics and gene expression. *Pharm. Res.* **18**, 1277–1283.

11

Design of Polyphosphoester-DNA Nanoparticles for Non-Viral Gene Delivery

Hai-Quan Mao* and Kam W. Leong[†]

*Department of Materials Science and Engineering
The Johns Hopkins University, Baltimore, Maryland 21205
[†]Department of Biomedical Engineering
The Johns Hopkins University, Baltimore, Maryland 21205

Advances in Genetics, Vol. 53
Copyright 2005, Elsevier Inc. All rights reserved.

0065-2660/05 $35.00
DOI: 10.1016/S0065-2660(05)53011-6

ABSTRACT

Development of safe and effective non-viral gene carriers is still critical to the
ultimate success of gene therapy. This review highlights our attempt to design
the gene carriers in a systematic manner. We have synthesized a series of
polymers with a phosphoester backbone containing different charge groups in
the sidechain connected to the backbone through a phosphate (P–O) or a
phosphoramide (P–N) bond. These gene carriers have different charge groups,
sidechain lengths, and branching structures, but they are structurally related to
allow a systematic investigation of the structure–property relationship, including
DNA binding capacity, cytotoxicity, DNA protection, biodegradability, DNA
release kinetics, and transfection efficiency.   © 2005, Elsevier Inc.

I. INTRODUCTION

Lingering concerns of viral vectors continue to fuel the development of non-
viral gene carriers (Cavazzana-Calvo et al., 2004; Kaiser, 2004; Williams and
Baum, 2003). The promising non-viral gene delivery systems, other than the
"gene gun" in DNA vaccine applications, are composed of ionic complexes
formed between DNA and polycationic liposomes or polycationic polymers
(Ewert et al., 2004; Niidome and Huang, 2002; Nishikawa and Huang, 2001;
Pedroso de Lima et al., 2003; Wagner, 2004). The latter have been increasingly
proposed as potential vectors because of their versatility. Rigidity, hydrophobi-
city/hydrophilicity, charge density, biodegradability, and molecular weight of the
polymer chain are all parameters that in principle can be adjusted to achieve an
optimal complexation with DNA. It is likely that for different cells or tissues, or

different routes of administration *in vivo*, the desirable characteristics of the DNA nanoparticles, or polyplexes, would differ. Polymeric carriers with their versatility are well positioned to meet the challenges.

Design of new polymeric gene carriers has therefore been an intensively pursued research area in recent years. Polymers spanning the spectrum of biodegradable, biostable, linear, branched, dendrimeric, and crosslinked have all been investigated. The promising candidates include different forms of polyethylenimines (PEI) (Ahn *et al.*, 2002; Blessing *et al.*, 2001; Brissault *et al.*, 2003; Brownlie *et al.*, 2004; Forrest *et al.*, 2003, 2004; Gautam *et al.*, 2003; Gosselin *et al.*, 2001; Kunath *et al.*, 2003; Lampela *et al.*, 2004; Wightman *et al.*, 2001; Zou *et al.*, 2000), modified polylysines (Ahn *et al.*, 2004; Jeong *et al.*, 2003; Kim *et al.*, 2004; Oupicky *et al.*, 2001; Putnam *et al.*, 2001), poly (α-(4-aminobutyl)-L-glycolic acid) (Lee *et al.*, 2003a; Lim *et al.*, 2000a; Maheshwari *et al.*, 2002), polyamidoamine dendrimers (Jones *et al.*, 2000; Kramer *et al.*, 2004; Lee *et al.*, 2003b; Zhang and Smith, 2000), poly((2-dimethylamino)ethyl methacrylate)s (Lim *et al.*, 2000b), poly(β-amino ester) s (Ahn *et al.*, 2002, 2004; Anderson *et al.*, 2003; Jon *et al.*, 2003), polyphosphazenes (Luten *et al.*, 2003), cyclodextrin-containing polycations (Bellocq *et al.*, 2003; Davis *et al.*, 2004; Hwang and Davis, 2001a,b; Popielarski *et al.*, 2003; Pun and Davis, 2002; Pun *et al.*, 2004; Reineke and Davis, 2003a,b), chitosans of different composition (Kiang *et al.*, 2004; Lee *et al.*, 1998; Leong *et al.*, 1998; MacLaughlin *et al.*, 1998; Mao *et al.*, 2001; Richardson *et al.*, 1999; Roy *et al.*, 1999; Thanou *et al.*, 2001), and modified collagen (Truong-Lee *et al.*, 1998, 1999; Wang *et al.*, 2004a), among others. It would be beyond the scope of this review to cover all the gene carrier designs. Instead we will concentrate on polyphosphoesters (PPE), a class of versatile cationic gene carriers that can highlight the different strategies of dealing with the barriers of non-viral gene transfer.

A. Barriers for non-viral gene transfer

Before the discussion of the different PPE designs, it would be informative to first highlight the potential barriers of non-viral gene transfer. We will also discuss the general principles of incorporating features into a polymer to overcome these barriers.

1. Condensation of the plasmid DNA

The DNA must be condensed to a size permitting cellular internalization. The charge density and the ionicity of the polymer is therefore an important parameter. A major theme of the PPE design is to systematically study how the charge structures of the polymers would influence their complexation with DNA and the subsequent transfection efficiency. The optimal size of the polymer–DNA complex (polyplex) remains debatable. Conventional wisdom favors smaller sizes, below 150 nm. This is also the general threshold below which receptor-mediated endocytosis is believed to be operative. Contact of the DNA nanoparticles with biological fluid would easily lead to aggregation, which may severely limit the ability of the nanoparticles to be endocytosed and reaching the final site, the nucleus. Minimizing aggregation of the nanoparticles is therefore an important design consideration. Incorporation of hydrophilic segments into the nanoparticles, such as conjugation of PEG to the polymer, which shows the benefit of reducing nonspecific interactions with plasma proteins and lessening complement system activation, is a popular and effective strategy.

2. Transport of polyplex to cell

The task of the nanoparticle is to reach the cell surface without being cleared from the system and to protect the DNA from degradation by nucleases. Most nanoparticles are formed by electrostatic interaction. They must remain stable in the extracellular fluid where a high concentration of polyelectrolytes might disrupt this interaction. Polymers that can bind the DNA tightly are therefore preferred. Excessive net positive charge on the surface however tends to lead to aggregation and coating by negatively charged proteins such as albumin. Strong cationicity of the polymer is nevertheless needed to produce a compact nanoparticle structure that can impede enzymatic degradation of the embedded DNA.

3. Internalization by cell

Once the nanoparticles reach the target cells, they bind with the negatively charged cell membrane and are subsequently endocytosed, by mechanisms specific or non-specific. Achieving gene transfer to the target cells is highly desirable. Cell-specific targeting ligands, such as monoclonal antibodies, peptides and sugars can be conjugated to the gene carriers to promote receptor-mediated endocytosis. Polymers that contain functional groups amenable to ligand conjugation would be desirable. The conjugation step can be executed pre-or post-complex formation. The former allows for fine control of the degree of substitution, but the presence of the ligand, such as a polypeptide, may interfere with the complexation. The ligand may also be buried in the nanoparticle and not presented on the surface properly. Post-complexation conjugation would unquestionably decorate the nanoparticle surface. The stability of the nanoparticle must survive the reaction, and often the need to isolate the reacted nanoparticle from unreacted ligands would significantly diminish the yield of the nanoparticle.

4. Escape from the endosome

After the non-specific or receptor-mediated endocytosis, most nanoparticles are sequestered in the endosomal compartment, which at some point fuses with the lysosomes. Potent enzymes and low pH in these compartments either degrade or swell the cationic polymer, freeing the DNA, and also breaking down the DNA. Some anionic lipids in the endosome may also compete with DNA to bind with the cationic polymer, dissociating the complex and somehow release the DNA into the cytoplasm (Xu and Szoka, 1996). Escape of the nanoparticle or the DNA into the cytoplasm is a significant step in the gene transfer process. Efficient destabilization of the endosomes/lysosomes by endosomolytic reagents such as chloroquine (Midoux et al., 1993), lipids (El Ouahabi et al., 1997), and peptides (Sosnowski et al., 1996) would enhance the gene transfer efficiency. Polymer that can possess this endosomal disruptive property has figured prominently in the design consideration (Jones et al., 2003; Kyriakides et al., 2002). This can take the form of lipophilic sidechain, or attachment of peptides to the sidechains to interact specifically with the vesicle membrane. Alternatively, polymers with excessive amino groups have been hypothesized to reduce the acidification of the endosome, eventually causing endosomal swelling and collapse (Boussif et al., 1995). This "proton sponge" effect has been first postulated for the potency of polyethylenimine in gene delivery. According to this hypothesis, PEI and PEI-DNA complexes sequestered into the endosome absorb protons that are pumped into this organelle. Protonation of the amine groups on PEI leads to swelling of the polymer, concomitant with an influx of chloride ions to neutralize the build-up of a charge gradient. The net effect is an increase in osmotic pressure to destabilize and rupture the endosome, leading to release of the endosomal content into the cytoplasm. PPEs have been synthesized to mimic the pH buffering capacity of PEI for this reason.

5. Transport to the peri-nuclear space

Nanoparticles escaped from the vesicles into the cytoplasm intact can diffuse to the nuclear membrane. There is also evidence the nanoparticle-containing vesicles can be transported to the nuclear membrane via the molecular motors. The property of the polymer that would accelerate this active and passive diffusion process is currently unknown. The nanoparticle can also be dissociated to release the DNA, which is degraded in both the endosome and the cytoplasm. Microinjected plasmid DNA is rapidly degraded in the cytoplasm with an apparent half-life of 50–90 min (Lechardeur et al., 1999). The translocation of unpacked DNA from cytoplasm to nucleus is probably through diffusion, a relatively slow process compared to cytoplasmic degradation, which leads to an inefficient nuclear transport. It is understandable that DNA would be more stable in the cytoplasm if it were still complexed with the cationic polymer. However, only certain synthetic polymers, such as PEI, show protection of DNA in the cytoplasm and promote nuclear transport. The pathway of the transport of nanoparticle from cytoplasm to nucleus is not well understood, with some hypotheses leaning toward the involvement of anionic phospholipids.

6. Translocation into nucleus

There is evidence that intact nanoparticle can be present in the nucleus of even non-dividing cells. Again this phenomenon remains poorly understood. It is generally believed that it is the unpacked DNA that translocates into the nucleus for ultimate gene expression. The inclusion of viral nuclear localization signals (NLS) has been demonstrated to be an efficient strategy to facilitate nuclear transport (Branden et al., 1999; Ziemienowicz et al., 1999). NLS are generally peptide or protein sequences that are recognized by the cellular machinery, which can complex with DNA to promote nuclear accumulation. It is not clear if attachment of NLS to the nanoparticle can efficiently transport the nanoparticle into the nucleus. The same characteristics of polymer that allows attachment of ligands can be used for the conjugation of NLS.

It becomes clear even with this over-simplified picture delineating the barriers for gene transfer that the correlations between physicochemical properties of the nanoparticle and their transfection potency would be multifactorial. It would be challenging to design polymers incorporating all the desirable features into a single structure, particularly in the absence of biological guidance. For instance, stable complexes would prevent DNA release while unstable ones would be vulnerable to rapid DNA degradation. Both are undesirable. Nanoparticle with an intermediate stability may exhibit the best transfection efficiency. But how to strike the balance remains a trial-and-error process. This review covers an attempt to study one molecular design parameter at a time, such as the type of charge group, or the length of the linker for the charge group, while keeping the rest of the polymer characteristics relatively constant. However, the gene transfer process being complicated and convoluted as it is, the mechanistic insight gained so far with the PPE studies has been limited. Nevertheless, these studies have revealed interesting phenomena and hopefully may serve to stimulate other researchers to come up with more rational and effective polymeric gene carrier designs.

B. Rationale of developing polyphosphoesters as gene carrier

The motivation for developing polyphosphoesters as gene carriers initially comes from the desire to overcome the barrier of releasing the DNA from the nanoparticle and to use this series of polymers for systematic mechanistic studies

(Huang *et al.*, 2004; Zhao *et al.*, 2003). In our earlier work we have observed through intracellular trafficking studies that chitosan nanoparticle may be too stable for releasing the DNA. A biodegradable polycation would ensure the "unpacking" of the nanoparticle. Similar in rationale to the design of other polycationic gene carriers, the biodegradability of the polyphosphoesters offers several advantages. It can provide a function of extracellular sustained release, and conceivably even intracellular sustained release. Such a gene delivery system can therefore substantially improve the bioavailability of DNA both inside and outside of cells. This controlled release property can be adjusted by varying the polycationic polymer and DNA ratio (N/P) ratio as well as the composition of the backbone and the sidechain structure, which in turn can also influence the transfection efficiency *in vivo*.

The pentavalency of a phosphorus atom in the backbone of polyphosphoesters makes it possible to conjugate functional groups, including charged groups through a phosphate (P–O) or a phosphoramide (P–N) bond as side chains. Starting from the parent polyphosphite, the P–H bonds can be readily converted for conjugation of different chemical structures. We have synthesized a series of cationic PPEs and PPAs from the precursor polymer poly(1,2-propylene-H-phosphonate). This backbone structure is chosen with consideration of biocompatibility. Comprised of propylene oxide in the backbone, the potential breakdown products should be relatively innocuous. Conjugation of charged groups to this parent polymer produces water-soluble PPEs or PPAs that can complex efficiently with DNA. In keeping the chemical structure of the backbone and the molecular weight constant, we can then systematically study the effect of the charge groups on toxicity and transfection efficiency. Table 1 lists a series of PPE and PPA gene carriers that have been investigated in our labs (Huang *et al.*, 2004; Zhao *et al.*, 2003). These gene carriers have different charge groups, sidechain lengths, and branching structures, but they are structurally related to allow a systematic investigation of the structure-transfection efficiency relationship.

II. SYNTHESIS AND STRUCTURES OF PPE AND PPA GENE CARRIERS

A. Synthesis of PPEs and PPAs

The synthetic schemes for these PPEs and PPAs are illustrated in Fig. 11.1. With the exception of PCEP (a PPE with charge-bearing backbone) (Wen *et al.*, 2004), all the carriers are synthesized from the same precursor polymer, poly(4-methyl-2-oxo-2-hydro-1,3,2-dioxaphospholane) [also termed poly(1,2-propylene H-phosphonate)], which is obtained by ring opening polymerization using

Table 11.1. Structures of PPA and PPE Gene Carriers

PPA	PPE	
Charged sidechains	Neutral sidechain	Charged backbone

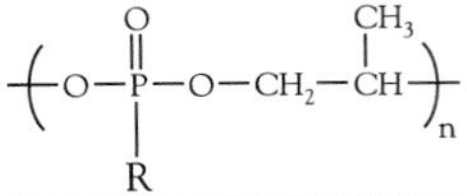

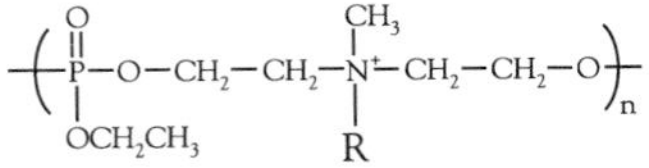

PPA backbone structure:

$$\left(\!-O-\overset{\displaystyle O}{\underset{\displaystyle R}{\overset{\|}{P}}}-O-CH_2-\overset{\displaystyle CH_3}{\underset{}{CH}}-\!\right)_n$$

PPE charged backbone structure:

$$\left(\!-\overset{\displaystyle O}{\underset{\displaystyle OCH_2CH_3}{\overset{\|}{P}}}-O-CH_2-\overset{\displaystyle CH_3}{\underset{\displaystyle R}{C}H}-N-CH_2-CH_2-O-\!\right)_n$$

R groups for:

PPA with different types of charge groups

PPA-EA	$-NH-CH_2CH_2NH_2$
PPA-MEA	$-NH-CH_2CH_2NH(CH_3)$
PPA-DMA	$-NH-CH_2CH_2N(CH_3)_2$
PPA-TMA	$-NH-CH_2CH_2N(CH_3)_3^+$

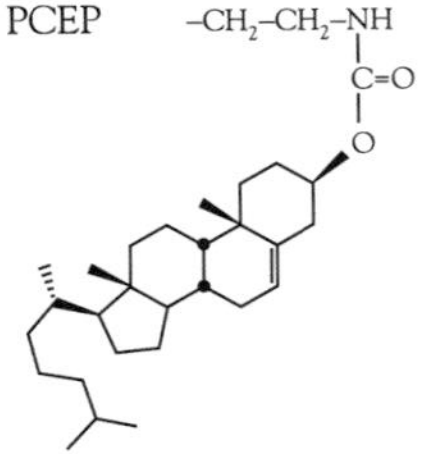

PPA with linear sidechains (different lengths)

PPA-EA	$-NH-CH_2CH_2NH_2$
PPA-EA	$-NH-CH_2CH_2CH_2NH_2$
PPA-EA	$-NH-CH_2CH_2CH_2CH_2NH_2$

PPA with branching sidechains (different lengths)

PPA-DEA	$-N\big(CH_2CH_2NH_2\big)\big(CH_2CH_2NH_2\big)$
PPA-EPA	$-N\big(CH_2CH_2NH_2\big)\big(CH_2CH_2CH_2NH_2\big)$
PPA-DPA	$-N\big(CH_2CH_2CH_2NH_2\big)\big(CH_2CH_2CH_2NH_2\big)$
PPA-BPA	$-N\big(CH_2CH_2CH_2NH_2\big)\big(CH_2CH_2CH_2CH_2NH_2\big)$

PPE with linear sidechains (different charge groups and chain lengths)

PPA-EA	$-O-CH_2CH_2NH_2$
PPE-MEA	$-O-CH_2CH_2NH(CH_3)$
PPE-HA	$-O-CH_2(CH_2)_4CH_2NH_2$

PPE with neutral sidechains

PPE-HE	$-O-CH_2CH_2OH$

PPE with charged backbone

PCEP	$-CH_2-CH_2-NH$

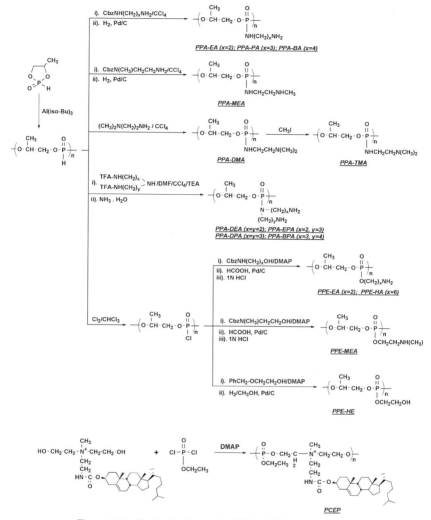

Figure 11.1. Synthetic schemes for PPE and PPA gene carriers.

4-methyl-2-oxo-2-hydro-1,3,2-dioxaphospholane (Wang et al., 2002a). Starting from this polyphosphite, various types of charge groups can be conveniently conjugated to yield PPAs (with P–N bonds) (Wang et al., 2002a,c, 2004b; Zhang et al., in press(a)) or PPEs (with P–O bonds) (Huang et al., 2004; Wang et al., 2001, 2002b, 2003). This synthetic scheme provides a common basis for a comparative study—polycationic PPEs and PPAs and electrostatically neutral PPEs share the same polymer backbone. The average molecular weight/average chain length of most of the polymer carriers in this series are similar, as they are prepared from the same parent polyphosphate, although in some cases the conjugation step leads to partial degradation of the polymer. In addition, the ring opening polymerization used to synthesize the precursor polymer can yield relatively high molecular weights.

1. Synthesis of PPAs from poly(1,2-propylene H-phosphonate)

The parent polymer, poly(1,2-propylene H-phosphonate) is synthesized by ring-opening polymerization of 4-methyl-2-oxo-2-hydro-1,3,2-dioxaphospholane using a cationic initiator triisobutyl aluminum (Wang et al., 2001, 2002a). Direct conversion of P–H bonds in poly(1,2-propylene H-phosphonate) to phosphor-amidyl group is achieved through Atherton-Todd reaction using CCl_4 as an oxidant (Wang et al., 2002a, 2004b). In the case of primary amino group side-chains (PPA-EA, PPA-PA and PPA-BA) and secondary amino group sidechains (PPA-MEA), the corresponding primary and secondary amino groups of the diamines are protected by benzyloxycarbonyl group before the diamines are conjugated to poly(1,2-propylene H-phosphonate). PPA carriers are obtained following deprotection by catalytic hydrogenation. Synthesis of PPA with ter-tiary amino group is straightforward—N,N-dimethylethylenediamine is reacted directly with poly(1,2-propylene H-phosphonate) to yield PPA-DMA. PPA with quaternary amino groups, PPA-DMA, is prepared by quaternization of PPA-DMA with methyl iodide. The weight average molecular weights of this series of PPAs range from 44 to 52 KDa.

In order to synthesize PPAs with branching side chains, alkylene triamines [(N-(2-aminoethyl)-1,2-ethylenediamine, (N-(2-aminoethyl)-1,3-propanedia-mine, N-(3-aminopropyl)-1,3-propanediamine, and N-(3-aminopropyl)-1,4-bu-tanediamine, respectively] are selectively protected by trifluoroacetate groups leaving only secondary amino groups for the conjugation to poly(1,2-propylene H-phosphonate) under similar conditions as described above (Wang et al., 2004b; Zhang et al., in press(b)). Due to the lower reactivity of secondary amino groups and the bulkier TFA-protected amino groups, the conversion efficiency from P–H groups to phosphoramidyl groups is lower, approximately 67%. The PPA carriers are obtained following deprotection in ammonia solution and purification by dialysis against distilled water. The weight average molecular weights of these PPAs range from 35 to 45 KDa.

2. Synthesis of PPEs from poly(1,2-propylene H-phosphonate)

The conversion from poly(1,2-propylene H-phosphonate) to PPE is achieved by chlorination of P–H bond first, followed by esterification with a protected hydroxyalkylamine (Fig. 11.1). Esterification of the P-Cl groups with, for example, benzyl *N*-(2-hydroxyethyl) carbamate (5–10% excess), is aided by 4-dimethylamino-pyridine (DMAP) as a catalyst yielding an intermediate polymer with protected sidechains (Wang *et al.*, 2001, 2002b, 2003). The protective groups are subsequently removed by catalytic hydrogenation using the formic acid-Pd/C method. This acidic hydrogenation step and chloric acid treatment in the purification step, however, would lead to significant degradation of the PPE backbone, particularly for the substituted amino ethyl sidechains. A PPE carrier with tertiary amino groups could not be prepared with a reasonable average molecular weight. The weight average molecular weights of PPE-EA, PPE-MEA and PPA-HA are 30 KDa, 13 KDa and 38 KDa, respectively.

PPE with electrostatically neutral sidechains can also be synthesized from poly(1,2-propylene chloro-phosphonate) using a similar method as described above (see Fig. 11.1). The esterification step is achieved using an excess amount of protected diol. Benzyloxy protecting groups are removed by Pd/C catalytic hydrogenation in methanol for about 16 h. Partial degradation is also observed during the catalytic hydrogenation process. PPE-HE has a weight average molecular weight of 9.1 KDa.

3. Synthesis of PCEP by polycondensation

PCEP is synthesized through polycondensation of ethyl dichlorophosphate with a diol carrying a positive charge and a cholesterol moiety, using triethylamine (TEA) as the acid acceptor (Wen *et al.*, 2004) (Fig. 11.1). Because of the bulky cholesterol moiety in the diol monomer, the degree of polymerization is low—the average molecular weight is only about 4000.

B. Cytotoxicity and tissue compatibility of PPAs and PPEs

Cytotoxicity of a polymeric gene carrier is important for practical considerations. It may also affect the gene expression by interfering with the transcription and translation processes in the cells. By carefully choosing the building blocks of polyphosphoesters, the degradation products of polyphosphoesters can be minimally toxic and have good biocompatibility. The backbone of this series of polyphosphoesters is polyphosphate based on α-propylene glycol, which has a favorable safety profile compared to other diols. The linkages within and between the repeating units are phosphate and phosphoramide bonds. These amino residues have also been carefully selected for toxicity consideration. For example, ethylene diol (for PPE-HE) and ethanolamine (for PPE-EA) both have minimal toxicity; and spermidine (for PPA-BPA) is present in cells at a concentration up to millimolar level (Tabor and Tabor, 1984).

The cytotoxicity of these carriers is assessed using cell viability assay (MTT assay). Cytotoxicity for PPAs with different types of charge groups

Table 11.2. LD_{50} Assessed in Cell Culture for PPA and PPE Carriers

PPA (1)[*]	PPA-EA	PPA-MEA	PPA-DMA	PPA-TMA		PEI
LD_{50} (μg/ml)	100	100	>500	>500		40
PPA (2)[*]	PPA-EA	PPA-PA	PPA-BA		PLL	PEI
LD50 (μg/ml)	104	117	110		45	24
PPA (3)[*]	PPA-DEA	PPA-EPA	PPA-DPA	PPA-BPA	PLL	PEI
LD50 (μg/ml)	122	96	103	105	45	24
PPE[*]	PPE-EA	PPE-MEA	PPE-HA	PPE-HE	PCEP	PEI
LD50 (μg/ml)	>1000	>500	>200	>10,000	90	12

[*]PPA (1): PPAs with different types of charge groups, tested in COS-7 cells; PPA (2): PPAs with primary amine charge groups but different lengths of linear sidechains, tested in HeLa cells; PPA (3): PPAs with primary amine charge groups but different lengths of branching sidechains, tested in HeLa cells; PPE: representative data from tests in various cell lines (HEK293 cells, HeLa cells and COS-7 cells).

decreases according to the order of sidechain amino groups: primary amine $\sim$ secondary amine > tertiary amine > quaternary amine (Wang et al., 2002c, 2004b). Carriers having primary amino groups exhibit similar levels of cytotoxicity with an LD_{50} of about 100 μg/ml (Table 11.2), which is lower than with branched polyethylenimine tested in the same cell lines (LD_{50} 24–40 μg/ml). PPEs in general show much lower cytotoxicity. PPE-HE (with a neutral sidechain) and PPE-EA are among the least toxic (Huang et al., 2004; Wang et al., 2002b). Cells could tolerate at least 10 mg/ml of PPE-HE or 1 mg/ml of PPE-EA without noticeable influence on cell proliferation. PCEP is an exception among all PPEs, showing a LD_{50} similar to that of PPAs with primary amino sidechains (Wen et al., 2004). This could be attributed to the amphiphilic nature of the carrier (it forms micelle).

Only mild acute tissue response to PPE-EA and PPE-HE is observed in the muscle of Balb/c mice (Huang et al., 2004; Wang et al., 2002b). On the other hand, a severe inflammatory response is observed in muscles with PEI injection at a dose that is 5.2-fold lower in mass to that of PPE-EA. Moreover, severe necrosis is noticed in all the muscle samples receiving PEI injection, with a large numbers of macrophages, histiocytes, and neutrophils present at the injection sites.

In a central nervous system (CNS) gene delivery model, intracisternal injection of PPE-EA does not provoke any detectable pathological changes in the CNS, whereas PEI induces apoptosis as revealed by TUNEL staining (Li et al., 2004). In addition, PPA-EA, PPA-MEA, PPA-DEA and PPA-TMA do not show any noticeable toxicity to tissue and cells in the spinal cord following intrathecal injections (Wang et al., 2004b).

III. POLYCATIONIC PPAS—STRUCTURE–PROPERTY RELATIONSHIP: DNA COMPACTION CAPACITY AND GENE TRANSFECTION EFFICIENCY

A. DNA compaction capacity and characterization of PPA/DNA nanoparticles

All PPAs listed in Table 11.1 exhibit high DNA compaction capability (Wang et al., 2002a, 2004b; Zhang et al., in press(a)). Plasmid DNA in these complexes is partially protected from enzyme degradation, as demonstrated using DNase I as a model enzyme. The PPAs, except PPA-BA, are able to condense plasmid DNA completely at an N/P ratio of less than or equal to 1, suggesting that these cationic PPA carriers have a high DNA compaction capacity. Among them, PPAs with tertiary amino groups, quaternary amino groups and those have branching sidechains have a slightly higher DNA compaction capacity. Similar to typical polycationic gene carriers, PPA/DNA nanoparticles aggregate severely at near neutral surface charge (at N/P ratio of ~ 1). The average size decreases as the N/P ratio increases and reaches a plateau at an N/P ratio of 8. All PPA/DNA particles show similar average size ranging from 120 to 140 nm at high N/P ratios (>8). Under this N/P ratio, surfaces of these particles are positively charged with a ζ potential of about 20 to 30 mV.

B. Effect of charge groups on transfection efficiency

Generally speaking, the transfection efficiency of polyphosphoester/DNA complexes varies as a function of multiple parameters including composition and surface charge of the nanoparticles (N/P ratios), cell culture condition, as well as polymer structure, namely the type of charged groups, length of the spacer, charge density, molecular weight, and likely the backbone structure.

Similar to most of polycationic gene carriers, transfection efficiencies of these polyphosphoester/DNA complexes are dependent on the charge ratio (N/P) of carrier to DNA. For example, the transfection efficiency of PPA-BPA/DNA complexes in HEK 293 cells increases with N/P ratio, reaching a maximum at N/P ratios between 15 and 20 (Wang et al., 2002a). In the presence of 100 μM of CQ, the optimal N/P ratios of PPA-BPA/DNA complexes shift to 5~10. Under these conditions, PPA-BPA/DNA nanoparticles transfect cells nearly as efficiently as PEI/DNA and Transfast®/DNA complexes in HEK 293 cells.

For applications that require more stable nanoparticles and/or intracellular delivery, PPA carriers are more suitable than PPEs. One such example is to deliver genes to CNS through a retrograde transport model. This is a non-invasive approach to deliver genes to the brain by injecting polycation/DNA complexes to peripheral muscle. Following intramuscular injection of PPA-BPA/DNA nanoparticles to the mouse tongue where the hypoglossal motor neurons locate, bcl-2 expression is detected on Day 2 in the brain stem of the injected mouse at a level similar to that obtained with PEI/DNA complexes. The nanoparticles are thought to be transported to the neuron body by the motor neuron in the injected muscle via a retrograde transport mechanism (Wang et al., 2002a).

PPA-mediated transfection is clearly charge group dependent. We first compared the gene expression mediated by four PPA carriers with an identical backbone, same side chain spacer, similar molecular weights but different charge groups containing primary to quaternary amino groups (Wang et al., 2004b) (PPA-EA, PPA-MEA, PPA-DMA, and PPA-TMA, see Table 11.1). Although the DNA-compaction capacity of these four PPAs increase in the order of PPA-EA < PPA-MEA < PPA-DMA ~ PPA-TMA,

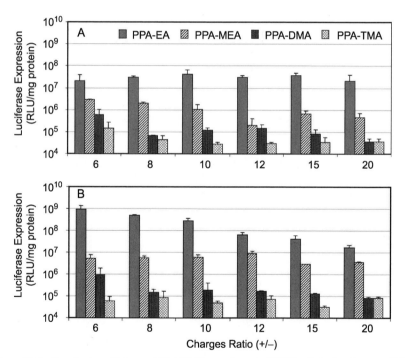

Figure 11.2. Transfection efficiency of PPA/DNA complexes in COS-7 cells in the absence of CQ (A) and in the presence of 100 μM CQ (B). Cells were incubated with PPA/DNA nanoparticles containing 3 μg of DNA at various N/P ratios for 3 h before refreshing the medium. Luciferase expression level was analyzed 48 h after transfection. Mean $\pm$ S.D. (n = 3). Reproduced with permission from Ref. (Wang et al., 2004b).

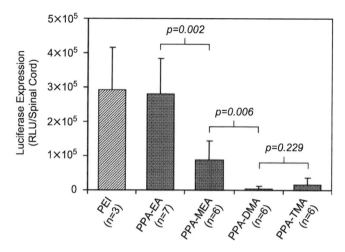

Figure 11.3. Luciferase activities in the spinal cord after intrathecal injection of PPA/DNA (N/P = 10) and PEI/DNA (N/P = 15) nanoparticles. Nanoparticles equivalent to 4 μg of pDNA were given to each rat. Transgene expression was analyzed 48 h after injection. Values were presented as mean ± S.D. (n = 3–7). The p-values were calculated using Student's t-test (two-tailed). Reproduced with permission from Ref. (Wang et al., 2004b).

their transfection efficiencies decrease in the same order. PPA-EA/DNA complexes gave the highest transfection efficiency in cell lines at charge ratios from 6 to 20 (Fig. 11.2).

The *in vivo* transfection efficiency of this series of PPA/DNA nanoparticles is evaluated in a rat spinal cord gene expression model following intrathecal injection of PPA/DNA nanoparticles at a charge ratio of 10 (Wang et al., 2004b). Matching the trend of transfection efficiency observed *in vitro*, gene expression level in the spinal cord is dependent upon the type of charge group. At a DNA dose of 4 μg, PPA-EA mediates the highest transgene expression among all four carriers. The luciferase expression activity is comparable to that of PEI, but with lower tissue toxicity. PPA-MEA with secondary amino group side chain follows with a 3.5-fold lower luciferase expression. PPA-DMA and PPA-TMA are not effective in this model, giving only a background level of gene expression (Fig. 11.3).

C. Effect of sidechain structure on transfection efficiency

Having established that primary amino group sidechain gives the highest transfection efficiency, we investigated the effect of sidechain structure (length and branching of PPA sidechains). We synthesized a series of seven PPA carriers:

PPA-EA, PPA-PA, PPA-BA, PPA-DEA, PPA-EPA, PPA-DPA, and PPA-BPA (Wang et al., 2002c; Zhang et al., in press(b)) (Table 11.1). These gene carriers all carry primary amino groups, but fall into two groups based on their branching structure of sidechains: those with linear side chains and those with branching side chains and higher charge density (2 positive charges per repeating unit). When PPA/DNA complexes are used to transfect HeLa cells, transfection efficiency correlates strongly with the branching structure of the sidechain, rather than the length of the sidechain. Transgene expression by PPAs with branching sidechains generally show about ten-fold higher efficiency than those with linear side chains (Wang et al., 2002c) (Fig. 11.4A). Of the four PPAs with branching side chains, PPA-DEA and PPA-DPA exhibit the highest transfection efficiencies. This transgene level is slightly higher than that mediated by PEI. The same trend is observed in transgene expression in primary rat hepatocytes (Fig. 11.4B). PPAs with branching side chains consistently show more than ten-fold higher gene expression than those with linear side chains.

D. Nanoparticles prepared with multiple PPAs with different types of charge groups: improving buffering capacity

It has been speculated that the transfection efficiency of PEI, which is correlated with the buffering capacity of PEI, is a result of the coexistence of primary (1°), secondary (2°) and tertiary (3°) amines in the PEI structure (Thomas and Klibanov, 2002). It has also been shown that copolymers with 1° and 3° and/or 2° amino groups exhibit higher transfection efficiency (Reschel et al., 2002; Thomas and Klibanov, 2002; Tang et al., 1996). Reschel et al. showed that, of all the polymethacrylamide carriers they synthesized, the highest transfection activity is found for a copolymer carrier containing both 1° and 3° amines (Reschel et al., 2002). Polyamidoamine (PAMAM) dendrimers also comprise a mixture of 1° amino groups (on the surface) and 3° amino groups (in the interior). Tang et al. reported that the fractured dendrimers (by partial hydrolysis to expose the interior 3° amino groups) show a 50-fold higher transfection efficiency than the intact dendrimers (Tang et al., 1996). These results suggest that the presence of 3° amine, and 2° amine in some cases, together with 1° amino groups could enhance transfection efficiency. The mechanism of enhancement is not clear at the current stage. Possible causes may include high buffering capacity of the carriers (for PEI) and increased carrier flexibility (for PAMAM dendrimer) that facilitate the compaction of DNA extracellularly or swelling of the complexes intracellularly to release the DNA (Reschel et al., 2002; Tang et al., 1996).

Our studies have demonstrated that the type of charge group significantly influences the transfection efficiency of the gene carrier. PPAs with

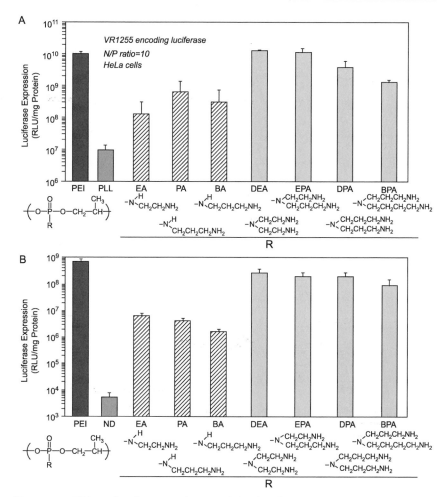

Figure 11.4. PPAs mediated gene transfection in (A) Hela cells and (B) primary rat hepatocytes at an N/P ratio of 10 using VR1255-Luciferase plasmid. Cells were incubated with PPA/DNA complexes containing 1 μg VR1255 plasmid DNA for 4 h before refreshing culture medium. Luciferase assay was performed 44 h after transfection. Data represent mean ± S.D. (n = 3).

primary amino group sidechains show high transfection efficiency in culture, whereas PPAs with 2°, 3° and quaternary (4°) amino groups are significantly less efficient (Wang *et al.*, 2004b). Based on the above hypothesis, a copolymer PPA carrier containing 1°, 2° and/or 3° amino groups, therefore, might be a superior carrier than the one with only 1° amine charge groups. Nevertheless, synthesizing a PPA with different amino groups in their sidechains in a controlled manner is challenging.

Alternatively, a multiple-component complex (for example, ternary complexes containing DNA with a carrier containing 1° amino groups and a carrier with 3° amino groups) may be able to mimic the function of the copolymer carrier design. We have evaluated a *ternary particle* system containing DNA and PPAs with 1° amino groups (PPA-BPA) and 3° amino groups (PPA-DMA), and a *quaternary particle* system containing DNA and PPAs with 1° and 2° and 3° amino groups (PPA-EA/PPA-MEA/PPA-DMA), respectively (Zhang *et al.*, in press(a)). Transfection of COS-7 cells using the ternary complexes (PPA-BPA/PPA-DMA/DNA) mediates significantly higher levels of gene expression than PPA-BPA or PPA-DMA carrier alone, and the transfection efficiency is dependent on the ratio of the two carriers. Under optimal conditions (at a PPA-BPA/PPA-DMA molar ratio of 4), the transfection efficiency achieved by the PPA-BPA/PPA-DMA mixture is 20 and 160 times higher than PPA-BPA and PPA-DMA mediated transfection, respectively (Zhang *et al.*, in press(a)) (Fig. 11.5). Characterization of the nanoparticles and cellular

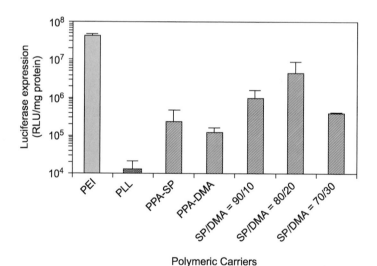

Figure 11.5. Effect of different ratios of PPA-BPA to PPA-DMA on the transfection efficiency of the ternary complexes at N/P ratio of 9 in COS-7 cells using VR1255-Luciferase plasmid. Cells were incubated with complexes containing 3 μg VR1255 plasmid for 4 h before refreshing culture medium. The transfection efficiency was expressed as mean relative light units ± S.D. (n = 3). Reproduced with permission from Ref. (Zhang *et al.*, in press(a)).

uptake studies indicate that the enhancement in transfection efficiency by these ternary and quaternary systems appears to be unrelated to their particle size, zeta potential, or DNA uptake. In addition, the titration assay and the transfection experiment using a proton pump inhibitor suggest that the enhancement effect is unlikely due to the slightly improved buffering capacity of the mixture over PPA-BPA.

Although the mechanism remains unclear, this study illustrates, from a different approach, that the charge group in the gene carrier, and in a general sense charge group in the nanoparticle formulation, is a crucial parameter determining the transfection efficiency of nanoparticles. This formulation method represents an alternative strategy to modulate the transfection efficiency of DNA nanoparticles.

E. PPA-DEA with imidazole moiety: improving buffering capacity

All PPA- and PPE-carriers we have tested show a significant enhancement of gene expression in cell culture, when chloroquine diphosphate (CQ) is present. This implies that these gene carriers, unlike PEI, have a low buffering capacity, a property that can facilitate endosomal escape of DNA or complexes into the cytosol.

In order to increase the buffering capacity of PPA carriers, we have conjugated imidazolyl acetic acid to the side chains of PPA-EA at two different grafting densities, 11% and 43% (Fig. 11.6). As expected, the DNA binding affinity of imPPA-EA (imidazole-modified PPA-EA) decreases as imidazolyl conjugation degree increases. The minimum N/P ratios to completely complex plasmid DNA for imPPA-EA-11% and imPPA-EA-43% are 1.5 and 6.0, respectively. The ζ-potential of complexes at an N/P ratio of 8 decreases significantly from 34 mV to about 4 mV. Most importantly, the titration curve (Fig. 11.6) shows that conjugation of imidazole groups markedly increases the buffering capacity of the carrier. The imPPA-EA-43% has a much higher buffering capacity than PPA-EA and even PEI. The cytotoxicity of the carrier is markedly reduced as well (LD_{50} increased from $\sim$100 μg/ml for PPA-EA to $\sim$720 μg/ml for imPPA-EA-43%).

However, this increased buffering capacity does not translate into enhancement in gene expression. The transfection efficiency decreases as the imidazole content increases. This is most likely due to the reduced DNA compaction capacity of imPPA-EA, leading to a "loose" complex. The average diameter of the complexes also increases substantially (from 78 nm for PPA-EA/DNA to 746 nm for imPPA-EA-43%). These data indicate that DNA compaction capacity is crucial in maintaining the nanoparticle integrity and transfection efficiency.

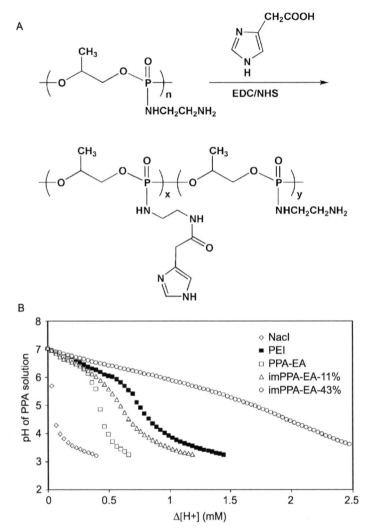

Figure 11.6. Synthesis of PPA-EA with imidazole groups conjugated to its sidechain amino groups (A), and the acid-base titration curve for PPA-EA and imidazole conjugated PPE-EA (B).

F. PPA-DPA with galactose ligand: hepatocyte targeting

We have synthesized galactosylated PPA carriers in an effort to increase the gene transfection efficiency to hepatocytes (Zhang *et al.*, in press(b)). Three Gal-PPA-DPA carriers are synthesized with different degrees of ligand conjugation (6.5%, 12.5%, and 21.8%) using a reduction amination scheme (Zhang *et al.*, in press(b)). The *in vitro* cytotoxicity of Gal-PPA decreases significantly with an increase in galactose substitution degree. The affinity of Gal-PPA/DNA nanoparticles to galactose-recognizing lectin increases with galactose substitution degree. However, decreased transfection efficiency is observed for these galactosylated PPAs in HepG2 cells. Based on the results of gel retardation and polyanion competition assays, we hypothesize that the reduced transfection efficiency of Gal-PPA/DNA nanoparticles is due to their decreased DNA-binding capacity and decreased particle stability. This is similar to the observation of reduced transfection efficiency for imidazole-modified PPA-EA carriers.

In order to overcome this reduced stability, we therefore prepared nanoparticles by precondensing DNA with PPA at a charge ratio of 0.5, yielding nanoparticles with negative surface charge, followed by coating with Gal-PPA, resulting in a Gal-PPA/DNA/PPA ternary complex (Zhang *et al.*, in press(b)). Such a ternary nanoparticle formulation leads to significant size reduction in comparison with binary nanoparticles, particularly at low N/P ratios (2 to 5). In HepG2 cells and primary rat hepatocytes, and at low N/P ratios (2 to 5), transfection efficiency mediated by ternary nanoparticles prepared with 6.5% Gal-PPA is 6–7200 times higher than PPA-DPA/DNA nanoparticles (Zhang *et al.*, in press(b)). Such an enhancement effect is not observed in HeLa cells that lack the asialoglycoprotein receptor (ASGPR). The improvement in transgene expression is most prominent at lower N/P ratios. Nevertheless, increasing the galactosylation degree of PPA carrier (higher than 12.5%) significantly reduces the stability of ternary particles, as well as the transfection efficiency.

This study demonstrates that ligand conjugation (in this case galactosylation) of PPA carrier significantly affects the physiochemical properties of PPA/DNA nanoparticles as a result of the lowered DNA compaction capacity, reduced stability and increased particle size. Despite the fact that galactosylated nanoparticles can efficiently recognize and bind to galactose-recognizing lectin, their transfection efficiency decreases significantly. This study accentuates the importance of keeping a balance between modifying the gene carrier (to improve the targeting and cell uptake) and maintaining its DNA compaction capacity and nanoparticle stability.

IV. POLYCATIONIC PPES—EFFECT OF SIDECHAIN STRUCTURE ON DEGRADATION OF CARRIER, DNA RELEASE RATE, AND TRANSFECTION EFFICIENCY

A. DNA compaction capacity of PPE/DNA

We will next review the polyphosphates (PPE) where the sidechains comprise the (P–O) instead of the (P–N) bond. Among the three cationic PPEs studied, PPE-EA, PPE-HA and PPE-MEA, DNA compaction capacity decreases in the sequence of PPE-EA > PPE-HA > PPE-MEA (Wang *et al.*, 2002b, 2003; Zhao *et al.*, 2003). PPE-EA, PPE-HA and PPE-MEA can compact DNA completely at an N/P ratio of 1.0, 1.5 and 4, respectively. Poor DNA compaction of PPE-MEA is due to the rapid degradation of PPE-MEA in PBS (degradation occurs in hours).

B. Degradation of PPEs and sustained release of DNA from PPE/DNA complexes

The biodegradability of polyphosphoesters not only offers a better biocompatibility and safety profile, but also additional advantage of extracellular sustained release, and conceivably even intracellular sustained release. Such a gene delivery system can therefore potentially address the DNA unpacking issue and improve the bioavailability of DNA both inside and outside of the cells, depending on the specific application.

In general, hydrolytic degradation of PPEs is substantially faster than that of PPAs, as phosphate bond is more labile than phosphoramidate bond (Wang *et al.*, 2003, 2002b; Zhao *et al.*, 2003). PPEs undergo hydrolytic degradation when incubated in PBS at 37 °C, which is due to the hydrolytic cleavage of the phosphoester bonds in the backbone as well as the side chains. The degradation rates are affected by the side chain length as well as the type of charge groups (Wang *et al.*, 2002b; 2003, Zhao *et al.*, 2003). Previous work by Penczek demonstrates that the degradation of polyphosphoester with a methoxy or ethoxy side chain is considerably slower (Baran and Penczek, 1995). However, when replacing the ethoxy side groups with ethoxyl amino groups, the M_w of PPE-EA dropped 12% in 24 h, and declines gradually to 10.5 KDa from 30.3 KDa (65% decrease) after 7 days. After 10 days of incubation, PPE-EA degrades to oligomers with a decreased net positive charge, which fail to condense plasmid DNA. Nevertheless, PPE-HA, with a longer hexylamino group, degrades much more slowly in PBS. The M_w of PPE-HA drops only 50% in 20 days from 37.6 KDa to 18.7 KDa, followed by another 20% decrease (to 14.8 KDa) in the next 30 days. PPE-MEA degrades at an even slower rate, with an M_w decrease of only 28% after 50 days of incubation under the same conditions (Fig. 11.7).

The relatively fast degradation characteristic of PPE-EA conveniently provides a sustained release system for plasmid DNA (Wang *et al.*, 2001, 2002b). Release of plasmid DNA from PPE-EA/DNA complexes can also be adjusted with the charge ratio. At an N/P ratio of 1.5, the DNA release kinetics is first-order in the first week, followed by a nearly constant release during the second week at a rate of 18 μg DNA/day per mg of complex (Wang *et al.*, 2001, 2002b) (Fig. 11.8). A faster release would be observed with PPE-MEA containing N-methylene groups in the side chain. The DNA release is complete in several hours.

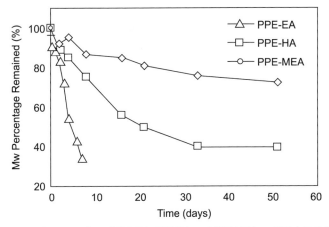

Figure 11.7. Degradation profiles of PPE-EA, PPE-HA and PPE-MEA in PBS (pH 7.4) at 37 °C. Reproduced with permission from Ref. (Zhao *et al.*, 2003).

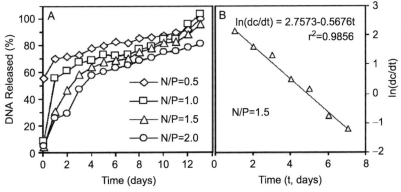

Figure 11.8. Plasmid DNA release profiles from PPE-EA/DNA complexes prepared at various N/P ratios. Complexes were incubated in PBS at 37 °C, and released DNA was quantified with UV absorption at 260 nm. (A) The DNA release rate from PPE-EA/DNA complexes followed first order kinetics in the first week (B). Reproduced with permission from (Wang *et al.*, 2002b).

The relatively rapid degradation of these polycations compared with that of polyphosphate with ethoxy groups in the sidechain suggests a self-catalytic degradation mechanism involving nucleophilic attack of the phosphate bonds by pendant amino groups (Wang *et al.*, 2003; Zhao *et al.*, 2003). This mechanism likely leads to a cleavage of the side chain, yielding negatively charged phosphate ions instead. Such a negatively charged backbone would shield further attack of OH⁻ on P=O, and result in a decreased degradation in subsequent hydrolysis. Following this reasoning, the slower degradation rate of PPE-MEA compared with PPE-EA and PPE-HA also suggests that the sidechains of PPE-MEA are cleaved at a faster rate.

The faster DNA release from PPE-MEA/DNA complexes is therefore a result of side-chain cleavage in PPE-MEA. Sidechain cleavage results in a less positively charged carrier, therefore reducing the binding ability of carrier to DNA. Taken together, degradation of the side chain appears to be a major factor affecting the DNA release from PPE/DNA complexes.

C. Effect of PPE sidechain structure on transfection efficiency

Due to the biodegradation of PPE, PPE/DNA particles can be used as a depot for sustained and local DNA release at the injection site, and potentially prolonging the gene expression in the muscle (Wang et al., 2001, 2002b, 2003; Zhao et al., 2003). Of the three carriers, PPE-MEA probably releases DNA too fast to provide a meaningful depot, whereas PPE-HA degrades too slowly. PPE-EA/DNA particle system could release DNA over a few days depending on the formulation, and represents a suitable carrier for further testing.

This hypothesis is evaluated in a murine model following intramuscular and intracisternal injections, respectively. The rationale of these applications is based on the fact that plasmid DNA could mediate gene expression in these tissues, and based on the hypothesis that sustained release of plasmid could enhance the gene expression (Li et al., 2004; Wang et al., 2002b). The success of this strategy is largely dependent upon the formulation (mainly N/P ratio). Intramuscular injection of PPE-EA/DNA complexes at N/P ratios of 0.5 and 1 leads to a higher transgene expression than naked DNA injection for both a β-galactosidase expression model (in anterior tibialis muscle, Fig. 11.9) and a systemic delivery model (interferon α2b expression in blood circulation) in Balb/c mice (Wang et al., 2002b). Nevertheless, as the PPE-EA/DNA particles are prepared at higher N/P ratios, the enhancement effect is less pronounced. There is also a dose limitation where this mechanism is operative. The enhancement effect decreases significantly as the DNA dose increases.

In a central nervous system (CNS) delivery model (Li et al., 2004), PPE-EA/DNA complexes at an N/P ratio of 2, injected intracisternally into the mouse cerebrospinal fluid, mediate a persistent level of transgene expression in the brain (mostly in cerebral cortex, basal ganglia, and diencephalons) for at least 4 weeks (Li et al., 2004) (Fig. 11.10). At week 4, PPE-EA/DNA complexes maintain a 15-fold higher luciferase expression than naked DNA, and about 3-fold higher than PEI/DNA complexes. The difference in optimal N/P ratio observed in these models could be the result of different stability of plasmid and

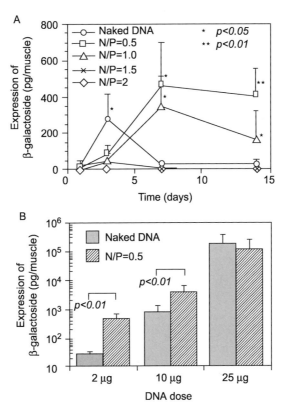

Figure 11.9. β-Galactosidase expression in mouse muscle after intramuscular injections of naked DNA and PPE-EA/DNA complexes with various N/P ratios. Mean ± S.D. ($n = 6$). (A) Effect of N/P ratio on β-gal expression level. Naked DNA and complexes (N/P = 0.5, 1.0, 1.5, 2.0) were given at a dose of 2 μg of DNA per muscle in 40 μl of saline. (B) Effect of DNA dose on β-gal expression level on Day 7. Each muscle received the same dose of naked DNA and complexes with N/P = 0.5 equivalent to 2, 10 and 25 μg of pcDNA. Reproduced with permission from Ref (Wang et al., 2002b).

polymeric carriers in different tissues. These results establish the potential of PPE-EA to achieve sustained gene expression in the CNS and in muscle.

V. PPE WITH A CHARGE CENTER IN THE BACKBONE

To further understand the effect of the charge group residing in the backbone of the polyphosphoester instead of the sidechain, a poly[[(cholesteryl oxocarbonylamido ethyl) methyl bis(ethylene) ammonium iodide] ethyl phosphate] (PCEP)

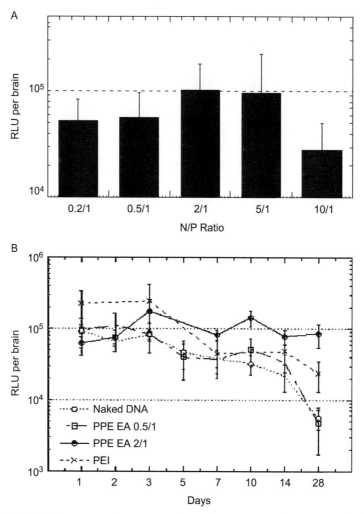

Figure 11.10. Luciferase expression in mouse brain after intracisternal injections of PPE-EA/
pCAG-Luc complexes: (A) effects of N/P ratios; (B) time course study. The
intracisternal injection in adult male Swiss mice (20–25 g) was carried out using a
10 μl Hamilton syringe connected with a 26-gauge needle. A plastic stopper was set
on the needle 4 mm from the tip. Ten μl of 5% glucose solution containing 1 μg of
DNA or equivalent PPE-EA/DNA or PEI/DNA nanoparticles were injected into
the cerebellomedullary cistern of each mouse. Mice were sacrificed either 3 days
after injection (A) or at different time points indicated (B). The supernatants from
the brain tissue homogenates were used for luciferase activity assay in a single-well
luminometer for 10 s. Data are presented as means ± S.D. (n = 6). Reproduced with
permission from Ref. (Li et al., 2004).

is synthesized as shown in Fig. 11.1 (Wen *et al.*, 2004). Carrying a positive charge in its backbone and a lipophilic cholesterol structure in the side chain, PCEP self-assembles into micelles in aqueous buffer at room temperature with an average size of 60–100 nm. It is analogous to the polymeric version of a cationic lipid. Intramuscular administration of the polyplexes shows a luciferase expression in muscle increasing with time during 3 months, although the expression level is lower than that by direct injection of naked DNA. In addition to biodegradability and lower toxicity than PEI, the PCEP micelle carrier offers structural versatility. The backbone charge density and the side chain lipophilicity are two parameters that can be varied through copolymerization and monomer variation to optimize the transfection efficiency.

VI. SUMMARY

Undoubtedly there remain significant challenges to improve non-viral transgene expression to the therapeutic levels. This review highlights our attempt to design the gene carriers in a systematic manner. Progress has been made on the issue of cytotoxicity, where PPE gene carriers can match the potency of PEI, for instance, but show lower toxicity both *in vitro* and *in vivo*. Effort to increase the buffering capacity of the gene carriers has been successful by including moieties such as imidazole, but produces overall disappointing results because of offsetting factors of poorer DNA protection and larger particle size. The issue of optimal unpacking of the DNA nanoparticles remains unresolved, which may be cell- and tissue-dependent. The PPE sidechain, and its hydrolysis, is a main factor determining the polymer degradation as well as the DNA release kinetics. The challenge is to strike a balance between releasing the DNA and protecting the DNA from degradation until it is available for transcription in the nucleus. Our studies on PPE-MEA, PPE-EA and PPE-HA, with a fast, intermediate and slow degradation profile, respectively, all fail to show any significant enhancement in transfection efficiency. This suggests that such an intracellular release mechanism alone probably would not be successful, unless combined with other approaches addressing other important steps in the gene transfer process. Our studies reinforce the notion that an integrative approach addressing all the barriers of the delivery process is crucial to the success of gene carrier development. Along this line, the versatility of the polyphosphoester gene carriers offers excellent opportunities for optimization.

Acknowledgment

The authors acknowledge the support of NIH/R01EB002849.

References

Ahn, C. H., Chae, S. Y., Bae, Y. H., and Kim, S. W. (2002). Biodegradable poly(ethylenimine) for plasmid DNA delivery. *J. Control Release* **80**(1–3), 273–282.

Ahn, C. H., Chae, S. Y., Bae, Y. H., and Kim, S. W. (2004). Synthesis of biodegradable multi-block copolymers of poly(L-lysine) and poly(ethylene glycol) as a non-viral gene carrier. *J. Control Release* **97**(3), 567–574.

Anderson, D. G., Lynn, D. M., and Langer, R. (2003). Semi-automated synthesis and screening of a large library of degradable cationic polymers for gene delivery. *Angew. Chem. Int. Ed. Engl.* **42**(27), 3153–3158.

Baran, J., and Penczek, S. (1995). Hydrolysis of Polyesters of Phosphoric-Acid.1. Kinetics and the Ph Profile. *Macromolecules* **28**(15), 5167–5176.

Bellocq, N. C., Pun, S. H., Jensen, G. S., and Davis, M. E. (2003). Transferrin-containing, cyclodextrin polymer-based particles for tumor-targeted gene delivery. *Bioconjug. Chem.* **14**(6), 1122–1132.

Blessing, T., Kursa, M., Holzhauser, R., Kircheis, R., and Wagner, E. (2001). Different strategies for formation of pegylated EGF-conjugated PEI/DNA complexes for targeted gene delivery. *Bioconjug. Chem.* **12**(4), 529–537.

Boussif, O., Lezoualc'h, F., Zanta, M. A., Mergny, M. D., Scherman, D., Demeneix, B., and Behr, J. P. (1995). A versatile vector for gene and oligonucleotide transfer into cells in culture and in vivo: Polyethylenimine. *Proc. Natl. Acad. Sci. USA* **92**(16), 7297–7301.

Branden, L. J., Mohamed, A. J., and Smith, C. I. (1999). A peptide nucleic acid-nuclear localization signal fusion that mediates nuclear transport of DNA. *Nat. Biotechnol.* **17**(8), 784–787.

Brissault, B., Kichler, A., Guis, C., Leborgne, C., Danos, O., and Cheradame, H. (2003). Synthesis of linear polyethylenimine derivatives for DNA transfection. *Bioconjug. Chem.* **14**(3), 581–587.

Brownlie, A., Uchegbu, I. F., and Schatzlein, A. G. (2004). PEI-based vesicle-polymer hybrid gene delivery system with improved biocompatibility. *Int. J. Pharm.* **274**(1–2), 41–52.

Cavazzana-Calvo, M., Thrasher, A., and Mavilio, F. (2004). The future of gene therapy. *Nature* **427**(6977), 779–781.

Davis, M. E., Pun, S. H., Bellocq, N. C., Reineke, T. M., Popielarski, S. R., Mishra, S., and Heidel, J. D. (2004). Self-assembling nucleic acid delivery vehicles via linear, water-soluble, cyclodextrin-containing polymers. *Curr. Med. Chem.* **11**(2), 179–197.

El Ouahabi, A., Thiry, M., Pector, V., Fuks, R., Ruysschaert, J. M., and Vandenbranden, M. (1997). The role of endosome destabilizing activity in the gene transfer process mediated by cationic lipids. *FEBS Lett.* **414**(2), 187–192.

Ewert, K., Slack, N. L., Ahmad, A., Evans, H. M., Lin, A. J., Samuel, C. E., and Safinya, C. R. (2004). Cationic lipid-DNA complexes for gene therapy: Understanding the relationship between complex structure and gene delivery pathways at the molecular level. *Curr. Med. Chem.* **11**(2), 133–149.

Forrest, M. L., Koerber, J. T., and Pack, D. W. (2003). A degradable polyethylenimine derivative with low toxicity for highly efficient gene delivery. *Bioconjug. Chem.* **14**(5), 934–940.

Forrest, M. L., Meister, G. E., Koerber, J. T., and Pack, D. W. (2004). Partial acetylation of polyethylenimine enhances in vitro gene delivery. *Pharm. Res.* **21**(2), 365–371.

Gautam, A., Waldrep, J. C., Orson, F. M., Kinsey, B. M., Xu, B., and Densmore, C. L. (2003). Topical gene therapy for pulmonary diseases with PEI-DNA aerosol complexes. *Methods Mol. Med.* **75**, 561–572.

Gosselin, M. A., Guo, W., and Lee, R. J. (2001). Efficient gene transfer using reversibly cross-linked low molecular weight polyethylenimine. *Bioconjug. Chem.* **12**(6), 989–994.

Huang, S. W., Wang, J., Zhang, P. C., Mao, H. Q., Zhuo, R. X., and Leong, K. W. (2004). Water-soluble and nonionic polyphosphoester: Synthesis, degradation, biocompatibility and enhancement of gene expression in mouse muscle. *Biomacromolecules* **5**(2), 306–311.

Hwang, S. J., and Davis, M. E. (2001). Cationic polymers for gene delivery: Designs for overcoming barriers to systemic administration. *Curr. Opin. Mol. Ther.* **3**(2), 183–191.

Jeong, J. H., Kim, S. W., and Park, T. G. (2003). Novel intracellular delivery system of antisense oligonucleotide by self-assembled hybrid micelles composed of DNA/PEG conjugate and cationic fusogenic peptide. *Bioconjug. Chem.* **14**(2), 473–479.

Jon, S., Anderson, D. G., and Langer, R. (2003). Degradable poly(amino alcohol esters) as potential DNA vectors with low cytotoxicity. *Biomacromolecules* **4**(6), 1759–1762.

Jones, N. A., Hill, I. R., Stolnik, S., Bignotti, F., Davis, S. S., and Garnett, M. C. (2000). Polymer chemical structure is a key determinant of physicochemical and colloidal properties of polymer-DNA complexes for gene delivery. *Biochim. Biophys. Acta* **1517**(1), 1–18.

Jones, R. A., Cheung, C. Y., Black, F. E., Zia, J. K., Stayton, P. S., Hoffman, A. S., and Wilson, M. R. (2003). Poly(2-alkylacrylic acid) polymers deliver molecules to the cytosol by pH-sensitive disruption of endosomal vesicles. *Biochem. J.* **372**(Pt. 1), 65–75.

Kaiser, J. (2004). Gene therapy. Side effects sideline hemophilia trial. *Science* **304**(5676), 1423–1425.

Kiang, T., Wen, J., Lim, H. W., and Leong, K. W. (2004). The effect of the degree of chitosan deacetylation on the efficiency of gene transfection. *Biomaterials* **25**(22), 5293–5301.

Kim, T. G., Kang, S. Y., Kang, J. H., Cho, M. Y., Kim, J. I., Kim, S. H., and Kim, J. S. (2004). Gene transfer into human hepatoma cells by receptor-associated protein/polylysine conjugates. *Bioconjug. Chem.* **15**(2), 326–332.

Kramer, M., Stumbe, J. F., Grimm, G., Kaufmann, B., Kruger, U., Weber, M., and Haag, R. (2004). Dendritic polyamines: Simple access to new materials with defined treelike structures for application in non-viral gene delivery. *Chembiochem.* **5**(8), 1081–1087.

Kunath, K., von Harpe, A., Fischer, D., and Kissel, T. (2003). Galactose-PEI-DNA complexes for targeted gene delivery: Degree of substitution affects complex size and transfection efficiency. *J. Control Release* **88**(1), 159–172.

Kyriakides, T. R., Cheung, C. Y., Murthy, N., Bornstein, P., Stayton, P. S., and Hoffman, A. S. (2002). pH-sensitive polymers that enhance intracellular drug delivery *in vivo. J. Control Release* **78**(1–3), 295–303.

Lampela, P., Soininen, P., Urtti, A., Mannisto, P. T., and A., Raasmaja (2004). Synergism in gene delivery by small PEIs and three different non-viral vectors. *Int. J. Pharm.* **270**(1–2), 175–184.

Lechardeur, D., Sohn, K. J., Haardt, M., Joshi, P. B., Monck, M., Graham, R. W., Beatty, B., Squire, J., O'Brodovich, H., and Lukacs, G. L. (1999). Metabolic instability of plasmid DNA in the cytosol: A potential barrier to gene transfer. *Gene. Ther.* **6**(4), 482–497.

Lee, M., Ko, K. S., Oh, S., and Kim, S. W. (2003a). Prevention of autoimmune insulitis by delivery of a chimeric plasmid encoding interleukin-4 and interleukin-10. *J. Control Release* **88**(2), 333–342.

Lee, J. H., Lim, Y. B., Choi, J. S., Lee, Y., Kim, T. I., Kim, H. J., Yoon, J. K., Kim, K., and Park, J. S. (2003b). Polyplexes assembled with internally quaternized PAMAM-OH dendrimer and plasmid DNA have a neutral surface and gene delivery potency. *Bioconjug. Chem.* **14**(6), 1214–1221.

Lee, K. Y., Kwon, I. C., Kim, Y. H., Jo, W. H., and Jeong, S. Y. (1998). Preparation of chitosan self-aggregates as a gene delivery system. *J. Controlled Release* **51**(2–3), 213–220.

Leong, K. W., Mao, H. Q., Truong-Le, V. L., Roy, K., Walsh, S. M., and August, J. T. (1998). DNA-polycation nanospheres as non-viral gene delivery vehicles. *J. Control Release* **53**(1–3), 183–193.

Li, Y., Wang, J., Lee, C. G., Wang, C. Y., Gao, S. J., Tang, G. P., Ma, Y. X., Yu, H., Mao, H. Q., Leong, K. W., and Wang, S. (2004). CNS gene transfer mediated by a novel controlled release system based on DNA complexes of degradable polycation PPE-EA: A comparison with poly-ethylenimine/DNA complexes. *Gene Ther.* **11**(1), 109–114.

Lim, Y. B., Han, S. O., Kong, H. U., Lee, Y., Park, J. S., Jeong, B., and Kim, S. W. (2000a). Biodegradable polyester, poly[alpha-(4-aminobutyl)-L-glycolic acid], as a non-toxic gene carrier. *Pharm. Res.* **17**(7), 811–816.

Lim, D. W., Yeom, Y. I., and Park, T. G. (2000b). Poly(DMAEMA-NVP)-b-PEG-galactose as gene delivery vector for hepatocytes. *Bioconjug. Chem.* **11**(5), 688–695.

Luten, J., van Steenis, J. H., van Someren, R., Kemmink, J., Schuurmans-Nieuwenbroek, N. M., Koning, G. A., Crommelin, D. J., van Nostrum, C. F., and Hennink, W. E. (2003). Water-soluble biodegradable cationic polyphosphazenes for gene delivery. *J. Control Release* **89**(3), 483–497.

MacLaughlin, F. C., Mumper, R. J., Wang, J. J., Tagliaferri, J. M., Gill, I., Hinchcliffe, M., and Rolland, A. P. (1998). Chitosan and depolymerized chitosan oligomers as condensing carriers for in vivo plasmid delivery. *J. Control. Release* **56**(1–3), 259–272.

Maheshwari, A., Han, S., Mahato, R. I., and Kim, S. W. (2002). Biodegradable polymer-based interleukin-12 gene delivery: Role of induced cytokines, tumor infiltrating cells and nitric oxide in anti-tumor activity. *Gene. Ther.* **9**(16), 1075–1084.

Mao, H. Q., Roy, K., Troung-Le, V. L., Janes, K. A., Lin, K. Y., Wang, Y., August, J. T., and Leong, K. W. (2001). Chitosan-DNA nanoparticles as gene carriers: Synthesis, characterization and transfection efficiency. *J. Control Release* **70**(3), 399–421.

Midoux, P., Mendes, C., Legrand, A., Raimond, J., Mayer, R., Monsigny, M., and Roche, A. C. (1993). Specific gene transfer mediated by lactosylated poly-L-lysine into hepatoma cells. *Nucleic Acids Res.* **21**(4), 871–878.

Niidome, T., and Huang, L. (2002). Gene therapy progress and prospects: Non-viral vectors. *Gene. Ther.* **9**(24), 1647–1652.

Nishikawa, M., and Huang, L. (2001). Non-viral vectors in the new millennium: Delivery barriers in gene transfer. *Hum. Gene. Ther.* **12**(8), 861–870.

Oupicky, D., Carlisle, R. C., and Seymour, L. W. (2001). Triggered intracellular activation of disulfide crosslinked polyelectrolyte gene delivery complexes with extended systemic circulation in vivo. *Gene Ther.* **8**(9), 713–724.

Pedroso de Lima, M. C., Neves, S., Filipe, A., Duzgunes, N., and Simoes, S. (2003). Cationic liposomes for gene delivery: From biophysics to biological applications. *Curr. Med. Chem.* **10**(14), 1221–1231.

Popielarski, S. R., Mishra, S., and Davis, M. E. (2003). Structural effects of carbohydrate-containing polycations on gene delivery. 3. Cyclodextrin type and functionalization. *Bioconjug. Chem.* **14**(3), 672–678.

Pun, S. H., and Davis, M. E. (2002). Development of a non-viral gene delivery vehicle for systemic application. *Bioconjug. Chem.* **13**(3), 630–639.

Pun, S. H., Bellocq, N. C., Liu, A., Jensen, G., Machemer, T., Quijano, E., Schluep, T., Wen, S., Engler, H., Heidel, J., and Davis, M. E. (2004). Cyclodextrin-modified polyethylenimine polymers for gene delivery. *Bioconjug. Chem.* **15**(4), 831–840.

Putnam, D., Gentry, C. A., Pack, D. W., and Langer, R. (2001). Polymer-based gene delivery with low cytotoxicity by a unique balance of side-chain termini. *Proc. Natl. Acad. Sci. USA* **98**(3), 1200–1205.

Reineke, T. M., and Davis, M. E. (2003a). Structural effects of carbohydrate-containing polycations on gene delivery. 2. Charge center type. *Bioconjug. Chem.* **14**(1), 255–261.

Reineke, T. M., and Davis, M. E. (2003b). Structural effects of carbohydrate-containing polycations on gene delivery. 1. Carbohydrate size and its distance from charge centers. *Bioconjug. Chem.* **14**(1), 247–254.

Reschel, T., Konak, C., Oupicky, D., Seymour, L. W., and Ulbrich, K. (2002). Physical properties and *in vitro* transfection efficiency of gene delivery vectors based on complexes of DNA with synthetic polycations. *J. Controlled Release* **81**(1–2), 201–217.

Richardson, S. C., Kolbe, H. V., and Duncan, R. (1999). Potential of low molecular mass chitosan as a DNA delivery system: Biocompatibility, body distribution and ability to complex and protect DNA. *Int. J. Pharm.* **178**(2), 231–243.

Roy, K., Mao, H. Q., Huang, S. K., and Leong, K. W. (1999). Oral gene delivery with chitosan— DNA nanoparticles generates immunologic protection in a murine model of peanut allergy. *Nat. Med.* **5**(4), 387–391.

Sosnowski, B. A., Gonzalez, A. M., Chandler, L. A., Buechler, Y. J., Pierce, G. F., and Baird, A. (1996). Targeting DNA to cells with basic fibroblast growth factor (FGF2). *J. Biol. Chem.* **271**(52), 33647–33653.

Tabor, C. W., and Tabor, H. (1984). Polyamines. *Ann. Rev. Biochemistr.* **53,** 749–790.

Tang, M. X., Redemann, C. T., and Szoka, F. C. (1996). *In vitro* gene delivery by degraded polyamidoamine dendrimers. *Bioconjug. Chem.* **7**(6), 703–714.

Thanou, M., Verhoef, J. C., and Junginger, H. E. (2001). Chitosan and its derivatives as intestinal absorption enhancers. *Adv. Drug Deliv. Rev.* **50**(Suppl. 1), S91–S101.

Thomas, M., and Klibanov, A. M. (2002). Enhancing polyethylenimine's delivery of plasmid DNA into mammalian cells. *Proc. Nat. Acad. Sci. USA* **99**(23), 14640–14645.

Truong-Le, V. L., August, J. T., and Leong, K. W. (1998). Controlled gene delivery by DNA-gelatin nanospheres. *Hum. Gene. Ther.* **9**(12), 1709–1717.

Truong-Le, V. L., Walsh, S. M., Schweibert, E., Mao, H. Q., Guggino, W. B., August, J. T., and Leong, K. W. (1999). Gene transfer by DNA-gelatin nanospheres. *Arch. Biochem. Biophys.* **361**(1), 47–56.

Wagner, E. (2004). Strategies to improve DNA polyplexes for *in vivo* gene transfer: Will "artificial viruses" be the answer? *Pharm. Res.* **21**(1), 8–14.

Wang, J., Lee, I. L., Lim, W. S., Chia, S. M., Yu, H., Leong, K. W., and Mao, H. Q. (2004a). Evaluation of collagen and methylated collagen as gene carriers. *Int. J. Pharm.* **279**(1–2), 115–126.

Wang, J., Gao, S. J., Zhang, P. C., Wang, S., Mao, M. Q., and Leong, K. W. (2004b). Polyphosphoramidate gene carriers: Effect of charge group on gene transfer efficiency. *Gene. Ther.* **11** (12), 1001–1010.

Wang, J., Huang, S. W., Zhang, P. C., Mao, H. Q., and Leong, K. W. (2003). Effect of side-chain structures on gene transfer efficiency of biodegradable cationic polyphosphoesters. *Int. J. Pharm.* **265**(1–2), 75–84.

Wang, J., Mao, H. Q., and Leong, K. W. (2001). A novel biodegradable gene carrier based on polyphosphoester. *J. Am. Chem. Soc.* **123**(38), 9480–9481.

Wang, J., Zhang, P. C., Lu, H. F., Ma, N., Wang, S., Mao, H. Q., and Leong, K. W. (2002a). New polyphosphoramidate with a spermidine side chain as a gene carrier. *J. Control. Release* **83**(1), 157–168.

Wang, J., Zhang, P. C., Mao, H. Q., and Leong, K. W. (2002b). Enhanced gene expression in mouse muscle by sustained release of plasmid DNA using PPE-EA as a carrier. *Gene. Ther.* **9**(18), 1254–1261.

Wang, X. L., Zhang, P. C., Dai, H., Leong, K. W., and Mao, H. Q. (2002c). Effect of pendant chain length of polyphosphoramidate gene carriers on their transfection efficiency. *In* "The 29th International Symposium on Controlled Release of Bioactive Materials" Controlled Release Society, Seoul, Korea.

Wen, J., Mao, H. Q., Li, W. P., Lin, K. Y., and Leong, K. W. (2004). Biodegradable polyphosphoester micelles for gene delivery. *J. Pharm. Sci.* **93**(8), 2142–2157.

Wightman, L., Kircheis, R., Rossler, V., Carotta, S., Ruzicka, R., Kursa, M., and Wagner, E. (2001). Different behavior of branched and linear polyethylenimine for gene delivery *in vitro* and *in vivo*. *J. Gene. Med.* 3(4), 362–372.

Williams, D. A., and Baum, C. (2003). Medicine. Gene therapy—new challenges ahead. *Science* 302(5644), 400–401.

Xu, Y., and Szoka, F. C., Jr. (1996). Mechanism of DNA release from cationic liposome/DNA complexes used in cell transfection. *Biochemistry* 35(18), 5616–5623.

Zhang, Z. Y., and Smith, B. D. (2000). High-generation polycationic dendrimers are unusually effective at disrupting anionic vesicles: Membrane bending model. *Bioconjug. Chem.* 11(6), 805–814.

Zhang, P. C., Wang, J., Leong, K. W., Mao, H. Q. (2005) Ternary complexes comprising polyphosphoramidate gene carriers with different types of charge groups improve transfection efficiency. *Biomacromolecules* 6, 54–60.

Zhang, X. Q., W. X. L., Huang, S. W., Zhuo, R. X., Liu, Z., Mao, H. Q., Leong, K. W. (2005) Galactosylated ternary DNA/polyphosphoramidate nanoparticles mediate high gene transfection efficiency in hepatocytes. *J. Control. Rel.* 102, 749–763.

Zhao, Z., Wang, J., Mao, H. Q., and Leong, K. W. (2003). Polyphosphoesters in drug and gene delivery. *Adv. Drug Deliv. Rev.* 55(4), 483–499.

Ziemienowicz, A., Gorlich, D., Lanka, E., Hohn, B., and Rossi, L. (1999). Import of DNA into mammalian nuclei by proteins originating from a plant pathogenic bacterium. *Proc. Natl. Acad. Sci. USA* 96(7), 3729–3733.

Zou, S. M., Erbacher, P., Remy, J. S., and Behr, J. P. (2000). Systemic linear polyethylenimine (L-PEI)-mediated gene delivery in the mouse. *J. Gene. Med.* 2(2), 128–134.

12

Development of HVJ Envelope Vector and Its Application to Gene Therapy

Yasufumi Kaneda,* Seiji Yamamoto,* and Toshihiro Nakajima†
*Division of Gene Therapy Science, Graduate School of Medicine
Osaka University, Suita, Osaka 565-0871, Japan
†GenomIdea Inc., 7-7-15 Saito-Asagi, Ibaragi
Osaka 567-0085, Japan

Advances in Genetics, Vol. 53
Copyright 2005, Elsevier Inc. All rights reserved.

0065-2660/05 $35.00
DOI: 10.1016/S0065-2660(05)53012-8

ABSTRACT

To create a highly efficient vector system that is minimally invasive, we initially developed liposomes that contained fusion proteins from the hemagglutinating virus of Japan (HVJ; Sendai virus). These HVJ-liposomes delivered genes and drugs to cultured cells and tissues. To simplify the vector system and develop more efficient vectors, the next approach was to convert viruses to non-viral vectors. Based on this concept, we recently developed the HVJ envelope vector. HVJ with robust fusion activity was inactivated, and exogenous DNA was incorporated into the viral envelope by detergent treatment and centrifugation. The resulting HVJ envelope vector introduced plasmid DNA efficiently and rapidly into both cultured cells *in vitro* and organs *in vivo*. Furthermore, proteins, synthetic oligonucleotides, and drugs have also been effectively introduced into cells using the HVJ envelope vector. The HVJ envelope vector is a promising tool for both *ex vivo* and *in vivo* gene therapy experiments. Hearing impairment in rats was prevented and treated by hepatocyte growth factor gene transfer to cerebrospinal fluid using HVJ envelope vector. For cancer treatment, tumor-associated antigen genes were delivered efficiently to mouse dendritic cells to evoke an anti-cancer immune response. HVJ envelope vector fused dendritic cells and tumor cells and simultaneously delivered cytokine genes, such as IL-12, to the hybrid cells. This strategy successfully prevented and treated cancers in mice by stimulating the presentation of tumor antigens and the maturation of T cells. For human gene therapy, a pilot plant to commercially produce clinical grade HVJ envelope vector has been established. © 2005, Elsevier Inc.

I. INTRODUCTION

Gene therapy is a promising treatment for intractable human diseases (Cavazzana-Calvo *et al.*, 2004; Marshall, 1995), but further development of effective gene transfer vector systems is required for the advancement of human gene therapy (Mulligan, 1993). Efficient and minimally invasive vector systems appear to be most appropriate for gene therapy. Numerous viral and non-viral (synthetic) methods for gene transfer have been developed (Lam and Brakefield, 2000; Ledley, 1995; Li and Huang, 2000; Mulligan, 1993). Viral methods are generally more efficient than non-viral methods for the delivery of genes to cells, but the safety of viral vectors is of concern due to the concomitant introduction of genetic elements from parent viruses, leaky expression of viral genes, immunogenicity, and changes in the host genome structure (Mulligan, 1995) as pointed out in the SCID-X1 gene therapy clinical trial (Cavazzana-Calvo *et al.*, 2004). Because non-viral vectors are less toxic and less immunogenic than viral vectors, the development of non-viral vectors has been pursued. Various modifications have been made to enhance the efficiency of gene delivery by non-viral vectors. Liposomes have been used to target and introduce macromolecules into cells. However, the gene transfer efficiency of liposomes was low and varied during the early days of liposome development. The synthesis of cationic lipids produced a revolutionary improvement in gene transfer efficiency by Felgner et al. (1987). They also developed a new type of liposome-DNA complex called a "lipoplex." Prior to this development, DNA had been incorporated into liposomes, but, with lipoplex, an electrostatic complex was made between negatively charged DNA and positively charged cationic liposomes. Numerous cationic lipids have been synthesized to further improve transfection efficiency and reduce the cytotoxicity of lipoplex (Li and Huang, 2000). Nevertheless, in lipoplex-mediated transfection, DNA is still delivered into cells by phagocytosis or endocytosis, not by fusion.

Because molecules that enter the cell by phagocytosis or endocytosis often become degraded before reaching the cytoplasm, fusion-mediated delivery systems have been developed. A fusigenic viral liposome with fusion proteins derived from hemagglutinating virus of Japan (HVJ; Sendai virus) was constructed (Kaneda et al., 1999). HVJ fuses with the cell membrane at a neutral pH, and the hemagglutinin-neuraminidase (HN) protein and fusion (F) protein of the virus contribute to cell fusion (Okada, 1993). For fusion-mediated gene transfer, DNA-loaded liposomes were fused with UV-inactivated HVJ to form the fusigenic viral-liposome called HVJ-liposome.

Fusion-mediated delivery protected the molecules in the endosomes and lysosomes from degradation (Dzau et al., 1996). When fluorescein isothiocyanate (FITC)-tagged oligodeoxynucleotide (ODN) was introduced into vascular smooth muscle cells using HVJ-liposomes, fluorescence was detected in the nuclei 5 min after transfer, and fluorescence was stable in the nucleus for at least 72 h In contrast, fluorescence was observed in cellular components (most likely, endosomes) and not in the nucleus when FITC-ODN was transferred directly in the absence of HVJ-liposomes, and no fluorescence was detected 24 h after transfer. Using a fluorescence resonance energy transfer system, we demonstrated that more than 80% of oligonucleotides labeled with two different fluorescent dyes at the 5' and 3' ends were intact in the nucleus, while less than 30% of the oligonucleotides were intact when Lipofectin was used. (Nakamura et al., 2001).

Another advantage of HVJ-liposomes is the ability to perform repeated injections. Gene transfer to rat liver cells was not inhibited by repeated injections. After repeated injections, the anti-HVJ antibodies generated in the rat were not sufficient to neutralize HVJ-liposomes. Cytotoxic T cells recognizing HVJ determinants were not detected in the rats transfected repeatedly with HVJ-liposomes (Hirano et al., 1998).

A similar approach has been used to enhance the gene transfer efficiency of a receptor-mediated gene delivery system by combining fusion peptide derived from influenza virus hemagglutinin (Wagner et al., 1992). A tissue-specific gene delivery system has been developed by binding tissue-specific molecules to a poly-L-lysine/DNA complex (Wu and Wu, 1988). Binding asialoglycoprotein and transferrin to a poly-L-lysine/DNA complex successfully targets DNA to hepatocytes and cancer cells, respectively (Wu and Wu, 1988; Zenke et al., 1990). However, the limitation of this system is the degradation of the DNA in the lysosomes. To avoid such degradation, a fusion-mediated gene delivery system has been investigated using influenza fusion proteins. Influenza virus fuses with cell membranes at an acidic pH, and hemagglutinin (HA) protein on the viral envelope is involved in the fusion between viral envelope and endosomal membrane. It has also been elucidated that a mutant N-terminal peptide of influenza HA subunit, HA-2, can fuse with cell membranes at neutral pH. The transferrin/poly-L-lysine/DNA complex bound with the HA-2 peptide increases gene transfer efficiency in cultured cancer cells more than 1,000 fold compared with that in the absence of the peptide (Wagner et al., 1992).

Reconstituted particles containing fusion proteins of HVJ have also been developed to promote fusion-mediated gene delivery (Bagai and Sarker, 1993; Ramani et al., 1997). HVJ virion was completely lysed with detergent, and the lysates were mixed with DNA solution. In some cases, several lipids were added to the mixture. By removing the detergent with dialysis or a column procedure, reconstituted HVJ particles containing DNA were constructed. Instead of the whole virion of HVJ, fusion proteins (F and HN) isolated from the virion were mixed with the lipid/DNA mixture in the presence or absence of detergent. Since F protein is recognized by the asialoglycoprotein receptor on hepatocytes, reconstituted HVJ particles containing only F protein have been constructed to specifically target hepatocytes in vivo (Ramani et al., 1998). However, DNA trapping efficiency of the reconstituted particles was not so high. To improve the limitation, another approach was that liposomes containing fusion proteins of HVJ and DNA-loaded liposomes were prepared separately and then both liposomes were fused together (Suzuki et al., 2000). These reconstituted fusion liposomes were as effective as conventional HVJ-liposomes, which contain the fully intact HVJ virion, in terms of the delivery of FITC-ODN and the luciferase gene to cultured cells. The LacZ gene was also transferred directly to mouse skeletal muscle in vivo using these reconstituted fusion particles.

A more direct and practical approach is the conversion of a fusigenic virion to a non-viral gene delivery particle. Numerous viruses such as influenza, vesicular stomatitis virus, and HVJ induce cell fusion. HVJ is the most abundantly produced in chick eggs. Therefore, we tried to construct an HVJ envelope vector system by incorporating plasmid DNA into inactivated HVJ particle (Kaneda *et al.*, 2002a). In this review article, we explain the development of the new vector system and its application to gene therapy studies. Effective expression of a transgene is also a big issue for gene therapy, but we do not refer to this issue here.

II. PREPARATION OF HVJ ENVELOPE VECTOR

There are some drawbacks to HVJ-liposomes, although they have been widely used for gene transfer both *in vitro* and *in vivo*. One disadvantage of HVJ-liposomes is the complicated procedure used to isolate and produce both inactivated HVJ and DNA-loaded liposomes. Additionally, the fusion activity of the HVJ-liposomes decreases to approximately 2% of that of native HVJ because of the reduced density of fusion proteins on the surface of HVJ-liposomes. To simplify the vector system and to develop a more effective gene delivery system, we attempted to incorporate plasmid DNA into inactivated HVJ particles without using liposomes (Fig. 12.1).

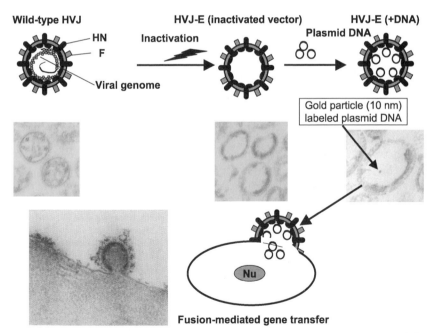

Figure 12.1. HVJ envelope vector system. For constructing HVJ envelope vector, gold-labeled plasmid DNA is mixed with inactivated HVJ particles purified through columns and the mixture is treated with mild detergent and centrifugation to incorporate DNA inside the particle. HVJ envelope vector can fuse with cell membrane to directly transfer DNA into cells.

HVJ is a mouse parainfluenza virus and is not a human pathogen (Okada, 1993). HVJ can fuse with cell membranes in both neutral and acidic conditions. Two distinct glycoproteins, HN and F, are required for cell fusion (Okada, 1993; Yeagle, 1993). HN is required for the binding of HVJ to cell surface sialic acid receptors and the subsequent degradation of the receptors by its sialidase activity. Then, F associates with lipids, such as cholesterol, in the cell membrane to induce cell fusion. The F glycoprotein is first synthesized as inactive F0 in cells infected with HVJ. F0 is then cleaved by host protease into the active F1 and F2 forms. F1 contains hydrophobic peptides with approximately 25 amino acids that induce cell fusion (Ghosh et al., 2000; Yeagle, 1993). F protein produced in chick eggs is converted to the active F1 form by the protease in chorioallantoic fluid, while the virus produced from cultured cells contains inactive F0 that needs to be cleaved by a protease to form active F1. Cells in the rodent airway contain enzymes to convert inactive F0 to active F1. Therefore, HVJ induces severe pneumonia in rodents but not in humans. Large amounts of viral proteins are produced in infected cells. Viral nucleocapsid protein induces cytotoxic T cells (CTLs) against infected cells (Chen et al., 1998). However, nucleocapsid protein is indispensable for virus production. Therefore, to develop highly efficient and minimally invasive vectors based on HVJ, our approach has been to use inactivated viral envelope in which the viral genome has been destroyed.

For this purpose, we have converted HVJ to a non-viral vector containing therapeutic genes instead of the viral genome (Kaneda et al., 2002a). HVJ amplified in the chorioallantoic fluid of 10- to 14-day-old chick eggs was inactivated with β-propiolactone (0.0075% to 0.001%) (Race et al., 1995) followed by UV irradiation (99 mjoule/cm^2) (Kaneda, 2002b). Then, inactivated HVJ envelope was purified by ion-exchange column chromatography and gel filtration (Nakajima et al., unpublished method). With this inactivation procedure, virus replication was completely destroyed, but hemagglutinating activity was not affected. HVJ particles in the chorioallantoic fluid were very heterogeneous with a diameter ranging from 150 to 600 nm. More homogeneous HVJ envelope was isolated with the improved purification method using the column procedure. The diameter of HVJ envelope obtained from the column procedure was 280 nm, and the zeta potential was approximately −5 mV.

Aliquots of the inactivated virus (3×10^{10} particles/1.5 ml tube) were centrifuged (10000 × g, 5 min), and the viral pellet was mixed with exogenous plasmid DNA. Exogenous plasmid DNA was incorporated into inactivated HVJ by treatment with mild detergent and centrifugation. First, inactivated HVJ of 3×10^{10} particles was mixed with 0.24% to 0.80% Triton X-100 in the presence of plasmid DNA (200 μg) in less than 100 μl of Tris-EDTA buffer for 5 min on ice and the mixture was centrifuged at 10000 g for 5 min to move the DNA into the HVJ particle. The DNA trapping efficiency of HVJ envelope vector was approximately 15% to 20%. Without centrifugation, the DNA trapping efficiency was approximately 3%. Different detergents were available for the preparation of the HVJ envelope vector including NP-40, CHAPS, octylglucoside, sodium cholate and dodecyl maltoside, although the optimal concentration of detergent for preparation of HVJ envelope vector is different. Without detergent treatment, DNA does not become incorporated into the viral particle. Electronmicroscopy confirms that DNA became incorporated into all of the particles of inactivated HVJ. The largest plasmid tested was 14 kb and its trapping efficiency was 18%.

The HVJ envelope vector differs from the reconstituted HVJ particles that are prepared by reassembling lipids and fusion proteins after solubilization of the virus particle. To prepare the HVJ envelope vector, plasmid DNA is incorporated into inactivated HVJ particles by treatment with mild detergent. The virion is not destroyed and not subjected to the dialysis, purification, and addition of lipids or proteins that occurs during the preparation of reconstituted HVJ particles (Bagai and Sarker, 1993; Ramani *et al.*, 1997; Suzuki *et al.*, 2000). Protein analysis of HVJ envelope vector using SDS-polyacrylamide gel electrophoresis indicate that the composition of the HVJ envelope vector is very similar to that of native HVJ (Kaneda, *et al.*, 2002a). Most native HVJ proteins are retained in the HVJ envelope vector. Fusion proteins HN and F1 are retained, and the molar ratio of these proteins in HVJ envelope vector was approximately 2 to 2.3, which is the same as the ratio in native HVJ (Okada, 1993). This ratio of F and HN is very important for fusion activity. Therefore, the fusion activity of HVJ envelope vector is as robust as wild-type HVJ. Electron microscopic observation confirmed that the fusion between HVJ envelope vector and cell membrane occurs only 3 to 5 sec after the attachment of the plasmid-containing HVJ envelope vector to a cell surface (Fig. 12.1).

In contrast to recombinant HVJ viral vector (Yonemitsu *et al.*, 2000), the HVJ envelope is a non-viral vector system that consists of an envelope derived from wild type HVJ virus by inactivation and purification. Since the viral genome is inactivated in the HVJ envelope vector, the virus does not replicate and viral genes are not expressed in the cells that are transfected with the HVJ envelope vector. However, cells infected with recombinant HVJ viral vector produce viral proteins. The recombinant HVJ vector produces a large amount of therapeutic products, but it may cause cellular toxicity and be highly immunogenic, which makes it less desirable for repeated administration.

III. GENE TRANSFER TO CULTURED CELLS USING HVJ ENVELOPE VECTOR

For *in vitro* transfection, the HVJ envelope vector containing luciferase expression plasmid was mixed with protamine sulfate and this mixture was added to cultured cells. Protamine sulfate enhanced luciferase gene expression 10- to

50-fold in all cell lines tested. For example, in a mouse colon cancer cell line, CT26, luciferase gene expression was enhanced approximately 20-fold when compared to the expression level in the absence of protamine sulfate. A 10 min incubation period was sufficient for high levels of luciferase expression. The optimum conditions for *in vitro* gene transfer have been previously summarized (Kaneda et al., 2002a). Mouse embryonal stem (ES) cells were also transfected using the mixture of the HVJ envelope vector and protamine sulfate. When HVJ envelope vector containing green fluorescent protein (GFP) expression plasmid was added to mouse ES cells, the efficiency of GFP expression (as determined by flow cytometry) was approximately 80%. The gene transfer efficiency was dependent on the amount of vector and was not inhibited by 10% fetal calf serum. Non-adherent floating cells are generally resistant to gene transfer methods, and the human T cell leukemia cell lines, NALM-6 and CCRF-CEM, have been particularly difficult targets. Luciferase gene expression in NALM-6 and CCRF-CEM was increased with protamine sulfate treatment, but the expression level was still very low. However, when protamine sulfate was used in conjunction with centrifugation, luciferase gene expression in NALM-6 cells and CCRF-CEM was enhanced 30- to 40-fold compared to the expression level without centrifugation. The optimal condition for gene transfer to these non-adherent cell lines was the centrifugation of the mixture of cells and vector (6×10^9 particles) at 10000 g for 30 min at 37 °C. Primary cells such as human aortic endothelial cells and rat neuronal cells were also effectively transfected with the HVJ envelope vector without significant cell damage.

Approximately 20 to 30 copies of plasmid DNA can be incorporated into one HVJ envelope vector particle when 200 μg of 7 kb plasmid is mixed with 3×10^{10} particles of inactivated HVJ. When the DNA concentration is increased, more copies of plasmid DNA can be incorporated into the vector. Using HVJ envelope vector, two different plasmids can be delivered to the same cell. For example, HVJ envelope vector can be prepared when the LacZ gene and GFP gene are mixed with inactivated HVJ and the mixture is treated with mild detergent and centrifuged. When this HVJ envelope vector is added to BHK-21 cells, both proteins are clearly expressed in the same cells (Fig. 12.2a).

Small interfering RNA (siRNA) is an attractive and effective tool for suppressing target protein by specifically digesting its mRNA (Dorsett and Tuschl, 2004; Tijsterman and Plasterk, 2004). SiRNA is superior to antisense oligonucleotides and ribozyme in terms of efficiency and specificity (Miyagishi et al., 2003; Yokota et al., 2004), but finding a suitable delivery system for siRNA has been problematic (Sioud, 2004). Drugs, synthetic oligonucleotides, proteins and peptides, as well as siRNA, can be incorporated into the HVJ envelope

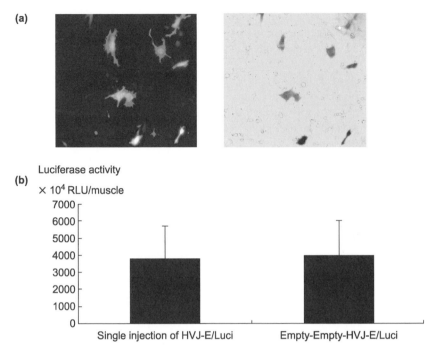

Figure 12.2. (a) Two different plasmids can be delivered to the same cell. HVJ envelope vector containing both GFP and LacZ gene is added to BHK-21 cells. At 24 h after gene transfer, both proteins clearly expressed in the same cells. (b) Gene expression is not inhibited after consecutive injection of HVJ envelope vector into mouse muscle. In both cases, luciferase gene expression in muscle is detected at 24 h after injection of HVJ envelope vector containing luciferase gene. In the mice shown by the blue bar, empty HVJ envelope vector has been injected into muscle twice in a 2-week interval before injection of luciferase gene-loaded vector. (See Color Insert.)

vector and delivered into cells. The HVJ envelope–mediated delivery efficiency of siRNA in cultured cells was 100%. Other researchers have also demonstrated that HVJ envelope vector efficiently delivers siRNA to islet cell lines (Itoh *et al.*, 2003) and Jurkat cells (Ishii *et al.*, 2003).

IV. GENE TRANSFER *IN VIVO* USING HVJ ENVELOPE VECTOR

HVJ envelope vector system can be used for *in vivo* gene transfer. The HVJ envelope vector has mediated gene transfer to a variety of tissues (lung, liver, uterus, eye, skin, muscle, brain, heart and cancerous tumors) in animals including mice, rats, rabbits and monkeys. In mouse liver, HVJ envelope vector-mediated luciferase gene expression was two times higher than the expression mediated by HVJ-liposomes prepared from inactivated HVJ (Kaneda *et al.*, 2002a).

In mouse skeletal muscle, consecutive injection of DNA-loaded HVJ envelope vector did not inhibit gene transfection. In this experiment, empty HVJ envelope vector was injected into mouse muscle tissue twice in a 2-week interval in one experimental group, while another group received saline injections. Two weeks later, HVJ envelope vector-containing luciferase gene was intramuscularly injected in all mice. As shown in Fig. 12.2b, similar luciferase gene expression was detected in the two groups. Thus, the HVJ envelope vector appears to be much less immunogenic than native HVJ which strongly induces CTLs against virus-infected cells.

We failed to get effective HVJ-liposome–mediated gene transfer in the mouse uterus. However, high levels of luciferase gene expression were obtained in the mouse uterus when the HVJ envelope vector was used (Nakamura *et al.*, 2003). Mice were anesthetized and subjected to laparotomy to expose the uterus. Twenty-five microliters of HVJ envelope vector (3×10^9 particles) containing plasmid DNA (16 μg) was slowly injected into the uterine cavity using a 30-gauge needle, and the cervix was clamped for 10 min. Then, the incision was closed to allow the mice to recover. After 24 h of transfection, the luciferase activity mediated by HVJ envelope vector was approximately 120 times higher than that obtained using Lipofectamine. A 5-fold increase in the amount of plasmid did not affect the level of luciferase activity when Lipofectamine was used. Luciferase activity was detected in the uterus for at least 3 days after transfection using the HVJ envelope. Transfer of the LacZ gene to the uterus with the HVJ envelope vector yielded gene expression mainly in the glandular epithelium of the endometrium. Few stroma cells were transfected with this procedure. Viral fusion proteins disappeared 3 days after gene transfer, but transferred DNA was detected in the uterus for 10 days. No transfer of injected DNA to mouse fetuses was detected in this experiment.

To safely transfer genes to brain tissue, we intrathecally injected HVJ envelope vector containing the LacZ gene into the cerebrospinal fluid (CSF) of male Wistar rats (270 to 300 g) (Fig. 12.3) (Shimamura *et al.*, 2003). The HVJ envelope vector was injected into the cisterna magna. Briefly, a stainless cannula (27 gauge) was introduced into the cisterna magna (subarachnoid space) of anesthetized rats. HVJ envelope vector (100 μl) containing the human hepatocyte growth factor (HGF) gene was infused at the rate of 50 μl/min after removing 100 μl of CSF. Then, the animals were placed head-down for 30 min. No behavioral changes, such as convulsions or abnormal movements, were observed. Cells that expressed β-galactosidase were present in the spiral ganglion

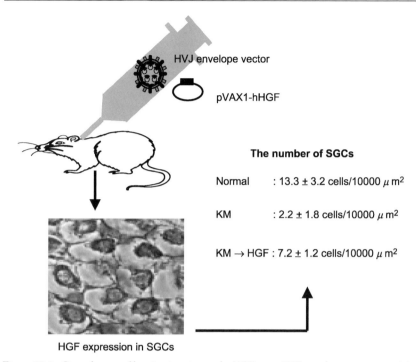

HGF expression in SGCs

Figure 12.3. Gene therapy of hearing impairment by HGF gene. HVJ envelope vector containing human HGF gene is administered to the cerebrospinal fluid of a rat. Human HGF gene expression is immunohistochemically detected at SGCs in the inner ear. By HGF gene expression, the number of SGCs is rescued after KM insult. (See Color Insert.)

cells (SGCs), cerebral cortex, and medulla. Luciferase expression was also examined in the brain, cochlea, lung, spleen and liver of the rats that were intrathecally injected with HVJ envelope vector containing the luciferase gene. One day after injection, strong transgene expression levels were observed in the cochleae and various areas of the brain. The highest luciferase activity was observed in the medulla, which is the area of the brain closest to the injection point. In contrast, on Day 1 and Day 5 after transfection, luciferase activity was not detected in the lung, spleen, or liver of the luciferase-injected rats. Luciferase activity was also not detected in any of the tested organs of the control-injected rats. To determine the optimal amount of HVJ envelope vector, we intrathecally administered 3×10^{10}, 4.5×10^{10}, and 6×10^{10} particles of HVJ envelope vector that contained 20 μg, 30 μg and 40 μg of luciferase gene, respectively. The highest expression level in the brain and cochlea resulted from the injection of 4.5×10^{10} particles of HVJ envelope vector (Oshima et al., 2004).

Next, we intravenously injected HVJ envelope vector. Luciferase gene–loaded HVJ envelope vector (6×10^9 particles) was injected into the mouse tail vein, and luciferase activity in several organs was measured 24 h after injection. Luciferase expression was detected in the spleen for at least 1 week. Intravenously administered naked plasmid (30 μg) that contained the luciferase gene resulted in luciferase expression in the liver, lung and spleen at much lower levels than those mediated by the HVJ envelope. When compared to the expression levels mediated by naked plasmid DNA transfer, luciferase gene expression mediated by HVJ envelope transfer was significant in the spleen, but not in the other organs (such as the lung and liver). When HVJ envelope vector containing FITC-ODNs was systemically injected, FITC-ODNs were observed in the marginal zone of the spleen. However, by intravenous injection of HVJ envelope vector, coagulation functions in the mouse blood is transiently suppressed, but the functions are recovered at 24 h after the injection. It is probably due to the hemagglutinating activity of HN protein. The protection of hemagglutinating activity of HN protein will be necessary for systemic injection of HVJ envelope vector to ensure the safety in gene therapy.

V. GENE THERAPY FOR HEARING IMPAIRMENT USING HVJ ENVELOPE VECTOR

By intrathecal injection of HVJ envelope vector, the vector reached the inner ear and gene expression was detected in SGCs which mediate auditory stimuli to the auditory nerve. Therefore, we tested treatment of hearing impairment with this approach (Oshima et al., 2004). Hearing impairment is caused by the loss of hair cells and the degeneration of SGCs. To recover hearing impairment, some neurotrophic factors, such as neurotrophin-3 (Ernfors et al., 1996), glial cell line-derived neurotrophic factor (Yagi et al., 2000) and brain-derived neurotrophic factor (Staecker et al., 1996) have been directly injected into inner ear tissue. However, we have reported that HGF also has pleiotropic effects on neural functions. HGF is a secretory protein that functions in an autocrine/paracrine manner on epithelial cells (Hayashi et al., 2001) and also on the nervous system (Miyazawa et al., 1998; Yoshimura et al., 2002). By analyzing the molecular interactions of HGF, we found that HGF activates the transcription factor, Ets-1. Many factors can be activated by Ets-1 (Tomita et al., 2003). HGF and its receptor, c-Met, are up-regulated by Ets-1, resulting in positive feedback of HGF activity. By the activation of VEGF and increases in metalloproteinases, angiogenesis is indirectly enhanced. Ets-1 also up-regulates itself. Therefore, even a small amount of HGF can induce pleiotropic effects on various cells expressing its receptor and the activity of HGF can be sustained. Therefore, we used the HGF gene as a therapeutic molecule to treat hearing impairment induced by aminoglycoside administration.

To transfer the human HGF gene into the SGCs and central nervous system, we injected 4.5×10^{10} particles of HVJ envelope containing the HGF gene into the CSF in the cisterna magna. We then measured the protein level of HGF in the CSF by ELISA. Five days after transfection, human HGF was readily detected in the CSF. An increase of rat HGF was also observed in the CSF of the rats transfected with human HGF gene. The expression of human HGF was detected in rats for 2 weeks after transfection, and endogenous rat HGF concentration was approximately 10 times that of human HGF (mean value: approximately 3 ng/ml on Day 5). Immunohistochemical detection of human HGF expression in the cytoplasm of SGCs indicated that the gene transfer efficiency was greater than 70% (Fig. 12.3). Additionally, the expression of c-Met, which is the tyrosine kinase receptor of HGF, was greatly enhanced on SGCs, suggesting that the increased c-Met and HGF expression synergistically affects the survival of SGCs.

We examined whether HGF could rescue the loss of SGCs induced by kanamycin (KM) insult (Oshima et al., 2004). A significant reduction of SGCs was induced by KM insult. The apoptosis was detected by TUNEL staining in SGCs. Pre-injection of HVJ envelope vector containing HGF gene completely prevented the apoptotic loss of SGCs. The apoptotic loss of hair cells in the inner ear was also rescued by HGF gene transfer using HVJ envelope vector.

To examine the potential for human gene therapy, we transferred the HGF gene into rats after hearing impairment was observed. Transfection of the human HGF gene into the subarachnoid space of the hearing-impaired rats significantly reduced the threshold shift in hearing function detected by auditory brain stem response when compared to rats transfected with the control vector. We also measured the number of SGCs in mid-modiolar sections of the cochleae from rats transfected with human HGF gene after kanamycin treatment. As shown in Fig. 12.3, by KM insult, the number of SGCs decreased to an undetectable level, but with HGF gene transfer, the number of SGCs are recovered to almost half of the normal level before KM insult. These results suggest that SGCs can be recovered by HGF gene transfection even after KM insult and that the recovery of SGCs results in the improvement of hearing impairment.

Similarly, cerebral infarction induced by the obstruction of the mid-cerebral artery in rats was prevented by the intrathecal injection of HVJ envelope vector containing HGF (Shimamura et al., 2004).

VI. APPROACHES TO CANCER GENE THERAPY USING HVJ ENVELOPE VECTOR

A. Transfection of dendritic cells (DCs) with melanoma-associated antigen (MAA) using HVJ envelope vector for immunotherapy of melanoma

Our first approach for developing cancer gene therapy was to construct cancer vaccines using the HVJ envelope vector system (Fig. 12.4). Polyvalent tumor antigen vaccines have been evaluated to increase the repertoire of anti-tumor T cells (Vilella *et al.*, 2003). In this study, we chose MAA genes, gp100 and TRP2, and transfected DCs with these two genes to evaluate *ex vivo* vaccination for prophylactic and therapeutic melanoma treatment.

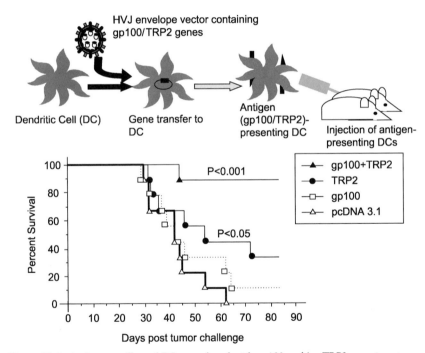

Figure 12.4. Anti-tumor effect of DCs transfected with gp100 and/or TRP2 gene in a tumor-dissemination model. At 24h after intravenous administration of BL6 melanoma cells (10^6), TAA-transfected DCs (10^6) were intradermally injected. Control groups consist of DCs transfected with pcDNA3.1 (△) (n = 9 mice). Experimental groups consist of vaccination with gp100-transfected DCs (□) (n = 9 mice), TRP2-transduced DCs (●) (n = 9 mice), and both gp100/TRP2-transfected DCs (▲) (n = 9 mice).

DCs are the most potent antigen-presenting cells. DCs, are capable of highly effective presentation of antigens to naïve T cells and they can initiate immune responses (Figdor et al., 2004). Several clinical studies have now been conducted in cancer patients with tumor-associated antigen (TAA)-loaded DCs (Nestle et al., 1998; Thurner et al., 1999). In several of these clinical studies the DCs were loaded by pulsing with protein antigens or with peptides derived from TAAs. Studies have demonstrated that ex vivo transfected DCs and adoptive therapy can be very effective in inducing antigen-specific immune responses. This type of strategy has been demonstrated in animals and pilot clinical studies for a variety of cancers (Gilboa et al., 1998; Engleman, 1997).

Transfected DCs with vectors expressing TAA genes have shown promising results. Studies have demonstrated that xenogeneic TAAs are very strong immunogens capable of cross-priming to host syngeneic TAAs. Human TRP2 and gp100 proteins have a high amino acid sequence homology to their mouse counterparts. We used xenogeneic TAAs for several reasons: to break self-antigen tolerance in the host, to induce strong antitumor immunity, and to enhance individual TAA immunogenicity. The approach provides a large expression of TAAs and activates DC presentation. Therefore, a highly efficient transfection system is absolutely necessary for efficient induction of tumor immunity. In recent studies, adenoviral vector was the original vector of choice for efficiently transfecting DCs (Arthur et al., 1997; Kaplan et al., 1999). However, adenovirus does have serious disadvantages as a vector for engineering DCs because of its high toxicity. Some studies have demonstrated that plasmid DNA in liposomes (Pecher et al., 2001), vaccinia virus (Yang et al., 2000; Prabakaran et al., 2002) and retroviruses (Akiyama et al., 2002; Bello-Fernandez et al., 1997) could be used to infect or transfect human DCs to present a variety of antigens. However, a highly efficient and minimally invasive vector system has not yet been achieved for transfection of DCs. To develop more efficient gene transfer to DCs that cause only minimal damage, we tested the potential of HVJ envelope vector for transfection of DCs with two different TAA genes (Fig. 12.4).

We used the luciferase gene to determine the optimal conditions for transfection of DCs. As previously reported, the optimum ratio of HVJ envelope vector to cultured cells is 6×10^3 to 1.2×10^4 (Kaneda et al., 2002a). We identified 1.2×10^4 as the optimal ratio and determined the most effective transfection conditions. When DCs were incubated with HVJ envelope vector containing luciferase gene, luciferase gene expression was not significant. We attempted to increase the luciferase gene expression by centrifuging the mixture of DCs and HVJ envelope vector. Centrifugation at $9000\,g$ was much more effective for gene transfection than $3000\,g$ and $13000\,g$. The viability of DCs after transfection was approximately 70%. The centrifugation time required for the most effective gene expression was 90 min. The highest luciferase gene expression was obtained in DCs that were centrifuged for 90 min at 9000 g and 37 °C. Six days after DCs were isolated was the best time to transfect the DCs for optimal luciferase gene expression. Thus, the optimal conditions for the most effective transfection of DCs was at 9000 g for 90 min at 37 °C in DCs cultured for 6 days after isolation from bone marrow. Under these optimal conditions, we assessed the expression of yellow fluorescent protein (YFP) expression to determine the transfected DCs population. Almost all DCs that were recognized with phycoerythrin (PE)-conjugated CD11c antibody expressed YFP, as determined by fluorescence microscopy. Flow cytometry analysis revealed that approximately 99% of DCs expressed YFP.

Surface markers of mature DCs were studied by flow cytometry analysis to confirm that differentiation was not inhibited by gene transfection using HVJ envelope vector. The LPS-induced expression of CD40, CD80, CD86 and HLA-DR was equivalent on non-transfected DCs and transfected DCs. The non-specific phagocytotic activity of DCs (evaluated by uptake of FITC-dextran) was suppressed in transfected DCs as well as in non-transfected DCs. Thus, gene transfer with HVJ envelope vector did not inhibit LPS-induced maturation of DCs.

Next, we used the HVJ envelope vector to transfect the glycoprotein 100 (gp100) and tyrosine-related protein 2 (TRP2) genes into DCs. The expression of each transfected antigen was detected by flow cytometry analysis of cells stained with antigen-specific antibody. On Day 1, gp100 and TRP2 were detected in 36% and 63% of DCs, respectively. Approximately 80% of DCs expressed both antigens on Day 7.

DCs transfected with either the gp100 or TRP2 gene or both genes after one day were injected i.d. into C57BL/6 mice. The delayed-type hypersensitivity (DTH) response was assessed in mice immunized with MAA-transfected DCs at 24, 48 and 72 h after TAA protein challenge. Significant cutaneous DTH was detected in mice immunized with MAA gene-transfected DCs compared to the control mice that were injected with DCs without MAA. The DTH response to gp100 was significant in the groups immunized with gp100-transfected DCs or gp100/TRP2-transfected DCs. The level of response was equivalent in the two groups. A similar response to TRP2 was obtained in the groups immunized with TRP2-transfected DCs or gp100/TRP2-transfected DCs.

We examined CTLs against BL6 tumor cells on Day 7 after DC administration. TAA-specific CTLs were generated with variable response rates. Mice immunized with DCs expressing gp100 and TRP-2 had the highest level of CTL activity against ^{51}Cr-labeled BL6 target cells. Mice immunized with single TAA-transfected DCs generated significant CTL activity against BL6 cells when compared to the control groups that received HVJ envelope vector containing pcDNA-3.1. However, these responses were lower than those generated by the cotransfer of gp100 and TRP2. Vaccination of DCs transfected with TAA genes induced TAA specific cellular immunity consisting of CTL activity and DTH-helper cell activity against melanoma cells.

We tested the effect of vaccination with MAA-transfected DCs on the inhibition of tumor growth in mice. No tumor growth was observed in mice immunized with gp100/TRP2 transfected DCs, although significant tumor masses were palpable in all control mice injected with DCs without TAA. Immunization with DCs transfected with single TAA gene was not effective, as small tumor masses were detected. However the tumor growth rate was slower than that of the control group. In this prophylactic study, vaccination with DCs transfected with both gp100 and TRP2 was much more effective for inhibition of tumor growth than vaccination with DCs transfected with TRP2 alone.

To investigate the therapeutic effect of vaccination with TAA-transfected DCs, we vaccinated the mice with various DC vaccines 24 h after intravenous inoculation with 10^4 BL6 cells. All control mice injected with PBS-treated DCs and pcDNA3.1-transfected DCs died of lung metastasis in 45 days and 62 days after inoculation, respectively. Vaccination with DCs that were transfected with a single TAA gene improved likelihood of survival. Vaccination with TRP2 transfected DCs was more effective than vaccination with gp100-transfected DCs. Vaccination with gp100/TRP2-transfected DCs significantly prolonged the survival rate. Eight of nine mice vaccinated with gp100/TRP2-transfected DCs survived more than 80 days after tumor inoculation (Fig. 12.4). Vaccination with immature DCs that were not treated with LPS was not effective for improving survival, and all of these mice died in 60 days (including the mice that received gp100/TRP2-transfected DCs).

B. Fusion of DC-Tumor cells and simultaneous gene transfer to the hybrid cells using HVJ envelope for the prevention and treatment of cancers

TAAs have been identified in some cancers such as melanoma (Boon et al., 1994; Kawakami and Rosenberg, 1997). However, TAAs in many cancers have not been identified. The identification of TAAs is required for the development of TAA-loaded DC vaccines.

To solve the problem, hybrid cell vaccines have been developed by fusing DCs with tumor cells (Gong et al., 1997; Wang et al., 1998). There is evidence that these DC-tumor-fused cells possess the properties of both tumor cells containing known and unknown TAAs and DCs containing high levels of MHC class I and II molecules and co-stimulatory molecules for priming and activating naïve CD4+ and CD8+ T cells. Therefore, even though tumor cells lose MHC class I molecules, TAAs can be presented on the surface of the fused cells by DC-derived MHC class I molecules. Polyethylene glycol and

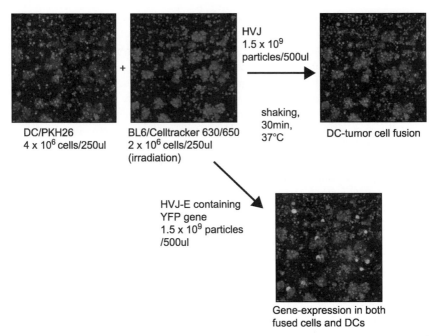

DC/PKH26   BL6/Celltracker 630/650   DC-tumor cell fusion
4 x 10⁶ cells/250ul   2 x 10⁶ cells/250ul
                        (irradiation)

HVJ
1.5 x 10⁹
particles/500ul

shaking,
30min,
37°C

HVJ-E containing
YFP gene
1.5 x 10⁹ particles
/500ul

Gene-expression in both
fused cells and DCs

Figure 12.5. Fusion of DCs and melanoma cells and simultaneous gene transfer to the fused cells. Mouse bone marrow-derived DCs stained with PKH26 and mouse melanoma BL6 labeled with Celltracker 630/650 are fused with inactivated HVJ (1.5×10^8 particles). When fused with HVJ envelope (HVJ-E) vector containing YFP gene, YFP gene expression is detected in both DCs and fused cells. (See Color Insert.)

electroporation have been used to induce tumor cell–DC fusion. HVJ has been identified as a powerful fusogen. We stained mouse DCs with red fluorescent reagent. Mouse melanoma cells were stained green and irradiated with gamma-rays. Both cells were mixed with inactivated HVJ. After 30 min, the fusion of DCs and tumor cells occurred with approximately 50% efficiency (Fig. 12.5).

We used HVJ envelope to generate powerful tumor-DC vaccines, because HVJ envelope can induce tumor cell–DC fusion and simultaneously transfer DNA or proteins to activate the immune response. To confirm simultaneous fusion and gene transfer mediated by HVJ envelope, DCs and tumor cells were mixed with HVJ envelope containing the GFP gene. As shown in Fig. 12.5, fusion between DCs and irradiated-tumor cells was observed, and YFP expression was detected in fused cells as well as in DCs.

Using this system, we attempted to generate anti-tumor immunity using HVJ envelope with or without the IL-12 gene. Ten days after the second immunization, spleen cells were isolated to assess cytolytic activity. The cytolytic activity of the effector cells obtained from the mice immunized with fused cells was significantly higher than that obtained from other vaccination protocols such as PBS, IL-12, Mix (mixture of DCs and tumor cells without fusion) and mix plus IL-12. Higher cytolytic activity was observed in the mice that received fused cells plus IL-12 as compared with that in the mice that received fused cells alone.

First, we examined the effect of this hybrid cell vaccine on the prevention of tumor generation. After two vaccinations, melanoma cells or renal cancer cells were intradermally injected. The mice vaccinated with fused cells plus IL-12 had significantly increased survival; all of these mice were alive 60 days after tumor challenge in the B16BL6 tumor model. The survival rate was 20% in the mice vaccinated with fused cells without IL-12. All mice in the other groups died. The effect of fused cells and IL-12 on the enhancement of tumor-specific immunity in mice was also observed against renal cancer. Therefore, these findings indicate that immunization with fused cells plus IL-12 strongly induces Th1 cytokines and activates tumor-specific CTLs, resulting in significant protection from melanoma, which has known TAAs or renal cancers which has unknown TAAs.

Next, we treated solid tumors by *in vivo* fusion and gene expression using HVJ envelope vector. Mouse DCs and HVJ envelope vector were injected into solid tumors. The effect of the transfer of a therapeutic gene by HVJ envelope vector was also evaluated. To detect *in vivo* fusion of DCs with tumor cells, red-stained DCs were injected into melanoma masses expressing YFP with inactivated HVJ. Megakaryocytes, which indicate tumor cell–tumor cell fusion, were detected, as well as large orange-stained cells which indicate DC-tumor cell fusion. However, the fusion efficiency *in vivo* was much less than that of *in vitro* cell fusion (Fig. 12.6).

For therapeutic experiments, melanoma cells were intradermally injected. When the tumor diameter was greater than 5 mm, an injection of DCs and HVJ envelope vector was administered. Spleen cells were isolated and assayed for anti-melanoma CTL activity. No therapeutic genes were incorporated into the HVJ envelope vector. The co-injection of DCs and empty HVJ envelope vector resulted in significant anti-melanoma CTL activity. DCs or HVJ envelope vector alone did not induce CTL activity. When tumor growth was observed, the co-injection of DCs and empty HVJ envelope vector inhibited tumor growth, but the tumor volume still gradually increased. When DCs were co-injected with HVJ envelope vector that contained the IL-12 gene, a greater inhibitory effect was obtained. The co-injection of DCs and HVJ envelope vector effectively prolonged mouse survival. The most effective therapeutic effect was obtained when HVJ envelope vector that contained IL-12 gene was co-injected with DCs into the tumor mass; approximately 60% of these mice survived 35 days after treatment.

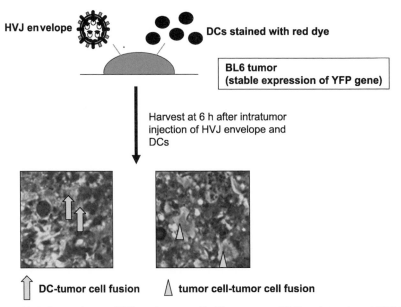

HVJ envelope   DCs stained with red dye

BL6 tumor
(stable expression of YFP gene)

Harvest at 6 h after intratumor
injection of HVJ envelope and
DCs

⇧ DC-tumor cell fusion   △ tumor cell-tumor cell fusion

Figure 12.6. *In vivo* fusion of DCs and tumor cells. The mixture of DCs and inactivated HVJ is injected into intradermal tumor mass derived from BL6 expressing YFP gene. When DCs are labeled with red dye (PKH26), orange-colored fused cells are detected, suggesting DC-tumor fusion. Tumor–tumor cell fusion is also observed.

Thus, we developed a novel cancer vaccine using HVJ envelope-mediated DC-tumor cell fusion. The DC-tumor cell fusion was induced by HVJ envelope vector at a high efficiency. Both known and unknown TAAs were presented on fused cells by MHC class I molecules from DCs. Furthermore, IL-12 gene transfer was achieved in the fused cells. The expression of IL-12 stimulates the maturation of naïve T cells and promotes a Th1 response (Xu *et al.*, 2003). It is believed that the presentation of TAAs, and the promotion of a Th1 response, together induce an efficient anti-tumor CTL response.

VII. TOWARD THE CLINICAL TRIAL

The HVJ envelope vector holds great promise for human gene therapy. We are improving the HVJ envelope vector system so that it can be used in clinical trials. So far, the virus has been produced in chick eggs (Okada, 1993), but egg-derived HVJ is difficult to use for clinical trials. It has been very difficult to produce large amounts of the virus in cultured cells. The production in cultured cells was less than 2% of that in chick eggs (Kaneda and Okada, unpublished

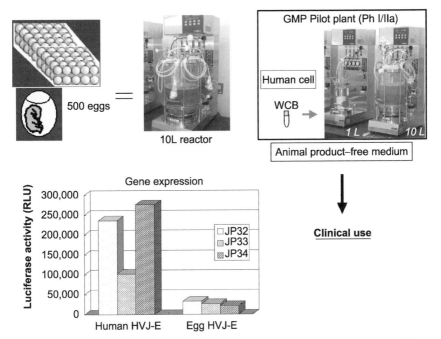

Figure 12.7. Development of clinical grade HVJ envelope vector. HVJ infects human cell from working cell bank (WCB), and human cell-derived HVJ is prepared using animal product-free medium in a pilot plant based on good manufacturing practice (GMP) standard. The envelope vector prepared from human cell-derived HVJ is more efficient in gene transfection compared with that from egg-derived HVJ. (See Color Insert.)

observation). We recently succeeded in producing HVJ in human cells. We cloned HEK 293 cells that produce a high titer of HVJ after infection, and determined the growth conditions for these cells in an animal product-free medium using a bioreactor. We are able to culture these cells in 10 liters of serum-free medium at a density of more than 10^6 cells/ml, and we can obtain HVJ at the efficiency of more than 3×10^9 particles/ml (Fig. 12.7). The production in a 10 liter culture is comparable to that in 500 chick eggs. A pilot plant to commercially produce clinical grade HVJ envelope vector was established in a venture company named GenomIdea, Inc. (7-7-15 Saito-Asagi, Ibaragi, Osaka 567-0085, Japan; http://www.anges-mg.com/news/030324.htm). GenomIdea created a working cell bank, master cell bank and master virus bank for producing clinical grade HVJ for Phase I and IIa trials. They also set up an apparatus to purify inactivated HVJ using two different column systems. The human cell-derived HVJ envelope vector was more effective for gene transfection than the egg-derived HVJ envelope vector (Fig. 12.7). Human cell-derived HVJ envelope vector is also available for drug delivery both *in vitro* and *in vivo*. This vector will soon be available for human gene therapy.

VIII. MAIN ABBREVIATIONS

HVJ	hemaggulutinating virus of Japan
HN	hemaggulutin-neuraminidase
F	fusion
FITC	fluorescein isothiocyanate
ODN	oligodeoxynucleotide
HA	hemagglutinin
CTLs	cytotoxic T cells
ES	embryonal stem
GFP	green fluorescent protein
siRNA	small interfering RNA
CSF	cerebrospinal fluid
HGF	hepatocyte growth factor
SGCs	spiral ganglion cells
KM	kanamycin
DCs	dendritic cells
MAA	melanoma-associated antigen
TAA	tumor-associated antigen
YFP	yellow fluorescent protein
PE	phycoerythrin
gp100	glycoprotein 100
TRP2	tyrosine-related protein 2
DTH	delayed-type hypersensitivity
WCB	working cell bank
GMP	good manufacturing practice
HVJ-E	HVJ envelope

References

Akiyama, Y., Maruyama, K., Watanabe, M., and Yamaguchi, K. (2002). Retroviral-mediated IL-12 gene transduction into human CD34+ cell-derived dendritic cells. *Int. J. Oncol.* **21**, 509–514.

Arthur, J. F., Butterfield, L. H., Roth, M. D., Bui, L. A., Kiertscher, S. M., Lau, R., Dubinett, S., Glaspy, J., McBride, W. H., and Economou, J. S. (1997). A comparison of gene transfer methods in human dendritic cells. *Cancer Gene Ther.* **4**, 17–25.

Bagai, S., and Sarker, D. P. (1993). Targeted delivery of hygromycin B using reconstituted Sendai viral envelopes lacking hemagglutinin-neuraminidase. *FEBS Lett.* **326**, 183–188.

Bello-Fernandez, C., Matyash, M., Strobl, H., Pickl, W. F., Majdic, O., Lyman, S. D., and Knapp, W. (1997). Efficient retrovirus-mediated gene transfer of dendritic cells generated from CD34+ cord blood cells under serum-free conditions. *Hum. Gene. Ther.* **8,** 1651–1658.

Boon, T., Cerottini, J. C., Van den Eynde, B., van der Bruggen, P., and Van Pel, A. (1994). Tumor antigens recognized by T lymphocytes. *Annu. Rev. Immunol.* **12,** 337–365.

Cavazzana-Calvo, M., Thrasher, A., and Mavillo, F. (2004). The future of gene therapy. *Nature* **427,** 779–781.

Chen, Y., Webster, R. G., and Woodland, D. L. (1998). Induction of CD8+ T cell responses to dominant and subdominant epitopes and protective immunity to Sendai virus infection by DNA vaccination. *J. Immunol.* **160,** 2425–2432.

Dorsett, Y., and Tuschl, T. (2004). siRNAs: Applications in functional genomics and potential as therapeutics. *Nat. Rev. Drug Discov.* **3,** 318–329.

Dzau, V., Mann, M. J., Morishita, R., and Kaneda, Y. (1996). Fusigenic viral liposome for gene therapy in cardiovascular diseases. *Proc. Natl. Acad. Sci. USA* **93,** 11421–11425.

Engleman, E. G. (1997). Dendritic cells: Potential role in cancer therapy. *Cytotechnology* **25,** 1–8.

Ernfors, P., Duan, M. L., ElShamy, W. M., and Canlon, B. (1996). Protection of auditory neurons from aminoglycoside toxicity by neurotrophin-3. *Nat. Med.* **2,** 463–467.

Felgner, P. L., Gadek, T. R., Holm, M., Roman, R., Chan, H. S., Wenz, M., Northrop, J. P., Ringold, G. M., and Danielsen, H. (1987). Lipofection; a highly efficient lipid-mediated DNA transfection procedure. *Proc. Natl. Acad. Sci. USA* **84,** 7413–7417.

Figdor, C. G., de Vries, I. J., Lesterhuis, W. J., and Melief, C. J. (2004). Dendritic cell immunotherapy: Mapping the way. *Nat. Med.* **10,** 475–480.

Ghosh, J. K., Peisajovich, S. G., and Shai, Y. (2000). Sendai virus internal fusion peptide: Structural and functional characterization and a plausible mode of viral entry inhibition. *Biochemistry* **39,** 1581–1592.

Gilboa, E., Nair, S. K., and Lyerly, H. K. (1998). Immunotherapy of cancer with dendritic-cell-based vaccines. *Cancer Immunol. Immunother.* **46,** 82–87.

Gong, J., Chen, D., Kashiwaba, M., and Kufe, D. (1997). Induction of antitumor activity by immunization with fusions of dendritic and carcinoma cells. *Nat. Med.* **3,** 558–561.

Hayashi, K., Morishita, R., Nakagami, H., Yoshimura, S., Hara, A., Matsumoto, K., Nakamura, T., Ogihara, T., Kaneda, Y., and Sakai, N. (2001). Gene therapy for preventing neuronal death using hepatocyte growth factor: In vivo gene transfer of HGF to subarachnoid space prevents delayed neuronal death in gerbil hippocampal CA1 neurons. *Gene Ther.* **8,** 1167–1173.

Hirano, T., Fujimoto, J., Ueki, T., Yamamoto, H., Iwasaki, T., Morishita, R., Kaneda, Y., Takahashi, H., and Okamoto, E. (1998). Persistent gene expression in rat liver in vivo by repetitive transfections using HVJ-liposome. *Gene Ther.* **5,** 459–464.

Ishii, T., Ohnuma, K., Murakami, A., Takasawa, N., Yamochi, T., Iwata, S., Uchiyama, M., Dang, N. H., Tanaka, H., and Morimoto, C. (2003). SS-A/Ro52, an autoantigen involved in CD28-mediated IL-2 production. *J. Immunol.* **170,** 3653–3661.

Itoh, Y., Kawamata, Y., Harada, M., Kobayashi, M., Fujii, R., Fukusumi, S., Ogi, K., Hosoya, M., Tanaka, Y., Uejima, H., Tanaka, H., Maruyama, M., Satoh, R., Okubo, S., Kizawa, H., Komatsu, H., Matsumura, F., Noguchi, Y., Shinohara, T., Hinuma, S., Fujisawa, Y., and M. Fujino, M. (2003). Free fatty acids regulate insulin secretion from pancreatic beta cells through GPR40. *Nature* **422,** 173–176.

Kaneda, Y., Saeki, Y., and Morishita, R. (1999). Gene therapy using HVJ-liposomes: The best of both worlds? *Mol. Med. Today* **5,** 298–303.

Kaneda, Y., Nakajima, T., Nishikawa, T., Yamamoto, S., Ikegami, H., Suzuki, Nakamura, H., Morishita, R., and Kotani, H. (2002a). Hemagglutinating virus of Japan (HVJ) envelope vector as a versatile gene delivery system. *Mol. Ther.* **6,** 219–226.

Kaneda, Y. (2002b). *In* "Gene Therapy Protocol – The second edition" (J. R. Morgan, ed.), pp. 63–72. Humana Press, New Jersey.

Kaplan, J. M., Yu, Q., Piraino, S. T., Pennington, S. E., Shankara, S., Woodworth, L. A., and Roberts, B. L. (1999). Induction of antitumor immunity with dendritic cells transduced with adenovirus vector-encoding endogenous tumor-associated antigens. *J. Immunol.* **163**, 699–707.

Kawakami, Y., and Rosenberg, S. A. (1997). Human tumor antigens recognized by T-cells. *Immunol. Res.* **16**, 313–339.

Lam, P. Y. P., and Brakefield, X. O. (2000). Hybrid vector designs to control the delivery, fate and expression of transgenes. *J. Gene Med.* **2**, 395–408.

Ledley, F. D. (1995). Non-viral gene therapy: The promise of genes as pharmaceutical products. *Hum. Gene Ther.* **6**, 1129–1144.

Li, S., and Huang, L. (2000). Non-viral gene therapy: Promises and challenges. *Gene Ther.* **7**, 31–34.

Marshall, E. (1995). Gene therapy's growing pains. *Science* **269**, 1052–1055.

Miyagishi, M., Hayashi, M., and Taira, K. (2003). Comparison of the suppressive effects of antisense oligonucleotides and siRNAs directed against the same targets in mammalian cells. *Antisense Nucleic Acid Drug Dev.* **13**, 1–7.

Miyazawa, T., Matsumoto, K., Ohmichi, H., Katoh, H., Yamashima, T., and Nakamura, T. (1998). Protection of hippocampal neurons from ischemia-induced delayed neuronal death by hepatocyte growth factor: A novel neurotrophic factor. *J. Cereb. Blood Flow Metab.* **18**, 345–348.

Mulligan, R. C. (1993). The basic science of gene therapy. *Science* **260**, 926–932.

Nakamura, H., Kimura, T., Ikegami, H., Ogita, K., Kohyama, S., Shimoya, K., Tsujie, T., Koyama, M., Kaneda, Y., and Murata, Y. (2003). Highly-efficient and minimally invasive *in vivo* gene transfer to the mouse uterus by Hemagglutinating Virus of Japan (HVJ) envelope vector. *Mol. Hum. Reprod.* **9**, 603–609.

Nakamura, N., Hart, D. A., Frank, C. B., Marchuk, L. L., Shrive, N. G., Ota, N., Taira, T., Yoshikawa, H., and Kaneda, Y. (2001). Efficient transfer of intact oligonucleotides into the nucleus of ligament scar fibroblasts by HVJ-cationic liposomes is correlated with effective antisense gene inhibition. *J. Biochem.* **129**, 755–759.

Nestle, F. O., Alijagic, S., Gilliet, M., Sun, Y., Grabbe, S., Dummer, R., Burg, G., and Schadendorf, D. (1998). Vaccination of melanoma patients with peptide- or tumor lysate-pulsed dendritic cells. *Nat. Med.* **4**, 328–332.

Okada, Y. (1993). *In* "Methods in Enzymology" (N. Duzgunes, ed.), Vol. 221, pp. 18–41. Academic Press, Inc., San Diego.

Oshima, K., Shimamura, M., Mizuno, S., Tamai, K., Doi, K., Morishita, R., Nakamura, T., Kubo, T., and Kaneda, Y. (2004). Intrathecal injection of HVJ-E containing HGF gene to cerebrospinal fluid can prevent and ameliorate hearing impairment in rats. *The FASEB J.* **18**, 212–214.

Pecher, G., Spahn, G., Schirrmann, T., Kulbe, H., Ziegner, M., Schenk, J. A., and Sandig, V. (2001). Mucin gene (MUC1) transfer into human dendritic cells by cationic liposomes and recombinant adenovirus. *Anticancer Res.* **21**, 2591–2596.

Prabakaran, I., Menon, C., Xu, S., Gomez-Yafal, A., Czerniecki, B. J., and Fraker, D. L. (2002). Mature CD83(+) dendritic cells infected with recombinant gp100 vaccinia virus stimulate potent antimelanoma T cells. *Ann. Surg. Oncol.* **9**, 411–418.

Race, E., Stein, C. A., Wigg, M. D., Baksh, A., Addawe, M., Frezza, P., and Oxford, J. S. (1995). A multistep procedure for the chemical inactivation of human immunodeficiency virus for use as an experimental vaccine. *Vaccine* **13**, 1567–1575.

Ramani, K., Bora, R. S., Kumar, M., Tyagi, S. K., and Sarkar, D. P. (1997). Novel gene delivery to liver cells using engineered virosomes. *FEBS Lett.* **404**, 164–168.

Ramani, K., Hassan, O., Venkaiah, B., Hasnain, S. E., and Sarkar, D. P. (1998). Site-specific gene delivery *in vivo* through engineered Sendai virus envelopes. *Proc. Natl. Acad. Sci. USA* **95**, 11886–11890.

Shimamura, M., Morishita, R., Endoh, M., Oshima, K., Aoki, M., Waguri, S., Uchiyama, Y., and Kaneda, Y. (2003). HVJ-envelope vector for gene transfer into central nervous system. *Biochem. Biophys. Res. Comm.* **300**, 464–471.

Shimamura, M., Sato, N., Oshima, K., Aoki, M., Kurinami, H., Waguri, S., Uchiyama, Y., Ogihara, T., Kaneda, Y., and Morishita, R. (2004). A novel therapeutic strategy to treat brain ischemia: Over-expression of hepatocyte growth factor gene reduced ischemic injury without cerebral edema in rat model. *Circulation* **109**, 424–431.

Sioud, M. (2004). Ribozyme- and siRNA-Mediated mRNA Degradation: A General Introduction. *Methods Mol. Biol.* **252**, 1–8.

Staecker, H., Kopke, R., Malgrange, B., Lefebvre, P., and Van de Water, T. R. (1996). NT-3 and/or BDNF therapy prevents loss of auditory neurons following loss of hair cells. *Neuroreport* **7**, 889–894.

Suzuki, K., Nakashima, H., Sawa, Y., Morishita, R., Matsuda, H., and Kaneda, Y. (2000). Reconstituted fusion liposomes for gene transfer *in vitro* and *in vivo*. *Gene Ther. Regulat.* **1**, 65–77.

Thurner, B., Haendle, I., Roder, C., Dieckmann, D., Keikavoussi, P., Jonuleit, H., Bender, A., Maczek, C., Schreiner, D., von den Driesch, P., Brocker, E. B., Steinman, R. M., Enk, A., Kampgen, E., and Schuler, G. (1999). Vaccination with mage-3A1 peptide-pulsed mature, monocyte-derived dendritic cells expands specific cytotoxic T cells and induces regression of some metastases in advanced stage IV melanoma. *J. Exp. Med.* **190**, 1669–1678.

Tijsterman, M., and Plasterk, R. H. (2004). Dicers at RISC; the mechanism of RNAi. *Cell* **117**, 1–3.

Tomita, N., Morishita, R., Taniyama, Y., Koike, H., Aoki, M., Shimizu, H., Mastumoto, K., Nakamura, T., Kaneda, Y., and Ogihara, T. (2003). Angiogenic property of hepatocyte growth factor is dependent on upregulation of essential transcription factor for angiogenesis, ets-1. *Circulation* **107**, 141–147.

Vilella, R., Benitez, D., Mila, J., Vilalta, A., Rull, R., Cuellar, F., Conill, C., Vidal-Sicart, S., Costa, J., Yachi, E., Palou, J., Malvehy, J., Puig, S., Marti, R., Mellado, B., and Castel, T. (2003). Treatment of patients with progressive unresectable metastatic melanoma with a heterologous polyvalent melanoma whole cell vaccine. *Int. J. Cancer* **106**, 626–631.

Wagner, E., Plank, C., Zatloukal, K., Cotten, M., and Birnstiel, M. L. (1992). Influenza virus hemagglutinin HA-2 N-terminal fusogenic peptides augment gene transfer by transferrin-polylysine-DNA complexes: Toward a synthetic virus-like gene-transfer vehicle. *Proc. Natl. Acad. Sci. USA* **89**, 7934–7938.

Wang, J., Saffold, S., Cao, X., Krauss, J., and Chen, W. (1998). Eliciting T cell immunity against poorly immunogenic tumors by immunization with dendritic cell-tumor fusion vaccines. *J. Immunol.* **161**, 5516–5524.

Wu, G. Y., and Wu, C. H. (1988). Receptor-mediated gene delivery and expression *in vivo*. *J. Biol. Chem.* **263**, 14621–14624.

Xu, S., Koski, G. K., Faries, M., Bedrosian, I., Mick, R., Maeurer, M., Cheever, M. A., Cohen, P. A., and Czerniecki, B. J. (2003). Rapid high efficiency sensitization of CD8+ T cells to tumor antigens by dendritic cells leads to enhanced functional avidity and direct tumor recognition through an IL-12-dependent mechanism. *J. Immunol.* **171**, 2251–2261.

Yagi, M., Kanzaki, S., Kawamoto, K., Shin, B., Shah, P. P., Magal, E., Sheng, J., and Raphael, Y. (2000). Spiral ganglion neurons are protected from degeneration by GDNF gene therapy. *J. Assoc. Res. Otolaryngol.* **1**, 315–325.

Yang, S., Kittlesen, D., Slingluff, C. L., Jr., Vervaert, C. E., Seigler, H. F., and Darrow, T. L. (2000). Dendritic cells infected with a vaccinia vector carrying the human gp100 gene simultaneously present multiple specificities and elicit high-affinity T cells reactive to multiple epitopes and restricted by HLA-A2 and -A3. *J. Immunol.* **164**, 4204–4211.

Yeagle, P. L. (1993). The fusion of Sendai virus. In "Viral fusion mechanisms" (J. Benty, ed.), pp. 313–334. CRC Press, Inc., London.

Yokota, T., Miyagishi, M., Hino, T., Matsumura, R., Tasinato, A., Urushitani, M., Rao, R. V., Takahashi, R., Bredesen, D. E., Taira, K., and Mizusawa, H. (2004). siRNA-based inhibition specific for mutant SOD1 with single nucleotide alternation in familial ALS, compared with ribozyme and DNA enzyme. *Biochem. Biophys. Res. Commun.* **314,** 283–291.

Yoshimura, S., Morishita, R., Hayashi, K., Kokuzawa, J., Aoki, M., Matsumoto, K., Nakamura, T., Ogihara, T., Sakai, N., and Kaneda, Y. (2002). Gene transfer of hepatocyte growth factor to subarachnoid space in cerebral hypoperfusion model. *Hypertension* **39,** 1028–1034.

Yonemitsu, Y., Kitson, C., Ferrari, S., Farley, R., Griesenbach, U., Juda, D., Steel, R., Scheid, P., Zhu, J., Jeffery, P., Kato, A., Hasan, M. K., Nagai, Y., Masaki, I., Fukumura, M., Hasegawa, M., Geddes, D. M., and Alton, E. F. W. (2000). Efficient gene transfer to airway epithelium using recombinant Sendai virus. *Nature Biotech.* **18,** 970–973.

Zenke, M., Steinlein, P., Wagner, E., Cotten, M., Beug, H., and Birnstiel, M. L. (1990). Receptor-mediated endocytosis of transferrin-polycation conjugates: An efficient way to introduce DNA into hematopoietic cells. *Proc. Natl. Acad. Sci. USA* **87,** 3655–3659.

Further Reading

Kaneda, Y. (1994). Virus (Sendai virus envelopes) mediated gene transfer. In "Cell Biology A Laboratory Handbook" (J. E. Celis, ed.), pp. 50–57. Academic Press, Orlando, FL.

13

Targeting of Polyplexes: Toward Synthetic Virus Vector Systems

Ernst Wagner, Carsten Culmsee, and Sabine Boeckle
Pharmaceutical Biology - Biotechnology, Department of Pharmacy
Ludwig-Maximilians-Universitaet Muenchen
Butenandtstr. 5-13, D-81377 Munich, Germany

ABSTRACT

Dominating issues in gene vector optimization are specific in recognizing the target cells and exploiting the proper intracellular trafficking routes. Any progress in this area will result in improved specific gene transfer, reduce the required therapeutic vector doses and, in consequence, lower the overall toxicity to the host. To provide polyplexes with the ability to distinguish between non-target

0065-2660/05 $35.00
DOI: 10.1016/S0065-2660(05)53013-X

and target cells, cell-binding ligands have been incorporated which recognize target-specific cellular receptors. In addition, polyplex domains with unspecific binding capacity (such as surface charges) have to be shielded or removed. Cell-binding ligands can be small molecules, vitamins, carbohydrates, peptides or proteins such as growth factors or antibodies. Such ligands have been incorporated into polyplexes after chemical conjugation to cationic polymers. The choice of the ligand and physical properties of the DNA formulation strongly influence extracellular routing (circulation in blood, tissue distribution), uptake and intracellular delivery of polyplexes. Recent efforts are discussed that aim at the development of polyplexes into virus-like supramolecular complexes; such particles should undergo structural changes compatible with extracellular and intracellular targeting. © 2005, Elsevier Inc.

I. INTRODUCTION

Specificity for the target tissue is a key issue in gene therapy and can be accomplished in several ways. Firstly, the therapeutic vector may be preferentially transported into the target tissue, where it is internalized by target cells and delivered into the nucleus ('targeted delivery'). Secondly, specific promotor/enhancer elements can be used in the expression cassette, which restrict the transcription to those cells that possess the required specific transcription factor combination ('transcriptional targeting').

To enhance targeted delivery, the concept of receptor-targeted gene transfer has been developed over the last two decades. Targeting ligands have been linked directly to DNA (Cheng *et al.*, 1983), to other non-viral vector formulations, or to viral vectors (Wickham, 2003). Many studies report successful ligand-mediated gene transfer to cultured cells which cannot be transfected by the unmodified parental vector. This review focuses on receptor-targeted polyplexes, the current stage of development and opportunities for further optimization.

II. EXTRACELLULAR DELIVERY: TISSUE AND CELL TARGETING

Many current vector formulations suffer from weak to non-existing targeting capacity. To provide non-viral vectors with the ability to distinguish between target and non-target tissue (Schätzlein, 2003), cell-binding ligands have been incorporated which recognize target-specific cellular receptors (see Section II.A). In addition, modifications have been made (see Section II.B) to shield vector domains which otherwise induce undesired binding to blood components or non-target cells (see Fig. 13.1).

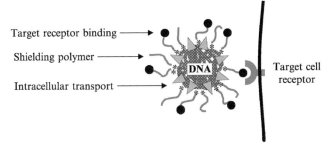

Figure 13.1. Receptor-targeted polyplexes. Gene transfer particles consisting of DNA, ligand-polycation conjugates, and- optionally, additional elements, such as shielding agents, or functions for intracellular transport such as endosome-destabilizing or nuclear targeting elements. Surface properties of the complexes (charge density, degree of coating with ligand and/or shielding agents) can influence cell binding. Unspecific interactions can be shielded by incorporation of PEG molecules; targeting to the cell surface is mediated by ligand-receptor interaction. Examples of targeting ligand/receptor pairs can be found in Table I.

A. Targeting ligands

A series of biological agents ('ligands') attach to cells via specific domains that interact with cell surface receptors. Such receptor-ligand interactions are involved in various processes, for example, entry of viruses, bacteria or toxins, provision of nutrients, cellular signalling through binding of growth factors and hormones, or clearance of metabolically aged proteins from the circulation.

Many reports describe the targeting of polyplexes by incorporation of various different cell targeting ligands that are covalently attached to a DNA-binding cationic polymer or protein. In some cases, recombinant chimeric proteins have been applied that present both a ligand and a non-polycationic DNA-binding domain (Fominaya and Wels, 1996; Uherek *et al.*, 1998). Numerous targeting ligands (Table 13.1), e.g., small chemical compounds such as vitamins, carbohydrates or synthetic peptide ligands and proteins such as growth factors or antibodies, have been evaluated. Polycations employed in gene transfer experiments include natural DNA-binding proteins such as histones; synthetic amino acid polymers such as polylysine, polyornithine, and histidinylated polylysine; other cationic polymers such as polyethylenimine, dendrimers, and poly (dimethylaminoethyl) methacrylates; or carbohydrate-based polymers such as chitosan. For review of the use of such polymers see (Brown *et al.*, 2001; De Smedt *et al.*, 2000; Han *et al.*, 2000).

Successful *in vitro* targeting in cell cultures has been described demonstrating up to 1000-fold enhanced gene expression in target cells in comparison to proper transfection controls like competition experiments with free ligands or

Table 13.1. Ligand-polycation Conjugates Used in Receptor-Targeted Polyplexes

Receptor	Ligand	Polycation used	Selected references
Airway cells	Surfactant proteins A and B	polylysine (pLL)	Baatz, 1994; Ross, 1995
Arterial wall	Artery wall binding peptide	PEG-pLL	Nah, 2002
ASGP receptor	Asialoglycoproteins	pLL	Cristiano, 1993; Wu, 1988, 1991
ASGP receptor	Synthetic galactosylated ligands	pLL, oligolysine, bisacridine, polyornithine, PEI	Haensler, 1993b; Merwin, 1994; Midoux, 1993; Morimoto, 2003; Nishikawa, 2000; Perales, 1994; Plank, 1992; Wadhwa, 1995
Carbohydrates	Lectins	pLL	Batra, 1994; Cotten, 1993; Yin, 1994
CD3	Anti-CD 3	pLL, PEI	Buschle, 1995; Kircheis, 1997
CD5	Anti-CD 5	pLL	Merwin, 1995
CD44	hyaluronic acid fragments	PPI	Uchegbu, 2004
CD117	Steel factor, Anti CD117	pLL	Schwarzenberger, 1996; Zauner, 1998
EGF-R	EGF, EGF peptide	pLL, PEI, pLL	Blessing, 2001; Chen, 1994; Cristiano, 1996; Fominaya, 1998; Liu, 2002, 2003; Wolschek, 2002
	Anti EGF-R, TGF-alpha	plus gal4 domain	
ErbB2	anti ErbB2	pLL plus gal4 domain	Fominaya, 1996; Uherek, 1998
FcR	IgG	pLL	Curiel, 1994; Rojanasakul, 1994
FGF2-R	basic FGF	pLL	Sosnowski, 1996
Folate receptor	Folate	pLL, PEI	Gottschalk, 1994; Guo, 1999; Mislick, 1995
Hepatocyte basolateral surface	Malarial circumsporozoite protein	pLL	Ding, 1995

Receptor/Target	Ligand	Carrier	References
Her2	Anti HER2	pLL, PEI	Chiu, 2004; Foster, 1997
Insulin receptor	Insulin	cation-modified albumin, pLL	Huckett, 1990; Rosenkranz, 1992
Integrin	RGD peptide	oligolysine, PEG-PEI	Harbottle, 1998; Hart, 1995; Kunath, 2003; Suh, 2002
LDL receptor family (hepatocytes)	Receptor associated protein (RAP)	pLL, poly(D)lysine	Kim, 2004
Mannose receptor (macrophages)	Synthetic ligands, mannosylated	pLL	Erbacher, 1996; Ferkol, 1996
Nerve growth factor (NGF) receptor TrkA	NGF serived synthetic peptide	oligolysine	Zeng, 2004
Neuroblastoma	Antibody ChCE7	pPEG-PEI	Coll, 1997
Ovarian carcinoma cell surface antigen OA3	Antibody OV-TL16 Fab' fragment		Merdan, 2003
PECAM (lung endothelium)	anti-PECAM antibody	PEI	Li, 2000
Poly-immunoglobulin receptor	Anti-secretory component	pLL	Ferkol, 1995
Serpin-enzyme receptor	peptide ligand	pLL	Ziady, 1999
Surface immunoglobulin	Anti-IgG, Anti-idiotype	pLL	Curiel, 1994; Schachtschabel, 1996
Thrombomodulin	Anti-thrombomodulin	pLL	Trubetskoy, 1992
Tn carbohydrate	Anti-Tn	pLL	Thurnher, 1994
Transferrin receptor	Transferrin	pLL, protamine, ethidium dimer, PEI, PEG-PEI,	Cotten, 1990; Kircheis, 1997; Kursa, 2003; Taxman, 1993; Wagner, 1990

ligand-free complexes. This apparent target specificity, however, can be misleading, because it does not regard interaction with other (non-target) cells. In fact, only few studies report successful *in vivo* targeting (Perales *et al.*, 1994; Wu *et al.*, 1991). In many cases the systemic administration of polyplexes through the tail vein in mice resulted in highest gene expression in the lung tissue; for example, see (Kircheis *et al.*, 1999). Non-specific interactions of polyplexes with blood components are supposed to be the major causes for the preferential lung transfection (see below).

B. Shielding from unspecific interactions

As outlined above, the presence of specific targeting ligands does not necessarily mean that the whole ligand-coated polyplex actually provides effective (tissue) targeting specificity. When using cationic polyplexes, systemic gene transfer at effective DNA doses is associated with acute toxicity. Plank *et al.* (1996) observed that positively charged polyplexes activate the alternative pathway of the complement system which is part of the innate immune system. Formation of erythrocyte agglomerates is another undesired side effect (Ogris *et al.*, 1999). Such non-specific interactions of polyplexes with blood components followed by trapping of polyplex aggregates in the lung capillaries are responsible for a preferential lung transfection and considerable toxicity (Chollet *et al.*, 2002; Verbaan *et al.*, 2001). One case was reported where severe acute toxicity systemic was observed after application of PEI polyplexes, with 50% of the mice dying with clinical signs of acute lung embolism (Kircheis *et al.*, 1999).

To make ligand-mediated targeting effective and more specific, polyplex domains with unspecific binding activity have to be masked to obtain the desired targeting specificity. For this purpose, hydrophilic polymers like polyethylene glycol (PEG) or hydroxypropyl methacrylate (pHPMA) have been attached to the polyplex surface. This shielding, however, markedly reduces gene transfer which can at least partly be restored by incorporating a targeting ligand. For the covalent attachment of ligand, and hydrophilic polymer to the polycation, several strategies have been developed, performing covalent coupling either before (Kursa *et al.*, 2003; Wolschek *et al.*, 2002) or after the polyplex formation (Fisher *et al.*, 2000; Ogris *et al.*, 1999). Another approach utilized the serum protein transferrin for both surface shielding and targeting (Kircheis *et al.*, 2001). Shielding with PEG not only improved circulation time but also reduced toxicity, increased solubility and provided stability for freeze-thawing (Ogris *et al.*, 2003). Applying such strategies, systemic targeting of tumors was demonstrated (see section on *in vivo* administration). Targeting and transfection efficiencies, however, were still poor. Apparently, the shielding process is a double-edged sword: the better particles are shielded and delivered to the target tissue, the lower is the efficiency of vector particles in the subsequent cellular uptake process and intracellular release.

III. CELLULAR UPTAKE AND INTRACELLULAR TRAFFICKING

Once the cell has been targeted the vector particle has to be internalized by the cellular uptake machinery via receptor-mediated endocytosis, macropinocytosis, phagocytosis or related processes. Several intracellular barriers then have to be overcome for successful transgene expression. The vector particle needs to survive in and escape from the endosome, traffic the cytoplasmic environment targeting the nucleus, enter the nucleus, and expose the carried nucleic acid to the cellular transcription machinery. Although these intracellular barriers present big hurdles for non-viral vectors, during the last years, parameters in vector formulations have been identified which strongly influence intracellular fate and efficiency of polyplexes.

Particle size, surface charge, and ligands all influence circulation time, cellular binding, uptake and intracellular trafficking of polyplexes (Goncalves et al., 2002; 2004; Kopatz et al., 2004; Ruponen et al., 2004). For example, large (several hundered nm) PEI polyplexes prepared in physiological salt solutions showed more than 100-fold higher transfection activity in comparison to 50 nm small complexes prepared in low salt glucose buffer (Ogris et al., 1998). Among several explanations (see below), limited transport and reduced binding of the small particles to the cell surfaces are possible reasons for this effect. Larger particles sediment rapidly onto the cultured cells, whereas small particles rather stay in solution which limits the contact with cells. Application of the small particles in more concentrated form and over extended periods of time improves transfection efficiency (Ogris et al., 1998). Rejman et al. (2004) report the size-dependent internalization of particles into non-phagocytic tumor cells via different pathways: irrespective of surface charge or ligands, small particles (<200 nm) internalize via clathrin-mediated endocytosis, whereas large particles (500 nm) internalize via caveolae-mediated endocytosis. In another study using PEI polyplexes, Kopatz et al. (2004) describe a model for non-viral entry of cationic polyplexes through adhesion to negatively charged transmembrane heparanproteoglycans, called syndecans, followed by clathrin-independent internalization of vesicles by the actin cytoskeleton. It remains unclear whether the latter pathway contributes to efficient gene delivery or rather represents a 'dead end'. As found by Goncalves et al. (2004) in studies with unmodified or histidinylated polylysine polyplexes and hepatoma cells, both clathrin-dependent pathways of polyplexes and clathrin-independent macropinocytosis are supposed to be the productive pathways mediating the delivery of genes. The macropinocytosis pathway, however, was found to impair transfection efficiency. Ruponen and colleagues (2004) describe that the uptake of DNA complexes depends on carrier, cell type and amounts of the polyanionic glycosaminoglycans (GAGs) heparan sulfate, chondroitin sulfate and hyaluronan on the cell-surface. However, all cell-surface GAGs inhibit the transgene expression and probably direct complexes into intracellular compartments that do not support gene transfer (Ruponen *et al.*, 2004).

After uptake of DNA complexes, escape from intracellular vesicles into the cytoplasm represents a major bottleneck. Entrapment in lysosomal or phagocytic vesicles is thought to be associated with degradation of the complexes in these compartments. The fate of the delivered DNA strongly depends on the selected polycationic carrier. For example, transferrin-polylysine DNA complexes were efficiently internalized into transferrin receptor-expressing cells, but persistent accumulation in intracellular vesicles prevented gene expression (Cotten *et al.*, 1990). The addition of the lysosomotropic agent chloroquine to the medium has been shown to increase transfection efficiency, probably by interfering with lysosomal degradation and enhancing the release of the DNA into the cytoplasm. The positive effect of chloroquine was especially strong in K562 cells, which can be explained by the unusually low pH in K562 early endosomes, resulting in preferential accumulation of the weak base in the endocytotic vesicles, swelling and subsequent destabilization of the endosomes (Cotten *et al.*, 1990).

Other polycationic carriers, such as polyethylenimine (Boussif *et al.*, 1995; Zou *et al.*, 2000) or dendrimers (Haensler and Szoka, 1993a; Tang and Szoka, 1997) can promote escape from intracellular vesicles to some degree. The high transfection efficiency of polyethylenimine (PEI) is thought to be based on a 'proton sponge' effect which contributes to the intrinsic ability of PEI to facilitate endosomal release. PEI can change its degree of protonation depending on the surrounding pH; in the absence of DNA only one out of six nitrogens is protonated at neutral pH. Complex-bound PEI is also only partially protonated at physiological pH. For complexation of DNA, usually a N/P (PEI nitrogen/ DNA phosphate) charge ratio of 5 or higher is applied, generating polyplexes with considerable buffering capacity at lower endosomal pH. Upon intracellular delivery of the DNA particle, the natural acidification within the endosome triggers protonation of complex-bound and free PEI, inducing chloride ion influx, osmotic swelling and destabilization of the vesicle (Sonawane *et al.*, 2003).

It has to be kept in mind that in addition to the cationic carrier, the actual formulation can have a large influence on gene transfer efficiency (Boussif *et al.*, 1996). While with large PEI/DNA polyplexes the endosomal escape is no major bottleneck (see above), inefficient endosomal release contributes to the lower efficacy of small PEI/DNA particles (Ogris *et al.*, 1998), and of purified PEI polyplexes devoid of free PEI (Boeckle *et al.*, 2004).

Additional approaches have been developed to enhance escape of DNA complexes into the cytoplasm. The inclusion of endosomolytic agents has been shown to dramatically enhance ligand-polylysine-based gene transfer. Addition of replication-defective adenovirus or rhinovirus in the transfection medium augments the levels of transferrin-mediated gene transfer more than thousand-fold in cell lines that express high levels of both virus and transferrin receptors (Curiel et al., 1991; Zauner et al., 1995). To broaden the applicability of these findings, endosomolytic agents were directly attached or incorporated into the DNA complexes (Cristiano et al., 1993; Fisher and Wilson, 1994, Wagner et al., 1992; Wu et al., 1994; Zauner et al., 1995). Apart from virus particles, also proteins such as adenovirus penton proteins (Fender et al., 1997), bacterial cytolysines (Gottschalk et al., 1995), or the transmembrane domain of diphtheria toxin (Fisher and Wilson, 1997), and synthetic peptides with virus-derived and other natural or artificial sequences (Gottschalk et al., 1996; Haensler and Szoka, 1993a; Mechtler and Wagner, 1997; Ogris et al., 2001; Plank et al., 1992) have been used.

Besides endosomal release agents, additional elements can be incorporated which modify the properties of the polyplex, such as elements promoting nuclear targeting. Because these elements mimic viral delivery functions, such complex systems have been termed "artificial viruses" (Plank et al., 1992; Remy et al., 1995; Wagner, 2004).

IV. *IN VIVO* ADMINISTRATION OF POLYPLEXES AND APPLICATIONS

Even if the polyplex has the desired specificity, targeting can only be successful if the DNA complex has the chance to reach the target cell and to bind to the cell surface receptor. This is no obvious step, as due to physical restrictions and biological barriers, including the reticuloendothelial system, vascular endothelium or the extracellular matrix, only a fraction of the applied material arrives at the target tissue. Moreover, successful targeting is a prerequisite, but not sufficient. Efficient intracellular delivery of the gene into the nucleus is required to obtain useful gene expression levels. Despite all of these obstacles, the first prototypes for receptor-targeted gene transfer have already shown encouraging evidence for in vivo (targeted) gene delivery and expression.

A. Gene transfer to the lung

Although gene transfer to the lung was already very efficient using 'simple' positively charged polyplexes with 22 kDa molecular weight linear PEI (L-PEI) (Goula et al., 2000; Zou et al., 2000), the application of such polyplexes remains restricted because of severe toxicity (Chollet et al., 2002) (see chapter II). Accordingly, polyplex targeting strategies appear as a preferable approach for systemic lung delivery. Davis and colleagues developed several strategies for polyplex-mediated gene transfer to the lung. For targeting to the polymeric immunoglobulin receptor, they generated small-sized anti-pIg Fab-polylysine polyplexes. Systemic delivery of anti-pIg polyplexes in rats resulted in reporter gene expression in cells of the airway epithelium and the submucosal glands (Ferkol *et al.*, 1995). In an alternative approach, using a synthetic peptide ligand (derived from alpha 1-antitrypsin) for the serpin-enzyme complex receptor, polyplexes were optimized regarding the ligand to polylysine ratio and molecular weight of polylysine (Ziady *et al.*, 1999). Long chain polylysines (54 kDa) generated smaller (25 nm) polyplexes than short chain polylysines (10 kDa; generating 40 nm polyplexes) and gave significantly higher and longer duration of expression *in vivo*.

Li *et al.* (2000) generated targeted PEI polyplexes by conjugation of PEI with anti-PECAM antibody; in the presence of the ligand, PEI to DNA charge ratios could be reduced without reducing gene transfer activity. This modification also significantly improved lung gene transfer in mice after i.v. administration and was associated with minor toxicity (as monitored by pro-inflammatory cytokine levels).

B. Gene transfer to the liver

The physical method of hydrodynamic delivery of naked DNA is a simple and very efficient method to obtain high gene expression levels in the liver of rodents (Liu *et al.*, 1999; Zhang *et al.*, 1999). The relatively harsh conditions and the administration of large volumes within a very short time, however, prevent the application for liver-directed gene transfer in larger species including humans. Systemic targeted *in vivo* gene transfer using polyplexes was reported for the first time by Wu and Wu (1988). For targeted delivery to the hepatocyte-specific asialoglycoprotein receptor, DNA/asialoorosomucoid-polylysine complexes were administered, which resulted in marker gene expression in rat livers. Expression was transient, but could be significantly prolonged by partial hepatectomy performed 30 min after injection of the polyplexes (Chowdhury *et al.*, 1993); this procedure induced hepatocyte proliferation and also increased persistence of plasmid DNA in hepatocytes of the treated animals. In another study, nagase analbuminemic rats were treated with a complexed cDNA expression construct for human albumin, followed by two-thirds partial hepatectomy (Wu *et al.*, 1991). Human albumin became detectable in the circulation after two days and concentration increased to a maximum by two weeks post-injection. Subsequent work applied the system for hepatocyte-specific gene transfer of the LDL receptor in a rabbit animal model of familial hypercholesterolemia, which resulted in a temporary amelioration of the disease phenotype (Wilson *et al.*, 1992).

As reported by Perales and colleagues (1994), prolonged (up to 140 days) gene expression of coagulation factor IX in the liver was obtained after application of DNA/galactose-polylysine complexes into the caudal vena cava of rats. The condensation of DNA into polyplexes of small particle size (of around 20 nm) was considered as an important factor.

Hashida and colleagues (Nishikawa et al., 2000) developed a synthetic multi-functional polyplex system consisting of polyornithine, which was modified first with galactose (to serve as ASGR ligand), then with a fusogenic peptide derived from the influenza virus HA2 (to serve as endosomal release domain), and complexed with a marker gene. Upon intravenous injection in mice, a large amount of transgene product was detected in the liver, and the hepatocytes contributed to more than 95% of total tissue gene expression. More recently, the same research group also evaluated galactosylated (Gal-PEI) or unmodified PEIs for gene transfer to mouse liver (Morimoto et al., 2003). In this work, they showed that different molecular weights (1.8, 10, or 70 kDa) of PEI greatly influence the polyplex activity. While the smallest polymer (PEI-1.8K) showed the highest but unspecific activity in cell culture, Gal-PEI-10K showed the highest receptor-specific activity in cell culture, and Gal-PEI-70K showed the highest liver expression levels upon portal vein injection of mice. Different polyplex sizes (>1000 nm for the small PEI, about 100 nm for the largest Gal-PEI) may contribute to the differences in transfection activity. In contrast to the previous studies, however, tail vein injection of this system did not result in liver expression.

Among other PEI-polyether conjugates, Nguyen et al. (2000) synthesized a conjugate of 2 kDa low molecular weight PEI with Pluronic 123 (P123, a polyoxyethylene-polyoxypropylene-polyoxyethylene block copolymer). In combination with free P123 and DNA, the conjugate formed 110 nm small and stable complexes that after i.v. injection into mice exhibited highest gene expression in liver, and also some expression in spleen, lung and heart. Targeting ligands have not yet been reported for this system.

In summary, targeted gene delivery to the liver is possible with polyplexes, but none of the systems reviewed in this section can compete in efficiency with the physical method of hydrodynamic delivery of naked DNA (Liu et al., 1999; Zhang et al., 1999).

C. Gene transfer to tumors

Non-targeted PEI polyplexes have been found to very effectively transfect tumor cell cultures. For this reason efforts have been made to apply these for in vivo cancer therapy in local applications. Direct intratumoral delivery of PEI polyplexes, however, resulted in expression levels which were far lower than in cell culture. Special routes of administration, such as local infusion of PEI polyplexes into the tumor mass by micropump (Coll et al., 1999) had to be applied to obtain satisfactory results. Gautam and colleagues (2002) have successfully delivered

PEI polyplexes as aerosol to established B16-F10 lung metastases. In their study, polyplex-mediated expression of p53 in combination with 9-nitrocamptothecin therapy resulted in significant inhibition of tumor growth.

Targeting tumors through the intravenous systemic route presents an interesting opportunity to attack tumor metastases. To reach this goal, interactions with blood components and healthy tissues have to be avoided, and targeting ligands have to be incorporated to improve the efficiency, as outlined in Section III.

Intravenous injection of targeted polyplexes shielded by transferrin or PEG resulted in gene transfer into distant subcutaneous tumors of mice (Kircheis et al., 2001; Kursa et al., 2003; Ogris et al., 1999). Marker gene expression levels in tumor tissues were 10- to 100-fold higher than in other organ tissues. The specificity of expression was also verified by *in vivo* imaging of luciferase reporter gene expression (Hildebrandt et al., 2003) and can be explained by the combination of several factors. First, the unique hyperpermeability of the tumor vasculature and inadequate lymphatic drainage results in enhanced permeability and retention (EPR) effect (Maeda, 2001). Therefore, shielded polyplexes with prolonged circulation half-life can accumulate in the tumor ('passive targeting'). Second, active targeting was achieved using the ligand transferrin, which binds to the transferrin receptors over-expressed on the cell surfaces of the target tumors. Third, replicating tumor cells are more accessible to transfection in comparison to non-dividing cells.

The therapeutic potential of such polyplexes was demonstrated by repeated systemic application of transferrin-coated polyplexes encoding tumor necrosis factor alpha (TNF-alpha) into tumor-bearing mice (Kircheis et al., 2002; Kursa et al., 2003). The treatment induced tumor necrosis and inhibition of tumor growth in four murine tumor models of different tissue origin. Since gene expression of TNF-alpha was localized within the tumor, no systemic TNF protein-related toxicities were observed.

Polypropylenimine (PPI) dendrimers (Zinselmeyer et al., 2002) have been applied for a related therapeutic strategy. Apparently, the low molecular weight of 1.7 kDa of PPI prevented undesired unspecific effects of the polymer such as blood aggregation and gene expression in the lung. Intravenous injection of TNF-alpha plasmid formulations delivered gene expression to solid tumors, and resulted, for example, in complete regression of A431 tumors (Uchegbu et al., 2004). Hyaluronic acid (a ligand for CD44) was conjugated to PPI dendrimers, and enhanced gene transfer to B16F10 tumors (which express CD44) was observed *in vitro*. CD44-targeted TNF-alpha DNA polyplexes showed the best effect in tumor growth retardation (Uchegbu et al., 2004). Passive tumor targeting was also observed with another dendrimer system; plasmid DNA complexed with dendritic poly(L-lysine) of the 6th generation (KG6) (Ohsaki et al., 2002; Okuda et al., 2004) circulated in the blood stream for

3 h after intravenous injection. At 60 min after administration, particles were observed in the tumor (Kawano et al., 2004). A combination of these dendrimer systems with targeting ligands has not yet been reported.

The epidermal growth factor receptor is over-expressed in several tumor types and therefore has been selected for tumor-targeting. For example, EGF-PEG-coated polyplexes were applied to human hepatoma-bearing SCID mice (Wolschek et al., 2002) and mediated enhanced gene expression in the EGF-receptor-rich tumor. Similar results were obtained with polylysine polyplexes linked with EGF-derived peptides and influenza-derived membrane-active peptides (Liu et al., 2002, 2003). Combined systemic administration of the p21 (WAF-1) and GM-CSF genes inhibited the growth of subcutaneous hepatoma cells and increased the survival rate of tumor-bearing mice (Liu et al., 2003).

V. OPTIMIZING POLYPLEXES TOWARD SYNTHETIC VIRUSES

Despite first proof-of-concept results demonstrating that polyplexes can mediate targeted in vivo gene delivery resulting in therapeutic effects in animal models, these studies also demonstrated clear limitations. The systemic targeting efficiencies are by far not perfect; only a few percent of the vector dose reach the target site; polyplex formulations often have significant toxic properties, a very heterogeneous composition, and low in vivo gene transfer activity.

For clinical use, defined and biocompatible systems will be necessary. Important aspects for optimization will include a defined assembly into monodisperse particle populations preferably of small size (Blessing et al., 1998). After polyplex formation, methods should be applied to purify polyplexes from toxic-free polymers and/or polyplex aggregates, such as demonstrated in (Boeckle et al., 2004). For improved storage stability, the generation of stabilized polyplexes is a major objective (Ogris et al., 2003). In addition, the use of biodegradable polymers would greatly improve the safety profile of polyplexes, since such polyplexes are less toxic and biodegradation allows excretion of polyplex compounds thereby avoiding long-term deposition of polymers within the organism; see references (Lim et al., 2002; Petersen et al., 2002; Wang et al., 2004; Zhao et al., 2003).

Current polyplexes are still very inefficient as compared to viral vectors. From this perspective, viruses might present ideal natural examples educating how to further optimize polyplexes into 'synthetic viruses' (Wagner 2004; Zuber et al., 2001). One unique property of viruses is their dynamic manner in responding to the biological micro-environment. Similar to viruses, polyplexes should alter their structure during the gene delivery process to make them most effective for the different subsequent gene delivery step. For example, the

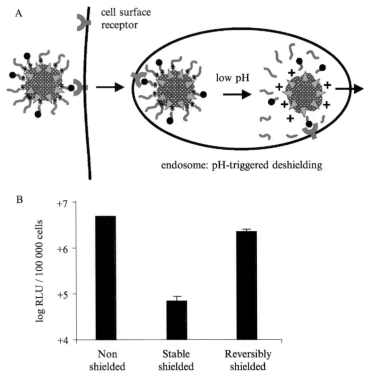

Figure 13.2. Bioreversible shielding of targeted polyplexes: toward artificial viruses. (A) The DNA particles have to be stabilized and to be shielded to avoid unspecific interactions in the blood circulation. Integrating PEG into the polyplex coat is one approach to mediating these characteristics. After binding to the cell by specific ligand-target receptor interaction, polyplexes are internalized into endosomal vesicles. The acidic environment triggers cleavage of bioreversibly linked PEG from the polyplex which enhances release from the vesicle to the cytosol. Incorporation of a membrane–active peptide (e.g., melittin analogs) further enhances gene transfer. (B) Transfection of HUH7 hepatoma cells using EGF receptor-targeted polyplexes containing 10% EGF-conjugated PEI plus 80% unmodified linear PEI 22 kDa, plus either 10% pLL (unshielded control polyplexes), or 10% PEG20kDa–pLL (stable shield) or 10% PEG20kDa-HZN-pLL (reversible shield). Note the pronounced increase in transfection efficiency by introducing a bioreversible PEG shield as compared to the stable PEG shield.

current PEG-shielded polyplexes mediate enhanced systemic delivery to the target cells, but reduce gene expression activity within the target cells; apparently the stable, irreversible PEG shield hampers intracellular uptake processes. In an optimized virus-like polyplex, PEG shielding should be presented in a bioresponsive fashion. Ideally, after entering the target cell and delivery into endosomal vesicles, the polyplex should strip off the PEG shield, and re-expose its cationic surface for efficient destabilization of the endosomal membrane. Such a dynamic change in character can be introduced into polyplexes by bioresponsive domains. For example, Walker *et al.* (2005) made use of the acidic milieu of the endosomes and introduced bioresponsive PEG-polycation conjugates with pH-labile linkages (Fig. 13.2). DNA particles shielded with these bioreversible PEG conjugates are supposed to lose their PEG shield at endosomal pH. They displayed up to 100-fold higher gene transfer activity compared to polyplexes with the analogous stable PEG shield. This example demonstrates that incorporation of delivery functions that are activated and presented in a bioresponsive fashion can strongly improve polyplex efficiency.

VI. CONCLUSIONS

As a result of twenty years of research, receptor-targeting has become an integral part in the development of non-viral gene transfer systems. Numerous examples of receptor-ligand pairs have already been tested *in vitro*. Several polyplex systems have been established, which display encouraging *in vivo* activity. In particular, specific targeting of organs such as the liver and the lung, and also tumors, was demonstrated, and therapeutic effects have been obtained in several animal models. At the current stage, transfection efficiency of these vectors is still very poor as compared to viral vectors or some physical methods of DNA delivery. The development of polyplexes into virus-like supramolecular complexes, which undergo programmed structural changes compatible with both extracellular and intracellular gene delivery steps, is a promising avenue to generate targeted and more efficient vector formulations applicable for *in vivo* gene therapies.

VII. MAIN ABBREVIATIONS

ASGP	asialoglycoprotein
EGF	epidermal growth factor
i.v.	intravenous
pHPMA	poly-N-2-hydroxypropyl-methacrylamide
PEG	polyethylene glycol
PEI	polyethylenimine
pLL	poly(L)lysine
PPI	polypropylenimine
TNF	tumor necrosis factor

Acknowledgments

The contributions to our research by Julia Fahrmeir, Carolin Fella, Lars Gädtke, Katharina von Gersdorff, Michael Günther, Melinda Kiss, Julia Kloeckner, Stefan Landshamer, Manfred Ogris, Jaroslav Pelisek, Silke van der Piepen, Wolfgang Roedl, Martina Rüffer, and Greg Walker are greatly appreciated. It is an exceptional pleasure to work within such an enthusiastic and very pleasant team. Many thanks also to Olga Brück for skillful assistance in preparing the manuscript.

References

Baatz, J. E., Bruno, M. D., Ciraolo, P. J., Glasser, S. W., Stripp, B. R., Smyth, K. L., and Korfhagen, T. R. (1994). Utilization of modified surfactant-associated protein B for delivery of DNA to airway cells in culture. *Proc. Natl. Acad. Sci. USA* **91,** 2547–2551.

Batra, R. K., Berschneider, H., and Curiel, D. T. (1994). Molecular conjugate vectors mediate efficient gene transfer into gastrointestinal epithelial cells. *Cancer Gene Ther.* **1,** 185–192.

Blessing, T., Kursa, M., Holzhauser, R., Kircheis, R., and Wagner, E. (2001). Different strategies for formation of pegylated EGF-conjugated PEI/DNA complexes for targeted gene delivery. *Bioconjug. Chem.* **12,** 529–537.

Blessing, T., Remy, J. S., and Behr, J. P. (1998). Monomolecular collapse of plasmid DNA into stable virus-like particles. *Proc. Natl. Acad. Sci. USA* **95,** 1427–1431.

Boeckle, S., von Gersdorff, K., van der Piepen, S., Culmsee, C., Wagner, E., and Ogris, M. (2004). Purification of polyethylenimine polyplexes highlights the role of free polycation in gene transfer. *J. Gene. Med.* **6,** May 5, Epub ahead of print.

Boussif, O., Lezoualc'h, F., Zanta, M. A., Mergny, M. D., Scherman, D., Demeneix, B., and Behr, J. P. (1995). A versatile vector for gene and oligonucleotide transfer into cells in culture and *in vivo*: Polyethylenimine. *Proc. Natl. Acad. Sci. USA* **92,** 7297–7301.

Boussif, O., Zanta, M. A., and Behr, J. P. (1996). Optimized galenics improve *in vitro* gene transfer with cationic molecules up to 1000–fold. *Gene Ther.* **3,** 1074–1080.

Brown, M. D., Schatzlein, A. G., and Uchegbu, I. F. (2001). Gene delivery with synthetic (non-viral) carriers. *Int. J. Pharm.* **229,** 1–21.

Buschle, M., Cotten, M., Kirlappos, H., Mechtler, K., Schaffner, G., Zauner, W., Birnstiel, M. L., and Wagner, E. (1995). Receptor-mediated gene transfer into human T lymphocytes via binding of DNA/CD3 antibody particles to the CD3 T cell receptor complex. *Hum. Gene Ther.* **6,** 753–761.

Chen, J., Gamou, S., Takayanagi, A., and Shimizu, N. (1994). A novel gene delivery system using EGF receptor-mediated endocytosis. *FEBS Lett.* **338,** 167–169.

Cheng, S., Merlino, G. T., and Pastan, I. H. (1983). A versatile method for the coupling of protein to DNA: synthesis of alpha 2-macroglobulin-DNA conjugates. *Nucleic Acids Res.* **11,** 659–669.

Chiu, S. J., Ueno, N. T., and Lee, R. J. (2004). Tumor-targeted gene delivery via anti-HER2 antibody (trastuzumab, Herceptin) conjugated polyethylenimine. *J. Control Release* **97,** 357–369.

Chollet, P., Favrot, M. C., Hurbin, A., and Coll, J. L. (2002). Side-effects of a systemic injection of linear polyethylenimine-DNA complexes. *J. Gene Med.* **4,** 84–91.

Chowdhury, N. R., Wu, C. H., Wu, G. Y., Yerneni, P. C., Bommineni, V. R., and Chowdhury, J. R. (1993). Fate of DNA targeted to the liver by asialoglycoprotein receptor-mediated endocytosis *in vivo*. Prolonged persistence in cytoplasmic vesicles after partial hepatectomy. *J. Biol. Chem.* **268,** 11265–11271.

Coll, J. L., Chollet, P., Brambilla, E., Desplanques, D., Behr, J. P., and Favrot, M. (1999). *In vivo* delivery to tumors of DNA complexed with linear polyethylenimine. *Hum. Gene Ther.* **10,** 1659–1666.

Coll, J. L., Wagner, E., Combaret, V., Metchler, K., Amstutz, H., Iacono-Di-Cacito, I., Simon, N., and Favrot, M. C. (1997). *In vitro* targeting and specific transfection of human neuroblastoma cells by chCE7 antibody-mediated gene transfer. *Gene Ther.* **4,** 156–161.

Cotten, M., Langle, R., Kirlappos, H., Wagner, E., Mechtler, K., Zenke, M., Beug, H., and Birnstiel, M. L. (1990). Transferrin-polycation-mediated introduction of DNA into human leukemic cells: Stimulation by agents that affect the survival of transfected DNA or modulate transferrin receptor levels. *Proc. Natl. Acad. Sci. USA* **87,** 4033–4037.

Cotten, M., Wagner, E., Zatloukal, K., and Birnstiel, M. L. (1993). Chicken adenovirus (CELO virus) particles augment receptor-mediated DNA delivery to mammalian cells and yield exceptional levels of stable transformants. *J. Virol.* **67,** 3777–3785.

Cristiano, R. J., and Roth, J. A. (1996). Epidermal growth factor mediated DNA delivery into lung cancer cells via the epidermal growth factor receptor. *Cancer Gene Ther.* **3,** 4–10.

Cristiano, R. J., Smith, L. C., Kay, M. A., Brinkley, B. R., and Woo, S. L. (1993). Hepatic gene therapy: Efficient gene delivery and expression in primary hepatocytes utilizing a conjugated adenovirus-DNA complex. *Proc. Natl. Acad. Sci. USA* **90,** 11548–11552.

Curiel, D. T., Agarwal, S., Wagner, E., and Cotten, M. (1991). Adenovirus enhancement of transferrin-polylysine-mediated gene delivery. *Proc. Natl. Acad. Sci. USA* **88,** 8850–8854.

Curiel, T. J., Cook, D. R., Bogedain, C., Jilg, W., Harrison, G. S., Cotten, M., Curiel, D. T., and Wagner, E. (1994). Efficient foreign gene expression in Epstein-Barr virus-transformed human B-cells. *Virology* **198,** 577–585.

De Smedt, S. C., Demeester, J., and Hennink, W. E. (2000). Cationic polymer based gene delivery systems. *Pharm. Res.* **17,** 113–126.

Ding, Z. M., Cristiano, R. J., Roth, J. A., Takacs, B., and Kuo, M. T. (1995). Malarial circumsporozoite protein is a novel gene delivery vehicle to primary hepatocyte cultures and cultured cells. *J. Biol. Chem.* **270,** 3667–3676.

Erbacher, P., Bousser, M. T., Raimond, J., Monsigny, M., Midoux, P., and Roche, A. C. (1996). Gene transfer by DNA/glycosylated polylysine complexes into human blood monocyte-derived macrophages. *Hum. Gene Ther.* **7,** 721–729.

Fender, P., Ruigrok, R. W., Gout, E., Buffet, S., and Chroboczek, J. (1997). Adenovirus dodecahedron, a new vector for human gene transfer. *Nat. Biotechnol.* **15,** 52–56.

Ferkol, T., Perales, J. C., Eckman, E., Kaetzel, C. S., Hanson, R. W., and Davis, P. B. (1995). Gene transfer into the airway epithelium of animals by targeting the polymeric immunoglobulin receptor. *J. Clin. Invest.* **95,** 493–502.

Ferkol, T., Perales, J. C., Mularo, F., and Hanson, R. W. (1996). Receptor-mediated gene transfer into macrophages. *Proc. Natl. Acad. Sci. USA* **93,** 101–105.

Fisher, K. D., Ulbrich, K., Subr, V., Ward, C. M., Mautner, V., Blakey, D., and Seymour, L. W. (2000). A versatile system for receptor-mediated gene delivery permits increased entry of DNA into target cells, enhanced delivery to the nucleus and elevated rates of transgene expression. *Gene Ther.* **7,** 1337–1343.

Fisher, K. J., and Wilson, J. M. (1994). Biochemical and functional analysis of an adenovirus-based ligand complex for gene transfer. *Biochem. J.* **299,** 49–58.

Fisher, K. J., and Wilson, J. M. (1997). The transmembrane domain of diphtheria toxin improves molecular conjugate gene transfer. *Biochem. J.* **321,** 49–58.

Fominaya, J., Uherek, C., and Wels, W. (1998). A chimeric fusion protein containing transforming growth factor-alpha mediates gene transfer via binding to the EGF receptor. *Gene Ther.* **5,** 521–530.

Fominaya, J., and Wels, W. (1996). Target cell-specific DNA transfer mediated by a chimeric multidomain protein. Novel non-viral gene delivery system. *J. Biol. Chem.* **271,** 10560–10568.

Foster, B. J., and Kern, J. A. (1997). HER2–targeted gene transfer. *Hum. Gene Ther.* **8,** 719–727.

Gautam, A., Waldrep, J. C., Densmore, C. L., Koshkina, N., Melton, S., Roberts, L., Gilbert, B., and Knight, V. (2002). Growth inhibition of established B16–F10 lung metastases by sequential aerosol delivery of p53 gene and 9-nitrocamptothecin. *Gene Ther.* **9,** 353–357.

Goncalves, C., Mennesson, E., Fuchs, R., Gorvel, J. P., Midoux, P., and Pichon, C. (2004). Macropinocytosis of polyplexes and recycling of plasmid via the clathrin-dependent pathway impair the transfection efficiency of human hepatocarcinoma cells. *Mol. Ther.* **10,** 373–385.

Goncalves, C., Pichon, C., Guerin, B., and Midoux, P. (2002). Intracellular processing and stability of DNA complexed with histidylated polylysine conjugates. *J. Gene. Med.* **4,** 271–281.

Gottschalk, S., Cristiano, R. J., Smith, L. C., and Woo, S. L. (1994). Folate receptor mediated DNA delivery into tumor cells: Potosomal. *Gene Ther.* **1,** 185–191.

Gottschalk, S., Sparrow, J. T., Hauer, J., Mims, M. P., Leland, F. E., Woo, S. L., and Smith, L. C. (1996). A novel DNA-peptide complex for efficient gene transfer and expression in mammalian cells. *Gene Ther.* **3,** 48–57.

Gottschalk, S., Tweten, R. K., Smith, L. C., and Woo, S. L. (1995). Efficient gene delivery and expression in mammalian cells using DNA coupled with perfringolysin O. *Gene Ther.* **2,** 498–503.

Goula, D., Becker, N., Lemkine, G. F., Normandie, P., Rodrigues, J., Mantero, S., Levi, G., and Demeneix, B. A. (2000). Rapid crossing of the pulmonary endothelial barrier by polyethyleni-mine/DNA complexes. *Gene Ther.* **7,** 499–504.

Guo, W., and Lee, R. J. (1999). Receptor-Targeted Gene Delivery Via Folate-Conjugated Poly-ethylenimine. *AAPS Pharmsci.* **1,** Article 19.

Haensler, J., and Szoka, F. C., Jr. (1993a). Polyamidoamine cascade polymers mediate efficient transfection of cells in culture. *Bioconjug. Chem.* **4,** 372–379.

Haensler, J., and Szoka, F. C., Jr. (1993b). Synthesis and characterization of a trigalactosylated bisacridine compound to target DNA to hepatocytes. *Bioconjug. Chem.* **4,** 85–93.

Han, S., Mahato, R. I., Sung, Y. K., and Kim, S. W. (2000). Development of biomaterials for gene therapy. *Mol. Ther.* **2,** 302–317.

Harbottle, R. P., Cooper, R. G., Hart, S. L., Ladhoff, A., McKay, T., Knight, A. M., Wagner, E., Miller, A. D., and Coutelle, C. (1998). An RGD-oligolysine peptide: A prototype construct for integrin-mediated gene delivery. *Hum. Gene Ther.* **9,** 1037–1047.

Hart, S. L., Harbottle, R. P., Cooper, R., Miller, A., Williamson, R., and Coutelle, C. (1995). Gene delivery and expression mediated by an integrin-binding peptide. *Gene Ther.* **2,** 552–554.

Hildebrandt, I. J., Iyer, M., Wagner, E., and Gambhir, S. S. (2003). Optical imaging of transferrin targeted PEI/DNA complexes in living subjects. *Gene Ther.* **10,** 758–764.

Huckett, B., Ariatti, M., and Hawtrey, A. O. (1990). Evidence for targeted gene transfer by receptor-mediated endocytosis. Stable expression following insulin-directed entry of NEO into HepG2 cells. *Biochem. Pharmacol.* **40,** 253–263.

Kawano, T., Okuda, T., Aoyagi, H., and Niidome, T. (2004). Long circulation of intravenously administered plasmid DNA delivered with dendritic poly(L-lysine) in the blood flow. *J. Control Release* **99,** 329–337.

Kim, T. G., Kang, S. Y., Kang, J. H., Cho, M. Y., Kim, J. I., Kim, S. H., and Kim, J. S. (2004). Gene transfer into human hepatoma cells by receptor-associated protein/polylysine conjugates. *Bioconjug. Chem.* **15,** 326–332.

Kircheis, R., Kichler, A., Wallner, G., Kursa, M., Ogris, M., Felzmann, T., Buchberger, M., and Wagner, E. (1997). Coupling of cell-binding ligands to polyethylenimine for targeted gene transfer. *Gene Ther.* **4,** 409–418.

Kircheis, R., Ostermann, E., Wolschek, M. F., Lichtenberger, C., Magin-Lachmann, C., Wightman, L., Kursa, M., and Wagner, E. (2002). Tumor-targeted gene delivery of tumor necrosis factor-alpha induces tumor necrosis and tumor regression without systemic toxicity. *Cancer Gene Ther.* **9,** 673–680.

Kircheis, R., Schuller, S., Brunner, S., Heider, K., Zauner, W., and Wagner, E. (1999). Polycation-Based DNA Complexes for Tumor-Targeted Gene Delivery in vivo. *J. Gene Med.* **1**, 111–120.

Kircheis, R., Wightman, L., Schreiber, A., Robitza, B., Rossler, V., Kursa, M., and Wagner, E. (2001). Polyethylenimine/DNA complexes shielded by transferrin target gene expression to tumors after systemic application. *Gene Ther.* **8**, 28–40.

Kopatz, I., Remy, J. S., and Behr, J. P. (2004). A model for non-viral gene delivery: Through syndecan adhesion molecules and powered by actin. *J. Gene Med.* **6**, 769–776.

Kunath, K., Merdan, T., Hegener, O., Haberlein, H., and Kissel, T. (2003). Integrin targeting using RGD-PEI conjugates for in vitro gene transfer. *J. Gene Med.* **5**, 588–599.

Kursa, M., Walker, G. F., Roessler, V., Ogris, M., Roedl, W., Kircheis, R., and Wagner, E. (2003). Novel Shielded Transferrin-Polyethylene Glycol-Polyethylenimine/DNA Complexes for Systemic Tumor-Targeted Gene Transfer. *Bioconjug. Chem.* **14**, 222–231.

Li, S., Tan, Y., Viroonchatapan, E., Pitt, B. R., and Huang, L. (2000). Targeted gene delivery to pulmonary endothelium by anti-PECAM antibody. *Am. J. Physiol. Lung Cell Mol. Physiol.* **278**, 504–511.

Lim, Y. B., Kim, S. M., Suh, H., and Park, J. S. (2002). Biodegradable, endosome disruptive, and cationic network-type polymer as a highly efficient and nontoxic gene delivery carrier. *Bioconjug. Chem.* **13**, 952–957.

Liu, F., Song, Y., and Liu, D. (1999). Hydrodynamics-based transfection in animals by systemic administration of plasmid DNA. *Gene Ther.* **6**, 1258–1266.

Liu, X., Tian, P., Yu, Y., Yao, M., Cao, X., and Gu, J. (2002). Enhanced antitumor effect of EGF R-targeted p21WAF-1 and GM-CSF gene transfer in the established murine hepatoma by peritumoral injection. *Cancer Gene Ther.* **9**, 100–108.

Liu, X., Tian, P. K., Ju, D. W., Zhang, M. H., Yao, M., Cao, X. T., and Gu, J. R. (2003). Systemic genetic transfer of p21WAF-1 and GM-CSF utilizing of a novel oligopeptide-based EGF receptor targeting polyplex. *Cancer Gene Ther.* **10**, 529–539.

Maeda, H. (2001). The enhanced permeability and retention (EPR) effect in tumor vasculature: The key role of tumor-selective macromolecular drug targeting. *Adv. Enzyme. Regul.* **41**, 189–207.

Mechtler, K., and Wagner, E. (1997). Gene transfer mediated by influenza virus peptides: The role of peptide sequence. *New J. Chem.* **21**, 105–111.

Merdan, T., Callahan, J., Petersen, H., Kunath, K., Bakowsky, U., Kopeckova, P., Kissel, T., and Kopecek, J. (2003). Pegylated polyethylenimine-fab' antibody fragment conjugates for targeted gene delivery to human ovarian carcinoma cells. *Bioconjug. Chem.* **14**, 989–996.

Merwin, J. R., Carmichael, E. P., Noell, G. S., DeRome, M. E., Thomas, W. L., Robert, N., Spitalny, G., and Chiou, H. C. (1995). CD5–mediated specific delivery of DNA to T lymphocytes: Compartmentalization augmented by adenovirus. *J. Immunol. Methods* **186**, 257–266.

Merwin, J. R., Noell, G. S., Thomas, W. L., Chiou, H. C., DeRome, M. E., McKee, T. D., Spitalny, G. L., and Findeis, M. A. (1994). Targeted delivery of DNA using YEE(GalNAcAH)3, a synthetic glycopeptide ligand for the asialoglycoprotein receptor. *Bioconjug. Chem.* **5**, 612–620.

Midoux, P., Mendes, C., Legrand, A., Raimond, J., Mayer, R., Monsigny, M., and Roche, A. C. (1993). Specific gene transfer mediated by lactosylated poly-L-lysine into hepatoma cells. *Nucleic Acids Res.* **21**, 871–878.

Mislick, K. A., Baldeschwieler, J. D., Kayyem, J. F., and Meade, T. J. (1995). Transfection of folate-polylysine DNA complexes: Evidence for lysosomal delivery. *Bioconjug. Chem.* **6**, 512–515.

Morimoto, K., Nishikawa, M., Kawakami, S., Nakano, T., Hattori, Y., Fumoto, S., Yamashita, F., and Hashida, M. (2003). Molecular weight-dependent gene transfection activity of unmodified and galactosylated polyethyleneimine on hepatoma cells and mouse liver. *Mol. Ther.* **7**, 254–261.

Nah, J. W., Yu, L., Han, S. O., Ahn, C. H., and Kim, S. W. (2002). Artery wall binding peptide-poly (ethylene glycol)-grafted-poly(L-lysine)-based gene delivery to artery wall cells. *J. Control Release* **78**, 273–284.

Nguyen, H. K., Lemieux, P., Vinogradov, S. V., Gebhart, C. L., Guérin, N., Paradis, G., Bronich, T. K., Alakhov, V. Y., and Kabanov, A. V. (2000). Evaluation of polyether-polyethyleneimine graft copolymers as gene transfer agents. *Gene Ther.* **7**, 126–138.

Nishikawa, M., Yamauchi, M., Morimoto, K., Ishida, E., Takakura, Y., and Hashida, M. (2000). Hepatocyte-targeted *in vivo* gene expression by intravenous injection of plasmid DNA complexed with synthetic multi-functional gene delivery system. *Gene Ther.* **7**, 548–555.

Ogris, M., Brunner, S., Schuller, S., Kircheis, R., and Wagner, E. (1999). PEGylated DNA/ transferrin-PEI complexes: Reduced interaction with blood components, extended circulation in blood and potential for systemic gene delivery. *Gene Ther.* **6**, 595–605.

Ogris, M., Carlisle, R. C., Bettinger, T., and Seymour, L. W. (2001). Melittin enables efficient vesicular escape and enhanced nuclear access of non-viral gene delivery vectors. *J. Biol. Chem.* **276**, 47550–47555.

Ogris, M., Steinlein, P., Kursa, M., Mechtler, K., Kircheis, R., and Wagner, E. (1998). The size of DNA/transferrin-PEI complexes is an important factor for gene expression in cultured cells. *Gene Ther.* **5**, 1425–1433.

Ogris, M., Walker, G., Blessing, T., Kircheis, R., Wolschek, M., and Wagner, E. (2003). Tumor-targeted gene therapy: Strategies for the preparation of ligand-polyethylene glycol-polyethylenimine/DNA complexes. *J. Control Release* **91**, 173–181.

Ohsaki, M., Okuda, T., Wada, A., Hirayama, T., Niidome, T., and Aoyagi, H. (2002). *In Vitro* Gene Transfection Using Dendritic Poly(L-lysine). *Bioconjug. Chem.* **13**, 510–517.

Okuda, T., Sugiyama, A., Niidome, T., and Aoyagi, H. (2004). Characters of dendritic poly (L-lysine) analogues with the terminal lysines replaced with arginines and histidines as gene carriers *in vitro. Biomaterials* **25**, 537–544.

Perales, J. C., Ferkol, T., Beegen, H., Ratnoff, O. D., and Hanson, R. W. (1994). Gene transfer *in vivo*: Sustained expression and regulation of genes introduced into the liver by receptor-targeted uptake. *Proc. Natl. Acad. Sci. USA* **91**, 4086–4090.

Petersen, H., Merdan, T., Kunath, K., Fischer, D., and Kissel, T. (2002). Poly(ethylenimine-co-L-lactamide-co-succinamide): A biodegradable polyethylenimine derivative with an advantageous pH-dependent hydrolytic degradation for gene delivery. *Bioconjug. Chem.* **13**, 812–821.

Plank, C., Mechtler, K., Szoka, F. J., and Wagner, E. (1996). Activation of the complement system by synthetic DNA complexes: A potential barrier for intravenous gene delivery. *Hum. Gene Ther.* **7**, 1437–1446.

Plank, C., Zatloukal, K., Cotten, M., Mechtler, K., and Wagner, E. (1992). Gene transfer into hepatocytes using asialoglycoprotein receptor mediated endocytosis of DNA complexed with an artificial tetra-antennary galactose ligand. *Bioconjug. Chem.* **3**, 533–539.

Rejman, J., Oberle, V., Zuhorn, I. S., and Hoekstra, D. (2004). Size-dependent internalization of particles via the pathways of clathrin- and caveolae-mediated endocytosis. *Biochem. J.* **377**, 159–169.

Remy, J. S., Kichler, A., Mordvinov, V., Schuber, F., and Behr, J. P. (1995). Targeted gene transfer into hepatoma cells with lipopolyamine-condensed DNA particles presenting galactose ligands: A stage toward artificial viruses. *Proc. Natl. Acad. Sci. USA* **92**, 1744–1748.

Rojanasakul, Y., Wang, L. Y., Malanga, C. J., Ma, J. K., and Liaw, J. (1994). Targeted gene delivery to alveolar macrophages via Fc receptor-mediated endocytosis. *Pharm. Res.* **11**, 1731–1736.

Rosenkranz, A. A., Yachmenev, S. V., Jans, D. A., Serebryakova, N. V., Murav'ev, V. I., Peters, R., and Sobolev, A. S. (1992). Receptor-mediated endocytosis and nuclear transport of a transfecting DNA construct. *Exp. Cell. Res.* **199**, 323–329.

Ross, G. F., Morris, R. E., Ciraolo, G., Huelsman, K., Bruno, M., Whitsett, J. A., Baatz, J. E., and Korfhagen, T. R. (1995). Surfactant protein A-polylysine conjugates for delivery of DNA to airway cells in culture. *Hum. Gene Ther.* **6**, 31–40.

Ruponen, M., Honkakoski, P., Tammi, M., and Urtti, A. (2004). Cell-surface glycosaminoglycans inhibit cation-mediated gene transfer. *J. Gene Med.* **6**, 405–414.

Schachtschabel, U., Pavlinkova, G., Lou, D., and Köhler, H. (1996). Antibody-mediated gene delivery for B-cell lymphoma *in vitro*. *Cancer Gene Ther.* **3**, 365–372.

Schätzlein, A. G. (2003). Targeting of Synthetic Gene Delivery Systems. *J. Biomed. Biotechnol.* **2003**, 149–158.

Schwarzenberger, P., Spence, S. E., Gooya, J. M., Michiel, D., Curiel, D. T., Ruscetti, F. W., and Keller, J. R. (1996). Targeted gene transfer to human hematopoietic progenitor cell lines through the c-kit receptor. *Blood* **87**, 472–478.

Sonawane, N. D., Szoka, F. C., Jr., and Verkman, A. S. (2003). Chloride Accumulation and Swelling in Endosomes Enhances DNA Transfer by Polyamine-DNA Polyplexes. *J. Biol. Chem.* **278**, 44826–44831.

Sosnowski, B. A., Gonzalez, A. M., Chandler, L. A., Buechler, Y. J., Pierce, G. F., and Baird, A. (1996). Targeting DNA to cells with basic fibroblast growth factor (FGF2). *J. Biol. Chem.* **271**, 33647–33653.

Suh, W., Han, S. O., Yu, L., and Kim, S. W. (2002). An angiogenic, endothelial-cell-targeted polymeric gene carrier. *Mol. Ther.* **6**, 664–672.

Tang, M. X., and Szoka, F. C. (1997). The influence of polymer structure on the interactions of cationic polymers with DNA and morphology of the resulting complexes. *Gene Ther.* **4**, 823–832.

Taxman, D. J., Lee, E. S., and Wojchowski, D. M. (1993). Receptor-targeted transfection using stable maleimido-transferrin/thio-poly-L-lysine conjugates. *Anal. Biochem.* **213**, 97–103.

Thurnher, M., Wagner, E., Clausen, H., Mechtler, K., Rusconi, S., Dinter, A., Birnstiel, M. L., Berger, E. G., and Cotten, M. (1994). Carbohydrate receptor-mediated gene transfer to human T leukaemic cells. *Glycobiology* **4**, 429–435.

Trubetskoy, V. S., Torchilin, V. P., Kennel, S. J., and Huang, L. (1992). Use of N-terminal modified poly(L-lysine)-antibody conjugate as a carrier for targeted gene delivery in mouse lung endothelial cells. *Bioconjug. Chem.* **3**, 323–327.

Uchegbu, I. F., Dufes, C., Elouzi, A., and Schatzlein, A. G. (2004). Gene therapy with polypropylenimine dendrimers. *Proceed 12th Inter. Pharm. Technol. Symp. 2004* 11–12.

Uherek, C., Fominaya, J., and Wels, W. (1998). A modular DNA carrier protein based on the structure of diphtheria toxin mediates target cell-specific gene delivery. *J. Biol. Chem.* **273**, 8835–8841.

Verbaan, F. J., Oussoren, C., van Dam, I. M., Takakura, Y., Hashida, M., Crommelin, D. J., Hennink, W. E., and Storm, G. (2001). The fate of poly(2-dimethyl amino ethyl)methacrylate-based polyplexes after intravenous administration. *Int. J. Pharm.* **214**, 99–101.

Wadhwa, M. S., Knoell, D. L., Young, A. P., and Rice, K. G. (1995). Targeted gene delivery with a low molecular weight glycopeptide carrier. *Bioconjug. Chem.* **6**, 283–291.

Wagner, E. (2004). Strategies to improve DNA polyplexes for *in vivo* gene transfer – will "artificial viruses" be the answer? *Pharm. Res.* **21**, 8–14.

Wagner, E., Kircheis, R., and Walker, G. F. (2004). Targeted nucleic acid delivery into tumors: New avenues for cancer therapy. *Biomed. Pharmacother.* **58**, 152–161.

Wagner, E., Zatloukal, K., Cotten, M., Kirlappos, H., Mechtler, K., Curiel, D. T., and Birnstiel, M. L. (1992). Coupling of adenovirus to transferrin-polylysine/DNA complexes greatly enhances receptor-mediated gene delivery and expression of transfected genes. *Proc. Natl. Acad. Sci. USA* **89**, 6099–6103.

Wagner, E., Zenke, M., Cotten, M., Beug, H., and Birnstiel, M. L. (1990). Transferrin-polycation conjugates as carriers for DNA uptake into cells. *Proc. Natl. Acad. Sci. USA* **87**, 3410–3414.

Walker, G. F., Fella, C., Pelisek, J., Fahrmeir, J., Boeckle, S., Ogris, M., and Wagner, E. (2005). Toward synthetic viruses: Endosomal pH triggered deshielding of targeted polyplexes greatly enhances gene transfer *in vitro* and *in vivo*. *Mol. Ther.* **11**, 418–425.

Wang, C., Ge, Q., Ting, D., Nguyen, D., Shen, H. R., Chen, J., Eisen, H. N., Heller, J., Langer, R., and Putnam, D. (2004). Molecularly engineered poly(ortho ester) microspheres for enhanced delivery of DNA vaccines. *Nat. Mater.* **3**, 190–196.

Wickham, T. J. (2003). Ligand-directed targeting of genes to the site of disease. *Nat. Med.* **9**, 135–139.

Wilson, J. M., Grossman, M., Wu, C. H., Chowdhury, N. R., Wu, G. Y., and Chowdhury, J. R. (1992). Hepatocyte-directed gene transfer *in vivo* leads to transient improvement of hypercholes-terolemia in low density lipoprotein receptor-deficient rabbits. *J. Biol. Chem.* **267**, 963–967.

Wolschek, M. F., Thallinger, C., Kursa, M., Rossler, V., Allen, M., Lichtenberger, C., Kircheis, R., Lucas, T., Willheim, M., Reinisch, W., Gangl, A., Wagner, E., and Jansen, B. (2002). Specific systemic non-viral gene delivery to human hepatocellular carcinoma xenografts in SCID mice. *Hepatology* **36**, 1106–1114.

Wu, G. Y., Wilson, J. M., Shalaby, F., Grossman, M., Shafritz, D. A., and Wu, C. H. (1991). Receptor-mediated gene delivery *in vivo*. Partial correction of genetic analbuminemia in Nagase rats. *J. Biol. Chem.* **266**, 14338–14342.

Wu, G. Y., and Wu, C. H. (1988). Receptor-mediated gene delivery and expression *in vivo*. *J. Biol. Chem.* **262**, 14621–14624.

Wu, G. Y., Zhan, P., Sze, L. L., Rosenberg, A. R., and Wu, C. H. (1994). Incorporation of adenovirus into a ligand-based DNA carrier system results in retention of original receptor specificity and enhances targeted gene expression. *J. Biol. Chem.* **269**, 11542–11546.

Yin, W., and Cheng, P. W. (1994). Lectin conjugate-directed gene transfer to airway epithelial cells. *Biochem. Biophys. Res. Commun.* **205**, 826–833.

Zauner, W., Blaas, D., Kuechler, E., and Wagner, E. (1995). Rhinovirus-mediated endosomal release of transfection complexes. *J. Virol.* **69**, 1085–1092.

Zauner, W., Ogris, M., and Wagner, E. (1998). Polylysine-based transfection systems utilizing receptor-mediated delivery. *Adv. Drug Deliv. Rev.* **30**, 97–113.

Zeng, J., Too, H. P., Ma, Y., Luo, E. S., and Wang, S. (2004). A synthetic peptide containing loop 4 of nerve growth factor for targeted gene delivery. *J. Gene. Med.* Sep. 14, Epub ahead of print..

Zhang, G., Budker, V., and Wolff, J. A. (1999). High levels of foreign gene expression in hepatocytes after tail vein injections of naked plasmid DNA. *Hum. Gene Ther.* **10**, 1735–1737.

Zhao, Z., Wang, J., Mao, H. Q., and Leong, K. W. (2003). Polyphosphoesters in drug and gene delivery. *Adv. Drug Deliv. Rev.* **55**, 483–499.

Ziady, A. G., Ferkol, T., Dawson, D. V., Perlmutter, D. H., and Davis, P. B. (1999). Chain length of the polylysine in receptor-targeted gene transfer complexes affects duration of reporter gene expression both *in vitro* and *in vivo*. *J. Biol. Chem.* **274**, 4908–4916.

Zinselmeyer, B. H., Mackay, S. P., Schatzlein, A. G., and Uchegbu, I. F. (2002). The lower-generation polypropylenimine dendrimers are effective gene-transfer agents. *Pharm. Res.* **19**, 960–967.

Zou, S. M., Erbacher, P., Remy, J. S., and Behr, J. P. (2000). Systemic linear polyethylenimine (L-PEI)-mediated gene delivery in the mouse. *J. Gene. Med.* **2**, 128–134.

Zuber, G., Dauty, E., Nothisen, M., Belguise, P., and Behr, J. P. (2001). Towards synthetic viruses. *Adv. Drug Deliv. Rev.* **52**, 245–253.

Index

Structural paradigm for self-assembly ABCD nanoparticles

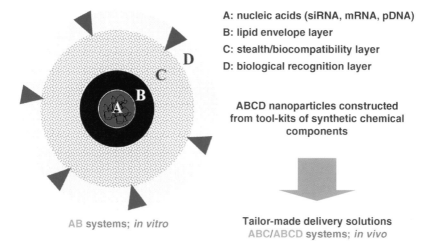

A: nucleic acids (siRNA, mRNA, pDNA)
B: lipid envelope layer
C: stealth/biocompatibility layer
D: biological recognition layer

ABCD nanoparticles constructed
from tool-kits of synthetic chemical
components

AB systems; *in vitro*

Tailor-made delivery solutions
ABC/ABCD systems; *in vivo*

Chapter 4, Figure 4.1. ABCD nanoparticle concept. Graphic illustration of ABCD nanoparticle structure to show how nucleic acids (A) are condensed in functional concentric layers of chemical components purpose designed for biological targeting (D), biological stability (C), and cellular entry (B).

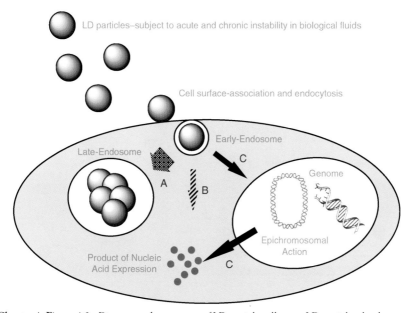

Chapter 4, Figure 4.2. Diagram to show process of LD particle cell entry. LD particles that have not succumbed to aggregation and/or serum-inactivation associate with the cell surface and enter usually by endocytosis. The majority in early endosomes become trapped in late endosomes (Path A) and the nucleic acids fail to reach the cytosol. A minority are able to release their bound nucleic acids into the cytosol. Path B is followed by RNA that acts directly in the cytosol. Path C is followed by DNA that enters the nucleus in order to act. The diagram is drawn making the assumption that plasmid DNA (pDNA) has been delivered which is expressed in an epichromosomal manner. Reprinted with permission of Bios. Scientific Publishers from (Miller, 1999).

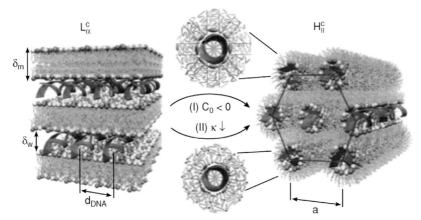

Chapter 4, Figure 4.5. LD particle internal structure and dynamics. *Left-hand side:* schematic of the lamellar $L_{\alpha I}$ phase of DNA molecules interacting with cationic bilayers forming a multilayered assembly typical of LD particle composition. DNA double helices are shown as ribbons (blue and purple), head groups of anionic/zwitterionic lipids are shown as white spheres while those of cytofectins are shown as grey spheres. The notation $\boldsymbol{\delta}_{m}$ refers to bilayer thickness, $\boldsymbol{\delta}_{w}$ to interbilayer separation and $\boldsymbol{d}_{DNA}$ to DNA interaxial spacing. *Right-hand side:* conversion from lamellar $L_{\alpha I}$ phase to the columnar, inverted hexagonal H_{II} phase thought to be typical of LD particle composition during the transfection process, takes place by two possible routes. The first involves pathway (**I**) typified by negative curvature $\boldsymbol{C}_{o}$ induced in each cationic monolayer due to the presence of DOPE 1. The second involves pathway (**II**) typified by loss in membrane rigidity $\boldsymbol{\kappa}$ thereby encouraging phase inversion. Reprinted with permission of AAAS from (Koltover *et al.*, 1998).

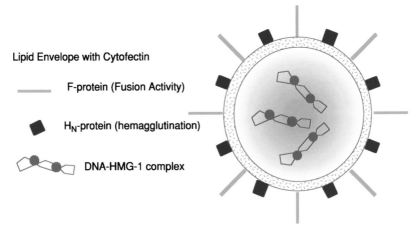

Chapter 4, Figure 4.8. HVJ-cationic liposome system. Cytofectins are incorporated in the lipid envelope Reprinted with permission of Bios. Scientific Publishers from (Miller, 1999).

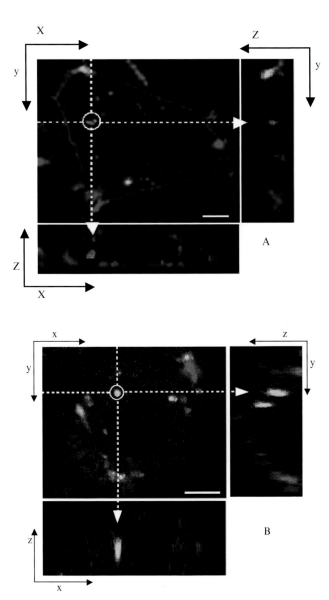

Chapter 5, Figure 5.6. Laser scanning confocal microscopy images of transfected mouse L cells, fixed six hours after incubation with complexes. For each set of images, the center image is the x–y (top) view at a given z; the right shows a y–z side view along the vertical dotted line; and the bottom a x–z side view along the horizontal dotted line. Red and green fluorescence corresponds to lipid and DNA labels, respectively; yellow, the overlap of the two, denotes CL–DNA complexes. Scale bars are 5 μm. (A): Cells transfected with H_{II}^{C} complexes ($M_{DOPE} = 0.69$) show transfer of fluorescent lipid to the cell plasma membrane and the release of DNA (green; in the circle) within the cell. (B): LSCM image of cells transfected with L_{α}^{C} complexes at $M_{DOPC} = 0.67$, where TE is low, as shown in Figure 5.5. The cell outline was observed in reflection mode, appearing in blue. No evidence for fusion is visible and intact CL–DNA complexes such as the one marked by a circle are observed inside the cells. This observation implies that DNA remains trapped within the complexes, consistent with the observed low transfection efficiency. Reprinted with permission from Lin *et al.* (2003). © 2003 Biophysical Society.

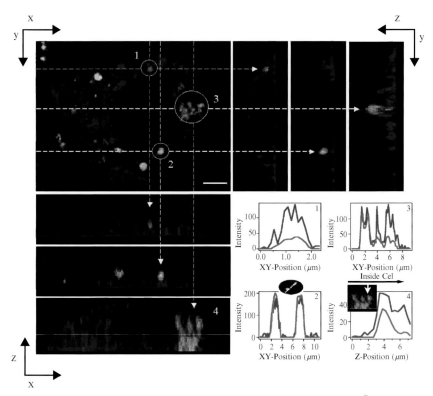

Chapter 5, Figure 5.8. A typical LSCM image of a mouse L cell transfected with L_α^C complexes at $m_{DOPC} = 0.18$, corresponding to cationic membranes with a high charge density $\sigma_M \approx 0.012$ e/$\overset{\circ}{A}^2$ and high TE (cf. Figure 5.7). Red and green fluorescence corresponds to lipid and DNA labels, respectively; yellow, the overlap of the two, denotes CL–DNA complexes. The cell outline was observed in reflection mode, appearing in blue. Scale bars are 5 μm. The center image is the x–y (top) view at a given z; on the right are y–z side views along the vertical dotted lines; at the bottom are x–z side views along the horizontal dotted lines. In the boxes in the lower right corner, plots of lipid and DNA fluorescence intensity along the x–y diagonal or z-axis are shown for objects labeled with numbers. Although the lamellar complexes used here show high TE, no lipid transfer to the cell plasma membrane is seen in contrast to high-transfecting H_{II}^C complexes (Figure 5.6A). Both released DNA (1) and intact complexes (2) are observed inside the cell. Labels (3) and (4): A complex in the process of releasing its DNA into the cytoplasm. Reprinted with permission from Lin *et al.* (2003). © 2003 Biophysical Society.

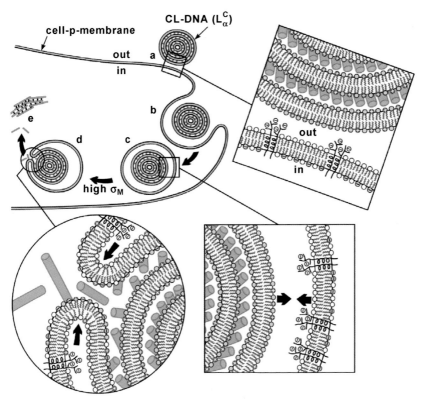

Chapter 5, Figure 5.12. Model of cellular uptake and endosomal release (through activated fusion) of L_α^C complexes. (a) Cationic complexes adhere to cells due to electrostatic attraction between positively charged CL–DNA complexes and negatively charged cell-surface proteoglycans (shown in expanded views). (b and c) After attachment, complexes enter through endocytosis. (d) Only complexes with sufficiently high membrane charge density escape from the endosome through activated fusion with the endosomal membrane. (e) Confocal microscopy shows that lipid-free DNA inside the cell exists primarily in the form of aggregates. These DNA aggregates must reside in the cytoplasm because oppositely charged cellular biomolecules able to condense DNA are not present in the endosome. Arrows in the expanded view of (c) indicate the electrostatic attraction between the oppositely charged membranes of the complex and endosome, which enhances adhesion and fusion. Arrows in the expanded view of (d) indicate the bending of the membranes required for fusion, which constitutes the main barrier for the process. Reprinted with permission from Lin *et al.* (2003). © 2003 Biophysical Society.

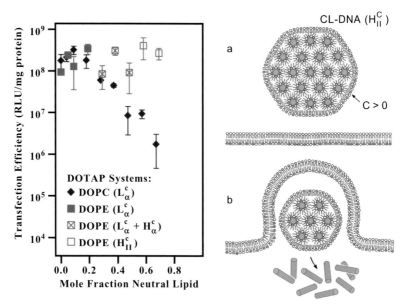

Chapter 5, Figure 5.13. Left: TE and complex structure as a function of molar fraction of neutral lipid for complexes prepared from DOTAP/DOPC and DOTAP/DOPE mixtures. Right: Schematic of the proposed mechanism of cell entry/endosomal fusion of inverted hexagonal complexes. An H_{II}^C CL–DNA complex is shown interacting with the plasma membrane or the endosomal membrane. The cell-surface proteoglycans of the cellular membrane have been omitted for clarity. The outer lipid monolayer covering the H_{II}^C CL–DNA complex has a positive curvature. However, the preferred curvature of the lipids forming the complex membrane is negative, as realized in the tubules coating DNA within the complex. Thus, the outer layer is energetically costly. This results in a driving force, independent of the cationic membrane charge density, for rapid fusion of the H_{II}^C complex with the bilayer of the cell plasma membrane or the endosomal membrane. Reprinted with permission from Lin *et al.* (2003). © 2003 Biophysical Society.

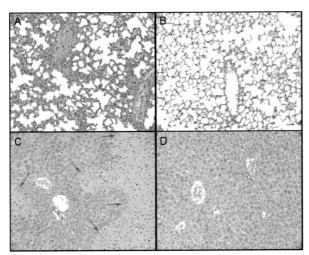

Chapter 7, Figure 7.4. Haemotoxylin and eosin stained sections of lungs (A and B) and livers (C and D) from mice 24 h after injecting linear PEI/pAA/DNA (400 μg/50 μg/50 μg) complexes without pAA "chaser" (B and D) and linear PEI/pAA/DNA complexes (240 μg/40 μg/40 μg) with 1.5 mg of pAA "chaser" (A and C) injected 30 min after injecting complex. Magnification for all panels is 100X. Although equivalent levels of luciferase were expressed from the mice, significantly reduced toxicity was observed in the mice that received the pAA chaser. Reprinted by permission from Gene Therapy, Trubetskoy *et al.* (2003), copyright 2003, Macmillan Publishers Ltd.

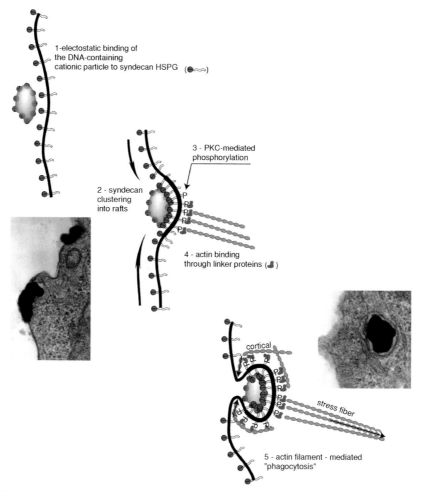

1 -electostatic binding of
the DNA-containing
cationic particle to syndecan HSPG

3 - PKC-mediated
phosphorylation

2 - syndecan
clustering
into rafts

4 - actin binding
through linker proteins

cortical

stress fiber

5 - actin filament - mediated
"phagocytosis"

Chapter 8, Figure 8.3. A model for the uptake of PEI/DNA complexes by adherent cells. Insets show TEM pictures of the complexes.

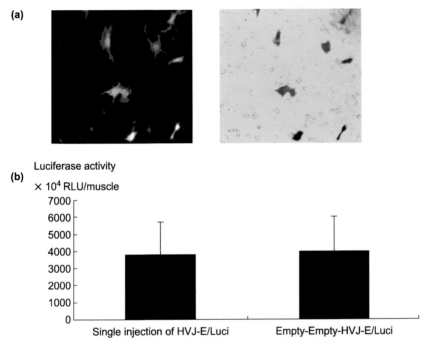

Chapter 12, Figure 12.2. (a) Two different plasmids can be delivered to the same cell. HVJ envelope vector containing both GFP and LacZ gene is added to BHK-21 cells. At 24 h after gene transfer, both proteins clearly expressed in the same cells. (b) Gene expression is not inhibited after consecutive injection of HVJ envelope vector into mouse muscle. In both cases, luciferase gene expression in muscle is detected at 24 h after injection of HVJ envelope vector containing luciferase gene. In the mice shown by the blue bar, empty HVJ envelope vector has been injected into muscle twice in a 2-week interval before injection of luciferase gene-loaded vector.

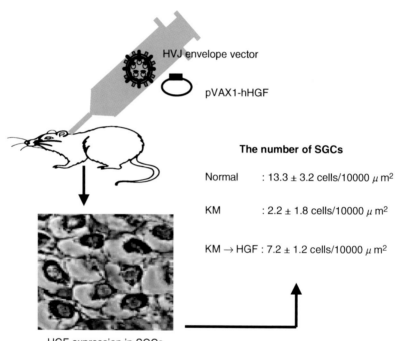

HVJ envelope vector

pVAX1-hHGF

The number of SGCs

Normal   : 13.3 ± 3.2 cells/10000 μ m^2

KM   : 2.2 ± 1.8 cells/10000 μ m^2

KM → HGF : 7.2 ± 1.2 cells/10000 μ m^2

HGF expression in SGCs

Chapter 12, Figure 12.3. Gene therapy of hearing impairment by HGF gene. HVJ envelope vector containing human HGF gene is administered to the cerebrospinal fluid of a rat. Human HGF gene expression is immunohistochemically detected at SGCs in the inner ear. By HGF gene expression, the number of SGCs is rescued after KM insult.

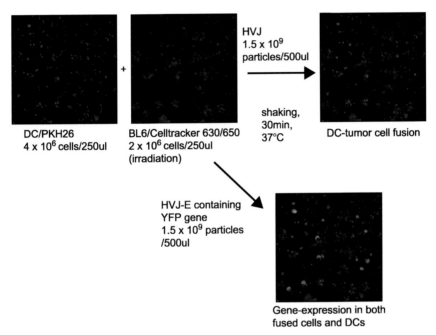

DC/PKH26
4 x 10^6 cells/250ul

+

BL6/Celltracker 630/650
2 x 10^6 cells/250ul
(irradiation)

HVJ
1.5 x 10^9
particles/500ul

shaking,
30min,
37°C

DC-tumor cell fusion

HVJ-E containing
YFP gene
1.5 x 10^9 particles
/500ul

Gene-expression in both
fused cells and DCs

Chapter 12, Figure 12.5. Fusion of DCs and melanoma cells and simultaneous gene transfer to the fused cells. Mouse bone marrow-derived DCs stained with PKH26 and mouse melanoma BL6 labeled with Celltracker 630/650 are fused with inactivated HVJ (1.5 × 10^8 particles). When fused with HVJ envelope (HVJ-E) vector containing YFP gene, YFP gene expression is detected in both DCs and fused cells.

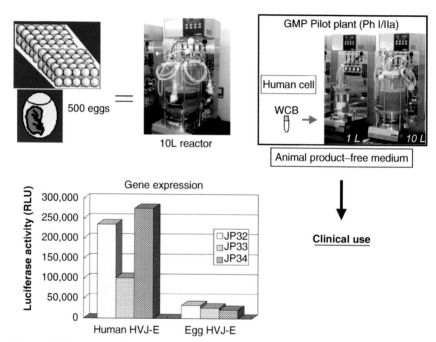

Chapter 12, Figure 12.7. Development of clinical grade HVJ envelope vector. HVJ infects human cell from working cell bank (WCB), and human cell-derived HVJ is prepared using animal product-free medium in a pilot plant based on good manufacturing practice (GMP) standard. The envelope vector prepared from human cell-derived HVJ is more efficient in gene transfection compared with that from egg-derived HVJ.